Recent Progress in Quantum Physics

Recent Progress in Quantum Physics

Edited by
Tom Gladstone

C WILLFORD PRESS

www.willfordpress.com

Published by Willford Press,
118-35 Queens Blvd., Suite 400,
Forest Hills, NY 11375, USA

ISBN: 978-1-68285-652-9

Cataloging-in-Publication Data

Recent progress in quantum physics / edited by Tom Gladstone.
    p. cm.
Includes bibliographical references and index.
ISBN 978-1-68285-652-9
1. Quantum theory. 2. Physics. 3. Mechanics. 4. Thermodynamics. I. Gladstone, Tom.
QC174.12 .R43 2019
530.12--dc23

For information on all Willford Press publications
visit our website at www.willfordpress.com

WILLFORD PRESS

Contents

Permissions

List of Contributors

Index

Preface

Quantum physics is a fundamental theory in physics that describes nature at the atomic and subatomic particle level. Wave-particle duality, uncertainty principle, quantization of energy and momentum are central aspects of quantum physics. It has diverse applications in quantum optics, quantum computing, quantum chemistry, medical and research imaging, etc. The behavior of subatomic particles that make up matter, such as electrons, protons, neutrons, etc. is governed by quantum mechanics. It also plays a foundational role in string theory and computational chemistry. The quantum field theory that studies the strong nuclear force is under the scope of quantum chromodynamics. It describes the interactions between quarks and gluons. The weak nuclear force is unified with electromagnetic force in their quantized forms through the approach of electroweak theory. The ever-growing need of advanced technology is the reason that has fueled the research in the field of quantum physics in recent times. This book brings forth some of the most innovative concepts and elucidates the unexplored aspects of this field. Coherent flow of topics, student-friendly language and extensive use of examples make this book an invaluable source of knowledge.

This book is a comprehensive compilation of works of different researchers from varied parts of the world. It includes valuable experiences of the researchers with the sole objective of providing the readers (learners) with a proper knowledge of the concerned field. This book will be beneficial in evoking inspiration and enhancing the knowledge of the interested readers.

In the end, I would like to extend my heartiest thanks to the authors who worked with great determination on their chapters. I also appreciate the publisher's support in the course of the book. I would also like to deeply acknowledge my family who stood by me as a source of inspiration during the project.

Editor

A Historical Survey of Sir Karl Popper's Contribution to Quantum Mechanics

William M. Shields

Worcester Polytechnic Institute, Worcester, Massachusetts, United States. E-mail: highc.king@verizon.net

Editors: *Ion C. Baianu, Christoph Lehner, Debajyoti Gangopadhyay & Danko Georgiev*

Sir Karl Popper (1902–1994), though not trained as a physicist and embarrassed early in his career by a physics error pointed out by Einstein and Bohr, ultimately made substantial contributions to the interpretation of quantum mechanics. As was often the case, Popper initially formulated his position by criticizing the views of others – in this case Niels Bohr and Werner Heisenberg. Underlying Popper's criticism was his belief that, first, the Copenhagen interpretation of quantum mechanics abandoned scientific realism and second, the assertion that quantum theory was complete (an assertion rejected by Einstein among others) amounted to an unfalsifiable claim. Popper insisted that the most basic predictions of quantum mechanics should continue to be tested, with an eye towards falsification rather than mere adding of decimal places to confirmatory experiments. His persistent attacks on the Copenhagen interpretation were aimed not at the uncertainty principle itself and the formalism from which it was derived, but at the acceptance by physicists of an unclear epistemology and ontology that left critical questions unanswered.
Quanta 2012; 1: 1–12.

1 Popper in the physics journals

Sir Karl Popper, by any measure one of the preeminent philosophers of the twentieth century, died in 1994 at the age of 92. He was productive to the end, publishing in the year of his death a criticism of Kuhn's incommensurability of paradigms [1]. That debate continues over his many and profound philosophical ideas and opinions is hardly surprising, almost two decades after his death. The proliferation of book-length biographies and scholarly philosophical articles is testimony to Popper's standing as a philosopher [2–5]. Major conferences are also regularly held on the thought of Karl Popper [6, 7].

It may come as a surprise, however, that beginning in the year 2000, Popper's name appears prominently in no less than a dozen papers in the journals of theoretical physics, in the majority of cases in the paper's title [8–20]. Several of these papers report the results of *Popper's experiment* carried out by physicists at the University of Maryland in 1999 [13–17]. Coincidentally, the final section of a recent text on the Einstein-Podolsky-Rosen (EPR) paradox in physics deals with this same experiment, proposed by Popper in the early 1980's [21, Sec. 5.4.3]. How is it that physicists in the new millennium are invoking Karl Popper's name, conducting experiments suggested by him, and arguing over the meaning of the results?

To suggest answers to this question, I review below Popper's fifty years of contributions to the interpretation of quantum mechanics. For reasons that will become clear in later sections of the paper, I trace a line of thought experiments that Popper proposed to test his own views against the views of the majority of theoretical physicists who created quantum mechanics. In the final section of the paper, I offer a possible explanation for why Popper's passionately-held opinions continue to attract the attention of physicists, and are worthy of that attention.

2 'Logik Der Forschung' and Einstein's refutation

The most exciting and fundamental discoveries of quantum mechanics were made while Popper was in college and graduate school. Though his dissertation was not in physics, he had studied the sciences and mathematics and was qualified to teach them on the secondary level. The thesis, completed in 1928, was titled *On the Problem of Method in the Psychology of Thinking* and it had more to do with the methodology of science than psychology. Popper characterized it as a "hasty last minute affair" [22, p. 78]. By the time he began work in earnest on *Logik der Forschung* in the early 1930's, Popper had educated himself on the new quantum mechanics, by then becoming accepted as a major advance in atomic physics.

Consistent with his interest in the logic and methodology of science, Popper focused his study on the philosophical underpinnings of the new theory. He was especially interested in the disputes that had arisen over how to interpret physically the mathematical formalism of the theory. Popper explains:

> At the time (1930) when ... I began writing my book, modern physics was in turmoil. Quantum mechanics had been created by Werner Heisenberg in 1925, but it was several more years before outsiders – including professional physicists – realized that a major breakthrough had been achieved. And from the very beginning there was dissension and confusion. The two greatest physicists, Einstein and Bohr, perhaps the two greatest thinkers of the twentieth century, disagreed with one another. [22, pp. 90-91]

Lacking doctoral-level knowledge of physics, Popper struggled to grasp the new theory:

> I was working on my own from books and from articles; the only physicist with whom I sometimes talked about my difficulties was my friend Franz Urbach. I tried to understand the theory and he had doubts whether it was understandable – at least by ordinary mortals. [22, p. 91].

Eventually Popper came to appreciate the core of the disagreement within physics and was able to sort out in his mind the various positions of Einstein, Bohr, Heisenberg, Schrödinger, and Born. By the time *Logik Der Forschung* was well underway, Popper felt qualified to address quantum theory in the book.

Never one to tackle a subject halfway, Popper devoted all of Chapter IX of *Logik Der Forschung* to *Some Observations on Quantum Theory* [23]. In the English translation, it runs to 35 densely-argued pages wherein Popper sets out the views he was to maintain, with some modifications, for the rest of his life [24]. He makes his purpose clear in the introductory section:

> What follows here might be described, perhaps, as an inquiry into the foundations of quantum theory. In this, I shall avoid all mathematical arguments and, with one single exception, all mathematical formulae. This is possible because I shall not question the correctness of the system of mathematical formulae of quantum theory. I shall only be concerned with the logical consequences of its physical interpretation which is due to [Max] Born. [24, p. 216]

Thereafter Popper lays out his criticism of what is commonly called the Copenhagen Interpretation of quantum theory (largely the work of Bohr) and the position of Heisenberg that the uncertainty relations must be viewed subjectively, as a "limitation of our knowledge" of physical systems [24, p. 220]. In the same passage, Popper notes that Moritz Schlick of the Vienna Circle had expressed strong support for Heisenberg's views (Heisenberg's explication of his own position can be found in [25]).

Popper devotes the next several sections to advocacy of a statistical interpretation of the uncertainty relations. He argues that, contra Heisenberg, it does indeed make sense to attribute well-defined positions and momenta to individual particles. Experimental results showing wave-like behavior of particles (as in slit experiments) can be explained as "statistical scatter relations." The scattering behavior is calculated using the mathematical machinery of quantum theory, but it does not imply anything about limitations on knowledge or an actual lack of a well-defined position and momentum at any moment in time [24, p. 225].

With this interpretation in hand, Popper wonders aloud whether anything has in fact been gained. His conclusion emphasizes his aim, which is in essence to succeed where Einstein had failed in his arguments with Bohr:

> The statistical elements of quantum theory must be inter-subjectively testable in the same way

as any other statements of physics. And my simple analysis preserves not only the possibility of spatio-temporal descriptions, but also the objective character of physics. [24, p. 234]

Events that took place two years after the publication suggested to Popper that he should have ended the chapter there. But he did not. In the next sections of the chapter, Popper describes an imaginary experiment (Gedankenexperiment) "which shows, in full agreement with quantum theory, that the precise measurements in question are possible" [24, p. 243]. I will omit details of the experimental design here – suffice to say that *it had the same idea as later versions, but was flawed in several key respects.*

These flaws were made apparent to Popper at a scientific conference held in Copenhagen in 1936. Following the conference, Popper was invited by Bohr at the urging of Victor Weiskopf, a leading theoretician, to stay on a few days to discuss quantum mechanics. Popper was already feeling uneasy about his Gedankenexperiment, which had been questioned by Einstein. After discussions with Bohr, Popper accepted that the experiment did not show what he intended, and he left Copenhagen quite upset over losing the argument [22, pp. 92-93]. While he did not stop thinking about quantum mechanics, he "remained for years greatly discouraged . . . I could not get over my mistaken thought experiment" [22, p. 94]. But following consultations with physicist Arthur March in the late 1940's, Popper returned to the problems of quantum mechanics with "something like renewed courage" [22, p. 94]. He began to revise and clarify his ideas while working on a set of appendices for the English version of *Logik Der Forschung* and the long-delayed *Postscript*. (The *Postscript* actually appears several years before the English edition of the book with its new appendices. Here I consider the appendices first in order to conclude the discussion of *Logik Der Forschung*.)

Appendix xi of the English version of *Logik Der Forschung* offers some general thoughts on the use of imaginary experiments in physics (see also [26, pp. 240-265]) before moving to a lengthy discussion of the Einstein-Podolsky-Rosen (EPR) experiment and Bohr's interpretation of it, which Popper predictably rejects. In Appendix xii, Popper bites the bullet and reprints in full Einstein's letter to him of 1935, in which Einstein describes the flaws in the Gedankenexperiment proposed in *Logik Der Forschung*. Popper admits that Einstein's letter "briefly and decisively disposes of my imaginary experiment in section 77" of *Logik Der Forschung* [24, p. 457]. By this time, however, Popper had already proposed a new experiment in the *Postscript* to *Logik Der Forschung*, and it is to that work that I now turn.

3 'Postscript to Logik Der Forschung': thoughts on Gedankenexperiments

Popper's *Postscript* evolved by the mid-1950s into a massive work that outsized the work it was a postscript to, so large that it had to be published in three volumes [27–29]. Popper's determination to have an impact on quantum theory is evidenced by his devoting the entire third volume to the subject. If *Quantum Theory and the Schism in Physics* [29] shows nothing else, it is proof that in the twenty years since his embarrassment before Bohr, he had devoted immense effort to mastering quantum mechanics. The book covers a wide range of quantum theoretical controversies and points of view, with a continual mixing of physical, mathematical, and philosophical ideas.

To explore this work fully is beyond my present scope. (This work appeared first in 1956, but again in a new (and more easily obtainable) version in 1982. The 1982 version, which in many ways represents Popper's final statement of views on quantum theory, opens with a 35-page *Preface* written in 1982, followed by a 62-page chapter written as a paper in 1966! The 1956 volume entitled *Schism* actually begins at page 97 of the 1982 edition. So the 1982 edition actually reads backwards from 1982 to 1956. In this section I confine the discussion to the 1956 material.) For the moment I wish to focus on Popper's revisiting of the idea of an experimentum crucis. As to the imaginary experiment proposed in *Logik Der Forschung*, Popper states bluntly that it was "invalid, and I wish to withdraw it" [29, p. 98]. In Chapter III, he discusses first the relatively simple experiment of collimated particles aimed at a small slit in a barrier. Classical mechanics would say that the particles will travel straight through, while quantum theory demands that if the slit is small enough, a scattering effect will be achieved. This scattering effect is due, according to Heisenberg, to the confinement of the wave packet representing the particle to a distance Δx (the width of the slit), causing according to the uncertainty relations a corresponding uncertainty in the momentum Δp, where $\Delta x \Delta p \geq \hbar$ [25, pp. 23-24]. But where Heisenberg (and Bohr) interpreted this result (which is easily observable) as amounting to a denial of a scattered particle's "particality," so to speak, so that retrodictive calculation of the particle's path is essentially meaningless, Popper argues that one can conclude no such thing. In his view, each scattered particle had a real path and had a well-defined position and momentum at all times. It was simply scattered by the slit: no epistemological conclusions can be drawn beyond that [29, pp. 144-147]. Popper then proceeds to examine the arguments based on the celebrated Gedankenexperiment proposed by

Einstein, Podolsky and Rosen, forever thereafter known as the EPR experiment [30].

After a discussion of the conflicting points of view of Einstein and Bohr – with interjections on why he agrees with Einstein – Popper does not mince words on what he believes is at stake:

> Reasonableness was the point at issue. The question is not whether by a subtle and highly scholastic argument we may continue to uphold an untenable position. The question is whether we should think critically and rationally in physics, or defensively and apologetically. [29, p. 150]

It is fairly apparent who Popper believes is doing the critical thinking and who is being defensive and apologetic.

Finally, Popper turns to an experiment analyzed (using classical wave theory) by Thomas Young in the early 1800's and, according to Popper, "discussed again and again by Bohr" [29, p. 151]. This is a two-slit experiment, in which the quantum scattering induced by one slit is then "projected" onto another barrier with two slits in it, neither of which aligns with the slit in the first barrier. What will then be observed, again assuming that all the slits are small enough to engage quantum phenomena, is an interference pattern just as if the whole experiment were conducted with macroscopic waves on a pond. But because we are using discrete particles, and each particle can go through only one of the two secondary slits, it appears that the two slits cooperate in producing the interference pattern. Indeed, closing one or the other of the paired slits changes the final pattern.

Popper uses this experiment as an opportunity to apply his propensity interpretation of probability calculations. Where Bohr would say that the result in each case (all slits open, two open, etc.) can be explained only be recourse to the complementary notions of wave and particle along with considering the active role of the observer, Popper argues that "it is the whole experimental arrangement which determines the propensities." By 'propensities' here, Popper refers to his interpretation of probability, which differs from the more commonly-held view that probability is statistical in nature. In Popper's view, probability is more akin to a field of force than it is to a mathematically-calculated frequency. It therefore acquires a 'reality' which can have physical effects: the probability fields can interact and interfere with one another. This approach, which Popper modified and refined over the years, responds to Einstein's dictum that 'God does not play dice', and tends to restore some measure of determinacy to physics that the standard versions of quantum theory deny. (Popper's approach to probability, especially as it is used in physics, is discussed at various

points throughout [29]; see also [31, pp. 59-60] [3, pp. 109-112]. A recent paper compares Popper's view on propensities with those originally suggested by the American philosopher Charles Sanders Peirce [32].) I mention the Young experiment, especially the aspect of closing one of the paired slits, because it appears to be a precursor in Popper's mind of a modified two-slit experiment he will propose and defend from 1981 to 1987. This new experimental proposal is the subject of the next section. I have found no evidence that Popper's 1956 efforts in the *Postscript* generated any interest in the physics community. Most likely, the *Postscript* was read by few physicists.

4 Gedankenexperiment refinements, 1981-1987

Popper's solitary efforts to offer a different view of quantum phenomena acquire a different status in the early 1980's. By this time, he had acquired colleagues in the theoretical physics community, one of whom, French physicist Jean-Pierre Vigier, had an international reputation. This collaboration led to the publication of a paper in a widely-read English-language journal, *Physics Letters*, in December of 1981 [33]. Italian physicist Augusto Garuccio, K. Popper and J.-P. Vigier (GPV) jointly proposed an experiment involving the interference of laser beams. The intent of the paper is clearly stated in the opening paragraph and rather obviously drafted by Popper:

> The present letter develops a gedanken experiment which leads to conflicting testable predictions of the Copenhagen ... and causal statistical ... interpretations of quantum theory. [33, p. 397]

It is not clear why the term "gedanken" is used here, since the experiment was clearly within the range of 1981 technology. Perhaps the term is used only to make clear that they had not built the apparatus and showed that it could perform as predicted. The GPV experimental design was based on work done by two other physicists, L. Mandel and R. L. Pfleegor, in the late 1960's, and modified an experiment first proposed by Garuccio and Vigier in 1980. Popper and his co-authors acknowledge helpful comments not only from Mandel, but also from John S. Bell, who derived the famous Bell's inequalities, and Alain Aspect, who was at that time (1981) conducting experiments of his own to test Bell's inequalities. The paper's concluding assertions include a comment on demarcation that must surely have been Popper's:

As one knows this typical Copenhagen retroactive action (which has been used to justify parapsychological phenomena) raises trouble with energy conservation and implies rejection of Feynman's quantum propagator D_c. The authors feel with Einstein that: (I) the flow of time is a real, irreversible and one-dimensional phenomenon, (II) only positive energies move in the forward time direction, and (III) the apparent microscopic time reversibility of the quantum mechanical wave equations only reflects the particle/anti-particle mixture of Einstein and Feynman which leads to correct perturbation theory. [33, p. 400]

More or less coincident with the publication of this paper, Popper re-issued *Quantum Theory and the Schism in Physics* with a new Preface, *On a Realistic and Commonsense Interpretation of Quantum Theory* and a new first section, *Quantum Mechanics Without 'The Observer'* [29, pp. 1-96]. Towards the end of the *Preface*, Popper proposes "a simple thought experiment which may be regarded as an extension of the Einstein-Podolsky-Rosen argument." Oddly, this experiment (though in this case admittedly a thought experiment, as no apparatus is actually described) is not at all the same as in the paper with Vigier and Garuccio, though the underlying purpose is the same. This experiment involved the placing of an emission source between two slits. The source emits particles with equal and opposite momenta (Figure 1). The particles pass through the small slits and scatter according to the uncertainty relations. But what would happen, Popper asks (and here is where the centuries-old Young optics experiment may have played a role), to the observed scattering on one side if the slit on the other side is widened so that by itself it would cause no quantum scattering? Will the "knowledge" imparted by the confinement of the particle in the still-small slit induce scattering on the other side so that the observed pattern does not change? As Popper explains:

> To sum up: if the Copenhagen interpretation is correct, then any increase in the precision of our *mere knowledge* of the position q_y of the particles going to the right should increase their scatter; and this prediction should be testable. [29, p. 29]

The paper in *Physics Letters* drew immediate attention in the physics community. First to respond was French physicist O. Costa de Beauregard, who offered in May of 1982 a brief letter entitled *Disagreement with Garuccio, Popper and Vigier* [34]. De Beauregard had no apparent objections to the experimental design of GPV but argued that the results would be fully consistent with the standard interpretations of quantum theory. He admits, however, that his remarks "contain no objection against the tentative theory of Garuccio et al." Just a few weeks later, Mandel criticized GPV's experimental arrangement and offered a modification of the design that would in principle address his objections [35]. The next month, June 1982, Garuccio and Vigier entered another paper in Physics Letters, but in this case Popper's place as a co-author was taken by one of Garuccio's colleagues, V. Rapisarda [36]. That Popper remains involved is made clear by note 4 in this paper, which cites a "private communication" from Popper. These authors noted that the GPV paper had "provoked a complex and heated discussion" in the physics community. Their present purpose was to

> ...present an experimental programme and experimental set-up which clearly falls outside of the field of the above-mentioned objections to the preceding discussion [GPV] and suppresses, as far as possible, experimental difficulties. [36, p. 17]

At the end of the paper, the authors asserted that this experiment will "escape all preceding objections" and, if conducted, "would really constitute a crucial distinction between CIQM [the Copenhagen Interpretation] and reality."

There the matter lay until 1985. In that year, a paper by Anthony Sudbery, a mathematician at the University of York, appeared in *Philosophy of Science* [37]. Sudbery's critical stance can be discerned from his lengthy title: *Popper's Variant of the EPR Experiment Does Not Test the Copenhagen Interpretation*. Sudbery analyzed the version of the experiment Popper presented in the 1982 *Preface* to *Quantum Theory and the Schism in Physics*, and in fact made no mention at all of the 1982 papers in *Physics Letters*. Sudbery criticizes two features of the thought experiment:

> The essential elements of [Popper's] deduction are: (i) the inverse relation between Δy and Δp_y, the uncertainties in position and momentum; and (ii) the correlation between the positions of two particles that have interacted in the past. Neither of these is universally true, whatever interpretation of quantum mechanics is in question; each of them holds only in special circumstances. According to the Copenhagen interpretation, they do not hold simultaneously; hence Popper's deduction of the effect E [increase in momentum spread] is not valid within this interpretation. [37, p. 472]

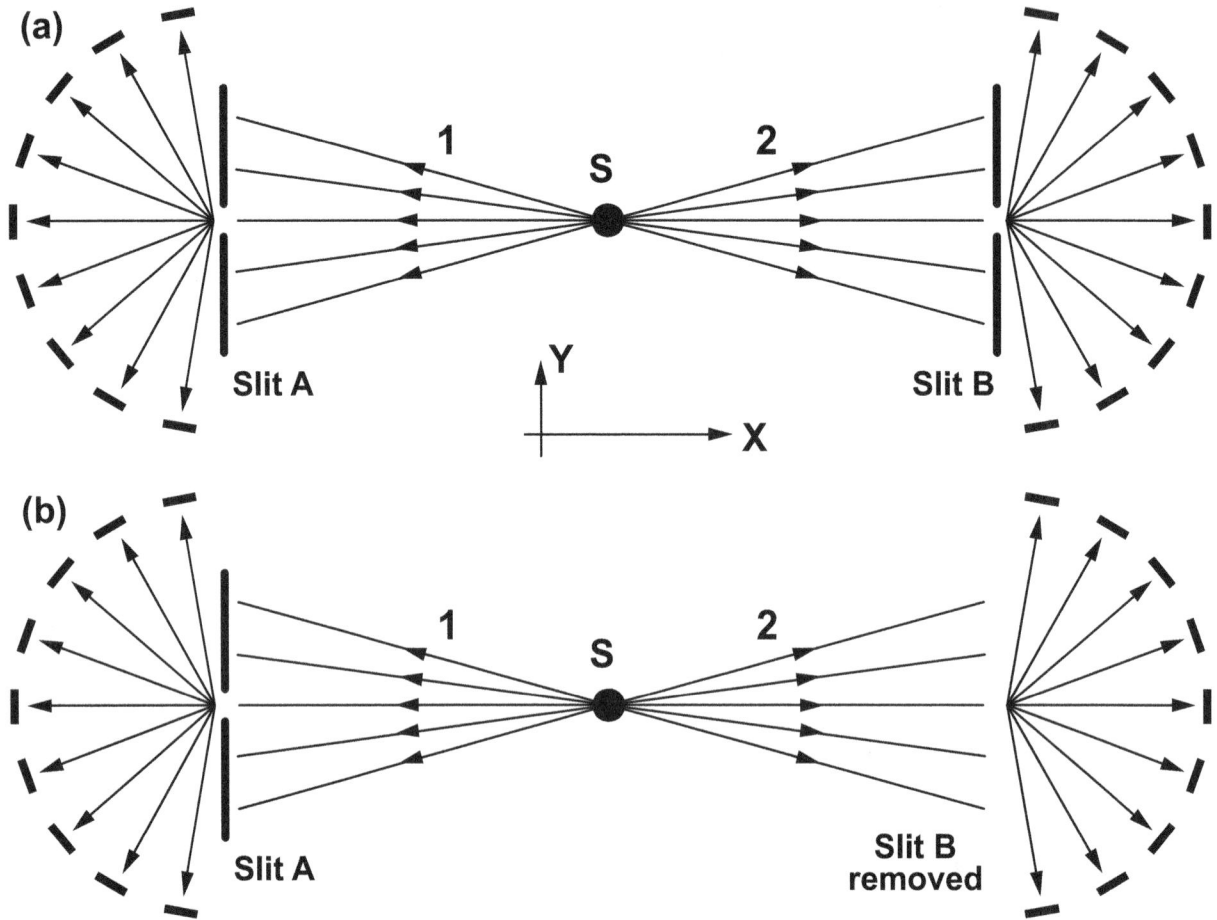

Figure 1: *Schematic diagram of Popper's thought experiment. (a) When both slits A and B are present, the particles are expected to show scatter in momentum. (b) Popper believed that by removing slit B one could test the Copenhagen interpretation.*

As to (i), Sudbery points out that the inverse relation (i.e., the Heisenberg relation) is not an equality but an inequality. Using the entangled wave function employed by Einstein, Podolsky and Rosen (the EPR wave function), application of the inequality leads to the possibility that the uncertainties in the positions and momenta of the entangled particles is in fact infinite. In such a case, Sudbery argues, "narrowing slit A has no effect on the range of momenta of the particles at either A or B, since these ranges cannot be increased any further". On element (ii), Sudbery argues that the experimental apparatus used to measure the entangled particles after separation, by the Copenhagen interpretation, disturbs the initial correlations between positions and momenta. Hence the wave function being measured *is not* the EPR function describing the initial state, which Sudbery writes in the form:

$$\psi(\mathbf{r}_A, \mathbf{r}_B) = \phi_1(x_A)\phi_2(x_B)\delta(y_A - y_B) \quad (1)$$

where ϕ_1 and ϕ_2 are separated localized wave packets, but a different wave function in which the correlations are no longer present:

$$\psi'(\mathbf{r}_A, \mathbf{r}_B) = \phi_1(x_A)\phi_2(x_B) \quad (2)$$

He summarizes as follows:

> If the particles approach the slits in the EPR wave function, so that the observation of one particle gives information about the other, then the spread of the counters that register particles do not depend on the width of the slit. Conversely, if the experiment is arranged so that the spread of counters does depend on the width of the slit, the observation of one particle gives no information about the other. [37, p. 473]

Although some researchers have rejected Sudbery's logic and conclusions [38], in his recent paper, Qureshi [20] finds that Sudbery's argument on (i) above "is the only robust criticism of Popper's experiment".

That Popper's views were not universally rejected, however, is made apparent in the 1985 text *Open Questions in Quantum Physics*, based on the proceedings of a conference held in 1983 at the University of Bari, Italy [39]. Popper's contribution, *Realism in Quantum Mechanics and a New Version of the EPR Experiment* [40] appears first in the book, and is followed by a discussion among Popper, Vigier, and nine other physicists [39, pp. 26-32].

Popper's essay in this book represents, in my view, the culmination of his thinking on quantum theory and its relationship to scientific realism and human knowledge. In the opening paragraphs, he comments:

> I am a realist, and I believe in the reality of matter, or energy, of particles, of fields of forces, of wavelike disturbances of these fields, and of propensity fields (de Broglie fields) ... and I suggest that quantum mechanics is misinterpreted when it is not interpreted realistically. I also suggest that quantum mechanics says nothing whatever about epistemology, about our knowledge and its limits, no more than Newtonian dynamics. [40, p. 4]

Contrast this view with that of Heisenberg, one of Popper's epistemological nemeses. In describing the meaning of the uncertainty relations as applied to the position and momentum of a free electron following a precise measurement of its velocity and no measurement at all of position:

> Then the principle states that every subsequent observation of the position will alter the momentum by an unknown and indeterminable amount such that after carrying out the experiment our knowledge of the electronic motion is restricted by the uncertainty relation ... It is a matter of personal belief whether [backcalculation of] the past history of the electron can be ascribed any physical reality or not. [25, p. 20]

Bohr, the other principal target of Popper's criticism, argued to the same effect:

> Indeed we have in each experimental arrangement suited for the study of proper quantum phenomena not merely to do with an ignorance of the value of certain physical quantities, but with the impossibility of defining these quantities in an unambiguous way. [41, p. 699]

A few pages into the essay, Popper characteristically joins the issue without hesitation; he has come a long way from the embarrassment of the *Logik* experiment:

> [This leads to] the doctrine that the Heisenberg formulae $\Delta p_x \Delta q_x \geq \frac{h}{2\pi}$ etc. are about limits to human knowledge or to the precision of possible measurements on particles. This Copenhagen thesis I deny. As a realist I assert that the formula is about the lower limits on the scatter of particles ... The particles themselves possess sharp positions and, at the same time, sharp momenta. [40, p. 4-5]

The discussion following Popper's essay was spirited and focused almost entirely on Popper's proposed experiment, which in this essay is the same as the experiment proposed in the Postscript, i.e., particles with correlated momentum states passing through slits. Some participants doubted it could be conducted, others argued over what the results (one way or the other) might mean. Gino Tarozzi, one of the conference organizers, was strongly in favor of conducting the experiment, commenting that in his view the experiment would be an effective test of "Einstein locality versus Heisenberg's indeterminacy relations" [39, p. 30].

Popper's final contributions to the interpretation of quantum mechanics appeared in the pages of *Letters to Nature* in 1987 [42, 43]. Two physicists at Essex University, M. J. Collett and R. Loudon, had taken the position that Popper's 1982 Postscript experiment "does not in fact provide a test" of the Copenhagen Interpretation [44, 45]. They argued that the source (say positronium) must be assigned a finite uncertainty in both position and momentum; calculation of this "source uncertainty effect" showed, in their view, that experiment was not a valid test of the Copenhagen interpretation. Popper, now 85 years old, replied to the letter a few months later. He began by thanking Collett and Loudon for "opening up a discussion of my 1982 proposal ... an experiment based upon Einstein, Podolsky and Rosen and a radical simplification of another proposal by myself," a reference to the ill-fated *Logik* experiment [42]. Popper answered the physicists' arguments point for point, and corrected their impression that his experiment was a test of quantum mechanics itself:

> My experiment was never intended as a crucial experiment of quantum mechanics but only of its (subjectivist) Copenhagen interpretation (which they call "the standard interpretation"). [42]

Popper pointed out that there "exist several interpretations of the formalism" and provided a citation to one developed by Jon Dorling. (Indeed, work continues along the lines Popper advocated, i.e., a statistical interpretation of quantum mechanics and especially of the uncertainty relations. See, for example, [46], the abstract of which begins: "I attempt to develop further the statistical interpretation of quantum mechanics proposed by Einstein, developed by Popper, Ballentine, etc.") Collett and Loudon's reply to this letter followed immediately after Popper's letter; they simply do not agree [45]. Popper insisted in characteristic fashion on having the last word. In a short note in Letters to Nature a month later, he first corrected a rather obvious error in the printing of his previous letter [43]. Then he took one last jab at Collett and Loudon's position:

I would point out that in their original criticism they speak of a "fixed source", whereas in their new criticism they replace this by a "massive source." To my mind this means a change of the problem: they never explain why a (non-massive positronium) source cannot be "fixed." [43]

Popper died in 1994, and thereafter, one might expect, his Gedankenexperiment would draw no further notice. Indeed, that was the case for more than a decade after 1987. But in 1999, Popper returned to the pages of the physics journals, and in a most surprising way.

5 Shih and Kim, 1999, and aftermath

In 1999, University of Maryland physicists Yanhua Shih and Yoon-Ho Kim reported the results of a realization of Popper's experiment [13–15]. Their experimental setup did not use Popper's point particle source (such as a decay of positronium) – it used entangled photons produced by a laser and refracted by lenses through slits (Figure 2). This arrangement avoided the "Sudbery problem" of the inability to eliminate uncertainty effects in the initial position of a particle. Shih and Kim point out that

> . . . a point source is not a necessary requirement for Popper's experiment. What is needed is the position entanglement of a two-particle system, i.e., if the position of particle 1 is precisely known, the position of particle 2 is also 100% determined. [14, p. 466]

Their results, taken at face value, "show that there appears to be a violation of the uncertainty principle" [14, p. 463]. This would mean, from Popper's point of view, that the Copenhagen interpretation is in error.

But Shih and Kim do not take that position. Instead, they argue that it is impermissible to apply the uncertainty relations to each of the entangled-state photons separately. These photons are, in their view, represented by a "non-factorizeable two-dimensional wave packet" such that "$\Delta y \Delta p_y \geq \hbar$ is not applicable to either photon 1 or photon 2 individually." They conclude:

> Our experimental demonstration of Popper's thought experiment call (*sic*) our attention to the important message: the physics of an entangled two-particle system is inherently different from that of two individual particles. [14, p. 470]

It is of interest that among the physicists whose assistance is acknowledged are none other than Jean-Pierre Vigier and Augusto Garuccio, Popper's 1981 collaborators [14, p. 470].

Shih and Kim's paper generated a cloudburst of responses, comments, criticisms, and suggestions for further work [8–12, 16, 17, 47–53]. The positions taken vary from Asher Peres's ungracious reference to "the absurdity of Popper's result" [49, p. 23] to Geoffrey Hunter's affirmation that

> Popper and EPR made no error – they agreed with Bohr, Heisenberg and other proponents of the Copenhagen interpretation that quantum theory predicts an instantaneous action at a distance . . . Popper and EPR's crucial point is that if such actions at a distance are not in fact observed (as in the Shih-Kim experiment), then quantum theory must be an incomplete (only statistical) theory of the physical world . . . [8, p. 248]

Rainer Plaga suggested an improvement ("Extension Step 1") in the Shih-Kim experiment that addresses a "conceptual flaw" having to do with the role of the observer [11]. In Plaga's opinion

> . . . it is of great importance to actually perform Popper's experiment with "Extension Step 1" . . . Should an experimental realization of "Extension Step 1" show that no virtual diffraction occurs, the relation between "quantum mechanical state" and "observed reality" . . . would be put into doubt. [11, p. 471]

A. J. Short agreed with Shih and Kim that their results do not suggest a violation of the uncertainty principle, but for different theoretical reasons [47, 48], while Brazilian physicist G. Rigolin disagreed with Short and claimed to "invalidate" his analysis [50]. A Korean group suggested a realization of Popper's experiment using a "dual measurement scheme" to achieve a modern version of Heisenberg's microscope thought experiment [10]. From Spain comes theoretician Pedro Sancho's application of Feynman's path integral methods to the Shih-Kim results [12]. And from one of India's leading theoreticians, C. S. Unnikrishnan, no less than three papers (all published since 2000) on many aspects of EPR, Popper's experiment, and the Shih-Kim results [51–53]. In addition to this spate of papers, Popper's experiment is analyzed in a recent textbook by Alexander Afriat and Franco Selleri devoted entirely to the EPR paradox [21] (published just slightly before Shih and Kim announced their results in 1999). The authors, both long-time contributors in the field, point

(a)

(b)

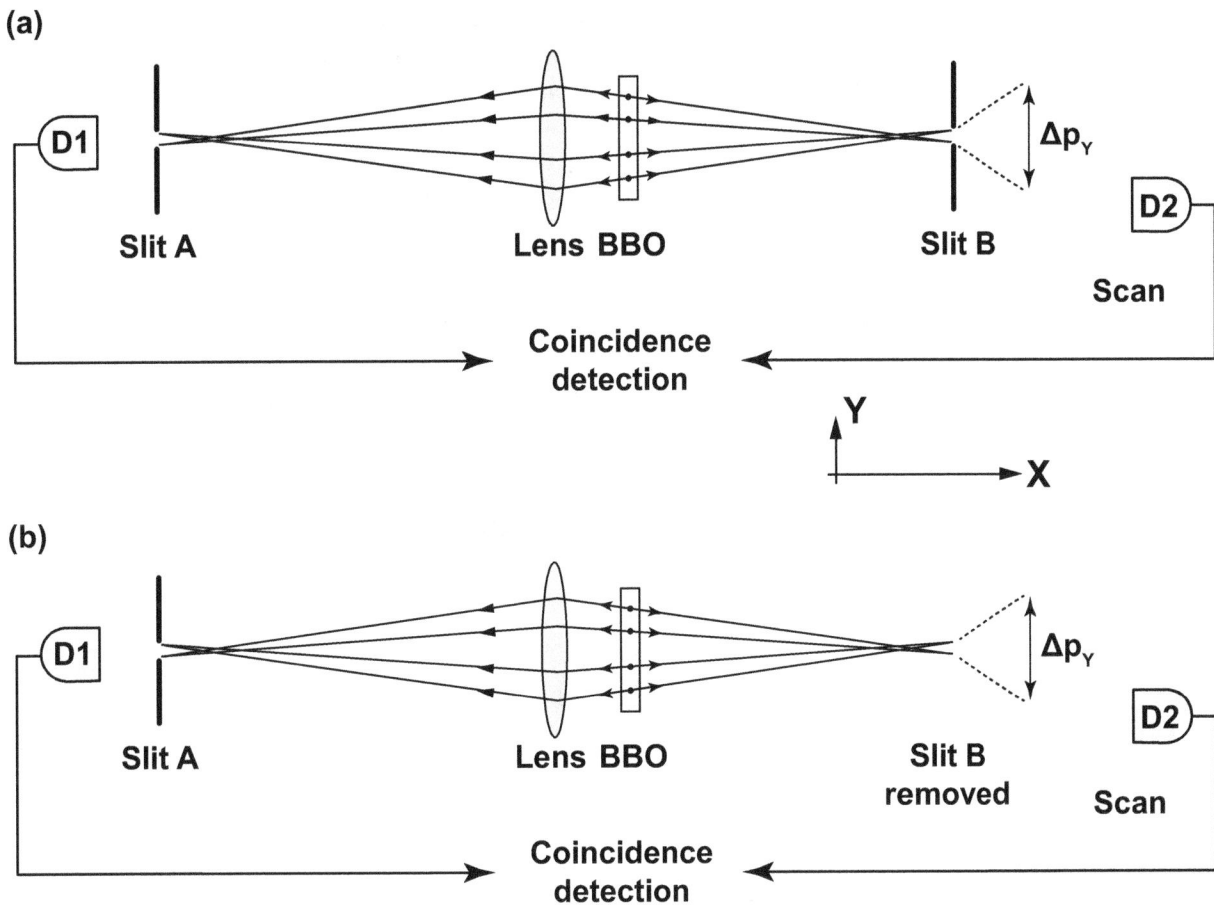

Figure 2: Modified version of Popper's experiment. An EPR photon pair is generated by spontaneous parametric down-conversion in a barium borate (BBO) crystal. A lens and a narrow slit A are placed in the path of photon 1 to provide the precise knowledge of its position on the y-axis and also determine the precise y-position of its twin, photon 2, on screen B due to a 'ghost image' effect. The distance between the lens and each of the slits is adjusted to 2f, where f is the focal length of the lens. Two detectors D1 and D2 are used to scan in the y-directions for coincidence counts. (a) Slits A and B are adjusted both very narrowly. (b) Slit A is kept very narrow and slit B is removed.

out the initial momentum problem in Popper's original proposal, but also note that the experiment could in principle be conducted by the use of collinear particles or photons [21, pp. 238-242].

6 Realism, uncertainty, knowledge

In this final section, I wish to explore the question: what nerve did Popper strike in his persistent challenge to the Copenhagen interpretation? What motivates physicists to devote time and resources to attempt Popper's experiment and battle over the meaning of the results?

To Popper, the interpretation of quantum mechanics represented a demarcational issue, not between science and non-science, but between the physical reality of things and human knowledge of those things. Popper read Bohr and Heisenberg as suggesting that "mere knowledge" of things had an observable physical effect, and to Popper this was nearing something like belief in the paranormal.

He could not accept mixing ontology and epistemology in this way [54]. Things exist, and we can come to know them by conjectures and refutations – but what we know (or think we know) and what is are not causally connected. *Ontology remains prior to epistemology.*

But more than realism was at stake. Popper tells us in the opening pages of Schism that his "strongest reason for my own opposition to the Copenhagen interpretation lies in its claim to finality and completeness" [29, pp. 5-6]. This statement calls to mind the 1935 dispute between Einstein, Podolsky and Rosen on the one hand, and Niels Bohr on the other. The Einstein paper concludes:

> While we have thus shown that the wave function does not provide a complete description of physical reality, we left open the question of whether or not such a description exists. [30, p. 780]

Bohr's responding paper certainly appears to lay a claim to finality and completeness:

Such an argumentation [EPR], however, would hardly seem suited to affect the soundness of quantum-mechanical description, which is based on a coherent formalism covering automatically any procedure of measurement like that indicated. [41, p. 696]

To Popper, this argument must have seemed like a contest between the critical, questioning posture he developed in *Logik Der Forschung*, and a position bordering on pure empiricism: we can know no more than what we see, and what we can know limits what exists. Philosopher Michael Redhead (who both met and corresponded with Popper) explains that in Popper's view, "probabilities in physics cannot, in general, be epistemic. How could human ignorance produce genuine physical effects?" [55, p. 172]. Popper designed and promoted "his experiment" not so much to prove quantum mechanics wrong, as to restore the "conjectures and refutations" attitude in quantum theory. To Popper, a physicist arguing that his theory is final and complete was anathema.

Perhaps most of the physicists who are once again engaged in a debate over 'Popper's experiment' are not overly concerned, as he was, with the nature of valid scientific inquiry. That is not terribly important. What is important is that Popper continues, from beyond the grave, to prick the "standard interpretation" of quantum mechanics and in so doing unsettles the field enough to generate renewed debate. One can hardly improve on Redhead's appreciation of Popper's contribution:

> Popper fought a lone battle against the Copenhagen interpretation at a time when anyone attempting to criticize orthodoxy was liable to be labeled at best an 'outsider' or at worst a crank. But Popper's carefully argued criticisms won the support of a number of admiring and influential physicists. He has done a great service to the philosophy of quantum mechanics by emphasizing the distinction between state preparation and measurement and trying to get a clearer understanding of the true significance of the uncertainty principle, but above all by spearheading the resistance to the dogmatic tranquilizing philosophy of the Copenhagenists. Because some detailed arguments are flawed, this does not mean that his overall influence has not been abundantly beneficial. [55, p. 176]

I believe that if Popper could once again weigh in on the debate – and weigh in he surely would! – he would demand above all else that the arguments continue, that more experiments be conducted, conjectures offered and refutations put forward. True, he would hope for a result

that showed human knowledge has nothing whatever to do with the position and momentum of particles. Yet even if the result seemed to show otherwise, he would continue to look for ways to preserve scientific realism and objectivity. What he could not abide was any hint of smugness, a complacency that our knowledge is complete and critical inquiry is at an end.

Acknowledgements

I am grateful to Hanspeter Fetz of the London School of Economics for pointing out Redhead's work to me.

References

[1] Popper KR. The Myth of the Framework. London: Routledge, 1994.

[2] O'Hear A. Karl Popper. London: Routledge, 1980.

[3] Stokes G. Popper: Philosophy, Politics and Scientific Method. Cambridge: Polity Press, 1998.

[4] Jarvie I, Pralong S. Popper's Open Society After 50 Years. London: Routledge, 1999.

[5] Hacohen MH. Karl Popper - The Formative Years, 1902-1945: Politics and Philosophy in Interwar Vienna. Cambridge: Cambridge University Press, 2000.

[6] Alai M, Tarozzi G. Karl Popper Philosopher of Science: Proceedings of the Conference, Cesena, 27-30/10/1994. Soveria Mannelli: Rubbettino, 2006.

[7] Amsterdamski S. Significance of Popper's Thought: Proceedings of the Conference Karl Popper: 1902-1994, March 10-12, 1995, Graduate School for Social Research, Warsaw. Amsterdam: Rodopi, 1996.

[8] Hunter G. Realism in the realized Popper's experiment. AIP Conference Proceedings 2002; 646 (1): 243-248. http://arxiv.org/abs/quant-ph/0507011

[9] Muller FA. Refutability revamped: how quantum mechanics saves the phenomena. Erkenntnis 2003; 58 (2): 189-211. http://philsci-archive.pitt.edu/1368/

[10] Nha H, Lee J-H, Chang J-S, An K. Atomic-position localization via dual measurement. Physical Review A 2002; 65 (3): 033827. http://arxiv.org/abs/quant-ph/0106053

[11] Plaga R. An extension of "Popper's experiment" can test interpretations of quantum mechanics. Foundations of Physics Letters 2000; 13 (5): 461-476. http://arxiv.org/abs/quant-ph/0010030

[12] Sancho P. Popper's experiment revisited. Foundations of Physics 2002; 32 (5): 789-805. http://dx.doi.org/10.1023/A:1016009127074

[13] Kim Y-H, Shih Y. Experimental realization of Popper's experiment: violation of the uncertainty principle? Foundations of Physics 1999; 29 (12): 1849-1861. http://arxiv.org/abs/quant-ph/9905039

[14] Shih Y, Kim Y-H. Experimental realization of Popper's experiment – violation of the uncertainty principle? Fortschritte der Physik 2000; 48 (5-7): 463-471. http://arxiv.org/abs/quant-ph/9905039

[15] Shih Y, Kim Y-H. Quantum entanglement: from Popper's experiment to quantum eraser. Optics Communications 2000; 179 (1-6): 357-369. http://dx.doi.org/10.1016/s0030-4018(99)00716-6

[16] Shih Y. Quantum imaging, quantum lithography and the uncertainty principle. European Physical Journal D 2003; 22 (3): 485-493. http://dx.doi.org/10.1140/epjd/e2003-00037-5

[17] Shih Y. Quantum imaging, quantum lithography and the uncertainty principle. Fortschritte der Physik 2003; 51 (4-5): 487-497. http://dx.doi.org/10.1002/prop.200310066

[18] Qureshi T. Popper's experiment, Copenhagen interpretation and nonlocality. International Journal of Quantum Information 2004; 2 (3): 407-418. http://arxiv.org/abs/quant-ph/0301123

[19] Qureshi T. Understanding Popper's experiment. American Journal of Physics 2005; 73 (6): 541-544. http://arxiv.org/abs/quant-ph/0405057

[20] Qureshi T. Analysis of Popper's experiment and its realization. Progress of Theoretical Physics 2012; 127 (4): 645-656. http://arxiv.org/abs/quant-ph/0505158

[21] Afriat A, Selleri F. The Einstein, Podolsky, and Rosen Paradox: In Atomic, Nuclear, and Particle Physics. New York: Plenum Press, 1999.

[22] Popper KR. Unended Quest: An Intellectual Autobiography. New York: Routledge, 1982.

[23] Popper KR. Logik der Forschung. Wien: Springer, 1934.

[24] Popper KR. The Logic of Scientific Discovery. New York: Basic Books, 1959.

[25] Heisenberg W. The Physical Principles of the Quantum Theory. New York: Dover, 1949 (original edition published in 1930).

[26] Kuhn TS. The Essential Tension: Selected Studies in Scientific Tradition and Change. Chicago: Chicago University Press, 1977.

[27] Popper KR. Postscript to the Logic of Scientific Discovery. Vol. 1: Realism and the Aim of Science. Totowa, New Jersey: Rowman and Littlefield, 1983.

[28] Popper KR. Postscript to the Logic of Scientific Discovery. Vol. 2: The Open Universe: An Argument for Indeterminism. Totowa, New Jersey: Rowman and Littlefield, 1982.

[29] Popper KR. Postscript to the Logic of Scientific Discovery. Vol. 3: Quantum Theory and the Schism in Physics. Totowa, New Jersey: Rowman and Littlefield, 1982.

[30] Einstein A, Podolsky B, Rosen N. Can quantum-mechanical description of physical reality be considered complete? Physical Review 1935; 47 (10): 777-780. http://dx.doi.org/10.1103/PhysRev.47.777

[31] Popper KR. Conjectures and Refutations: The Growth of Scientific Knowledge. New York: Routledge, 1963.

[32] Suárez M. Propensities and pragmatism, 2011. http://philsci-archive.pitt.edu/8957/

[33] Garuccio A, Popper KR, Vigier J-P. Possible direct physical detection of de Broglie waves. Physics Letters A 1981; 86 (8): 397-400. http://dx.doi.org/10.1016/0375-9601(81)90346-7

[34] Costa De Beauregard O. Disagreement with Garuccio, Popper and Vigier. Physics Letters A 1982; 89 (4): 171-172. http://dx.doi.org/10.1016/0375-9601(82)90200-6

[35] Mandel L. Tests of quantum mechanics based on interference of photons. Physics Letters A 1982; 89 (7): 325-326. http://dx.doi.org/10.1016/0375-9601(82)90183-9

[36] Garuccio A, Rapisarda V, Vigier J-P. New experimental set-up for the detection of de Broglie waves. Physics Letters A 1982; 90 (1-2): 17-19. http://dx.doi.org/10.1016/0375-9601(82)90038-X

[37] Sudbery A. Popper's variant of the EPR experiment does not test the Copenhagen interpretation. Philosophy of Science 1985; 52 (3): 470-476.

[38] Angelidis T. On some implications of the local theory Th(g) and of Popper's experiment. In: Gravitation and Cosmology: From the Hubble Radius to the Planck Scale, Amoroso R, Hunter G, Kafatos M, Vigier J-P (editors), Springer Netherlands, 2002, pp.525-536. http://dx.doi.org/10.1007/0-306-48052-2_56

[39] Tarozzi G, van der Merwe A. Open Questions in Quantum Physics: Invited Papers on the Foundations of Microphysics. Boston: D. Reidel, 1985.

[40] Popper KR. Realism in quantum mechanics and a new version of the EPR experiment. In: Open Questions in Quantum Physics: Invited Papers on the Foundations of Microphysics, Tarozzi G, van der Merwe A (editors), Boston: D. Reidel, 1985, pp.3-25.

[41] Bohr N. Can quantum-mechanical description of physical reality be considered complete? Physical Review 1935; 48 (8): 696-702. http://dx.doi.org/10.1103/PhysRev.48.696

[42] Popper KR. Popper versus Copenhagen. Nature 1987; 328 (6132): 675. http://dx.doi.org/10.1038/328675a0

[43] Popper KR. Correction needed. Nature 1987; 329 (6135): 112. http://dx.doi.org/10.1038/329112b0

[44] Collett MJ, Loudon R. Analysis of a proposed crucial test of quantum mechanics. Nature 1987; 326 (6114): 671-672. http://dx.doi.org/10.1038/326671a0

[45] Collett MJ, Loudon R. Popper versus Copenhagen. Nature 1987; 328 (6132): 675-676. http://dx.doi.org/10.1038/328675b0

[46] Shirai H. Reinterpretation of quantum mechanics based on the statistical interpretation. Foundations of Physics 1998; 28 (11): 1633-1662. http://dx.doi.org/10.1023/A:1018841625620

[47] Short AJ. Popper's experiment and conditional uncertainty relations. Foundations of Physics Letters 2001; 14 (3): 275-284. http://arxiv.org/abs/quant-ph/0005063

[48] Short AJ. Momentum changes due to quantum localization. Fortschritte der Physik 2003; 51 (4-5): 498-503. http://arxiv.org/abs/quant-ph/0112121

[49] Peres A. Karl Popper and the Copenhagen interpretation. Studies In History and Philosophy of Science Part B 1999; 33 (1): 23-34. http://arxiv.org/abs/quant-ph/9910078

[50] Rigolin GG. Uncertainty Relations for Entangled States. Foundations of Physics Letters 2002; 15 (3): 293-298. http://arxiv.org/abs/quant-ph/0008100

[51] Unnikrishnan CS. Popper's experiment, uncertainty principle, signal locality and momentum conservation. Foundations of Physics Letters 2000; 13 (2): 197-200. http://dx.doi.org/10.1023/A:1007839718507

[52] Unnikrishnan CS. Is the quantum mechanical description of physical reality complete? Proposed resolution of the EPR puzzle. Foundations of Physics Letters 2002; 15 (1): 1-25. http://dx.doi.org/10.1023/A:1015823125892

[53] Unnikrishnan CS. Proof of absence of spooky action at a distance in quantum correlations. Pramana 2002; 59 (2): 295-301. http://www.ias.ac.in/pramana/aug2002/qt24.htm

[54] Popper KR. Objective Knowledge. London: Oxford University Press, 1972.

[55] Redhead M. Popper and the quantum theory. In: Karl Popper: Philosophy and Problems, O'Hear A (editor), Cambridge: Cambridge University Press, 1995, pp.163-176. http://dx.doi.org/10.1017/S1358246100005488

2

Exploring Quantum, Classical and Semi-Classical Chaos in the Stadium Billiard

Chris C. King

Department of Mathematics, University of Auckland, Auckland, New Zealand. E-mail: dhushara@gmail.com

Editors: *James F. Glazebrook & Danko Georgiev*

his paper explores quantum and classical chaos in the stadium billiard using Matlab simulations to investigate the behavior of wave functions in the stadium and the corresponding classical orbits believed to underlie wave function scarring. The simulations use three complementary methods. The quantum wave functions are modeled using a cellular automaton simulating a Hamiltonian wave function with discrete (square pixel) boundary conditions approaching the stadium in the classical limit. The classical orbits are computed by solving the reflection equations at the classical boundary thus giving direct insights into the wave functions and eigenstates of the quantum stadium. Finally, a simplified semi-classical algorithm is developed to show the comparison between this and the quantum wave function method. Quanta 2014; 3: 16–31.

1 Introduction

Estimation of the quantum wave functions in chaotic systems, from the nucleus of atoms from helium to uranium to the eigenfunctions of a wave-particle in the Bunimovich stadium [1, 2] (the classical Coliseum stadium shape consisting of a rectangle or square capped off with semi-circular discs) proved to be initially intractable and even in the post-modern era of revived semi-classical methods computationally complex to the point of being counter-intuitive. Part of the purpose of this paper is to simplify understanding these quantum solutions so their dynamics can be more directly appreciated and understood. A full quantum solution to a quantum chaotic wave function, given its boundary constraints proved illusive for the founders of quantum theory in the case of the helium nucleus. In 1970 Gutzwiller developed a semi-classical method using a trace formula, which integrates the Green's function of wave propagation over all coordinates so that it can be expressed in terms of the repelling orbits hidden within the classically chaotic version of the quantum system [3, 4].

The trace formula for a particle of mass m and momentum $p = \sqrt{2mE}$ inside a box with arbitrarily shaped walls is

$$
\begin{aligned}
g(E) &= \sum_n \frac{1}{E - E_n} \\
&= \bar{g}(E) + \frac{1}{\iota\hbar} \sum_\gamma \sum_{k=1}^\infty \frac{ml_\gamma}{p} \frac{e^{\iota kpl_\gamma/\hbar - \iota\pi k\nu_\gamma/2}}{\left|2 - \mathrm{Tr}\left(M_\gamma\right)^k\right|^{\frac{1}{2}}} \quad (1)
\end{aligned}
$$

where $\bar{g}(e)$ is a smooth function giving the mean density of states, summed over all orbits γ of length l_γ with ν_γ the phase shift counting the focal points and twice the number of reflections off the walls and over all retracings k of these orbits [5]. The stability matrix M_γ records the sen-

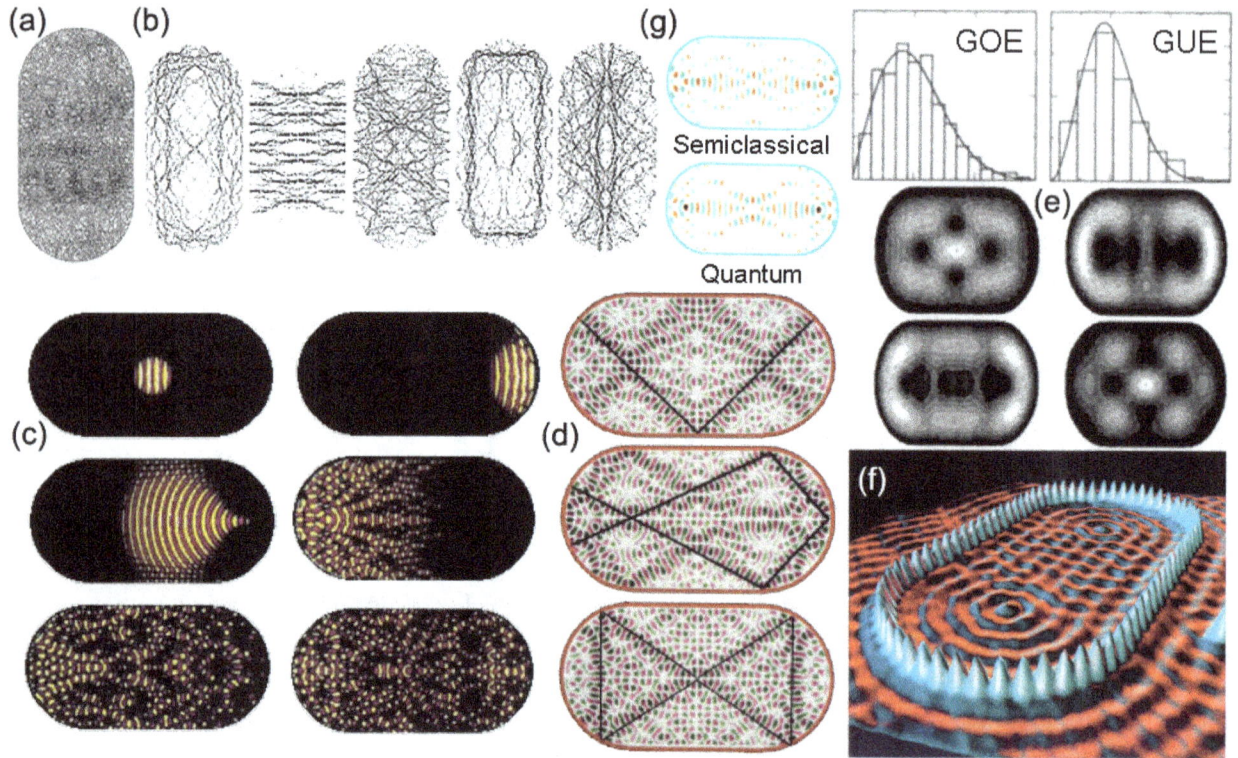

Figure 1: *Quantum chaos: (a) the stadium is densely filled with repelling periodic orbits [9], three of which are shown in black in (d). The quantum solution of the stadium potential well (b) [4] and (d) [5] shows scarring of the wave function along these repelling orbits, thus repressing the classical chaos, through probabilities clumping on the repelling orbits. A semi-classical simulation (c) shows why this is so. The quantum solution is scarred on precisely these orbits (d). This causes it to coincide with the eigenfunctions of the repelling periodic orbits, just as the orbital waves of an atom constructively interfere with themselves, in completing an orbit to form a standing wave, like that of a plucked string. Over time, although the behavior may be transiently chaotic, the quantum system eventually settles into a periodic solution or a linear combination of these. Experimental realizations, such as a magnetized electron in a quantum dot (e) [10] [Gaussian orthogonal ensemble (GOE) and Gaussian unitary ensemble (GUE) statistics], or the scanning tunneling view of an electron on a copper sheet bounded by a stadium of carefully-placed iron atoms (f) [11], confirm the general picture, although quantum tunneling in the latter case leaked the wave function outside too much to demonstrate proper scarring. The semi-classical approach matches closely the quantum calculation (g).*

sitivity of the orbit to changes in initial conditions. From the left hand side of the equation it can be seen that if the sum converges (this is generally possible only with some difficulty and depends on reordering the sum terms), the eigenvalues will appear as singularities in the sum. The general similarity of the trace formula to the Riemann zeta function and this function's Gaussian unitary ensemble (GUE) type statistics shared by quantum chaotic systems (see panel (e) in Figure 1) has led to extensive attempts to solve quantum chaotic systems through the paradigm of the zeta zeros (for a detailed discussion on the quantum chaos connection with the Riemann zeta function see section 7).

Both quantum and semi-classical approaches have been used to model systems such as the stadium billiard. In the quantum approach a wave packet is propagated throughout the potential well using a time dependent Schrödinger equation via Fourier transform techniques. This wave propagation approach is discretely modeled in the cellular automaton described in section 2. Alternatively the semi-classical approach uses the least repelling periodic orbits in the classical stadium to generate amplitudes based on the path lengths of all classical trajectories connecting a given location to another in the region.

Heller and Tomsovic [5–7] among others have used the semi-classical orbit-based approach to develop the theory of scarring of the quantum wave [8]. The authors comment:

> Accurate excited-state eigenvalues have been computed from knowledge of relatively few periodic orbits. However, deep questions about the convergence of the trace formula and its modifications remain unanswered. [5, p.40]

A particularly complete expose of the methods involved is provided in their *Physical Review E* paper [7].

2 The Quantum Cellular Automaton

To get a more direct picture of the process of wave spreading and wave function scarring [see panels (a-d) in Figure 2], this simulation depends on a simple wave propagation formula for a cellular automaton which preserves the essential nature of Hamiltonian dynamics and the Green's function governing transmission of wave amplitudes from point to point. Somewhat paradoxically, although the local rules are defined on a rectangular grind with only N, S, E and W neighbors, the process well models wave transmission at a variety of angles and under situations where the curved surfaces of the potential well boundaries refocus the wave fronts continuously in various ways. Again this is possible although the local rules at the boundary provide for exclusively horizontal and vertical reflection.

In effect, the wave period and implicit frequency of the generating wave gives the energy and momentum of the transmitted wave-particle and the individual pixels on the boundary provide a simulation of the smaller quantum features at the atomic level occurring in any real physical implementation of the quantum stadium, as exemplified in even more rudimentary discrete form, in the atomic sized stadium shown in panel (f) of Figure 1.

The cellular automation iterates the wave function according to the rule:

$$\psi_C(t + \Delta t) = \frac{1}{2}\left(\psi_E(t) + \psi_W(t) + \psi_N(t) + \psi_S(t) - 2\psi_C(t)\right) \tag{2}$$

As in panel (e) of Figure 2, black positions whose four neighbors are white are all iterated in one step, followed by the white positions iterated in terms of their newly updated black neighbors. This provides a discrete iteration modeling a Hamiltonian process resembling action-angle variables with an angle shift of $\frac{\pi}{2}$. In keeping with the quantized nature of the system, the boundary of the region consists of a discrete series of vertical and horizontal edges corresponding to a discrete region of square cells approaching the curved stadium shape in the infinite limit. Vertical and horizontal edges have a *reflecting rule* according to which the relevant internal boundary cells are doubled and the external out of region boundary cells are omitted. The rule is illustrated by the following edge and corner situations:

$$\text{W edge: } \psi_C(t + \Delta t) = \frac{1}{2}\left(2\psi_E(t) + \psi_N(t) + \psi_S(t) - 2\psi_C(t)\right)$$

$$\text{SW corner: } \psi_C(t + \Delta t) = \frac{1}{2}\left(2\psi_E(t) + 2\psi_N(t) - 2\psi_C(t)\right)$$

The process is also capable of handling applied forces by integrating their effect:

$$\Delta a_i(t + \Delta t) = \frac{1}{m_i}\left(F_i(t + \Delta t) - F_i(t)\right)$$
$$v_i(t + \Delta t) = v_i(t) + a_i(t)\Delta t \tag{3}$$
$$\psi_i(t + \Delta t) = \psi_i(t) + v_i(t + \Delta t)\Delta t$$

The simulation uses a variety of generating functions to set up excitations whose period corresponds to a fraction of the orbit length of a classical repelling orbit in the stadium. Broadly they are of the type:

$$\psi(x, y) = Ae^{-k_1(x^2+y^2)}\cos 2\pi k_2(ux \pm vy), \ u^2 + v^2 = 1 \tag{4}$$

combining one or more plane cosine waves of appropriate period lying along an orbit path clipped by being multiplied by a circular Gaussian, although circular and unidirectional traveling wave packets can also be established by clipping a sector of an emerging circular wave.

Six of the scars resulting from superimposing successive probabilities measured as the square of the amplitude are shown in panels (b-g) of Figure 3. The predominant repelling orbits clearly show as probability peaks coinciding with the corresponding repelling orbits in just the manner of the semi-classical simulations in Figure 3 (see also [5]). In panels (a*-h*) are shown the corresponding generating functions for each of the images (a-h) in Figure 3.

The images in Figure 3 were generated using the first 3300 iterations of the generating function, however when the scars (b) and (f) were iterated on from 3300 to 6600 iterations the scar remained stable, as shown in Figure 4, verifying that it is not just a transient feature of the expanding generating functions.

These patterns of scarring can also be compared with other potential wells such as a rectangular one which leads to ordered wave functions and the lemon shaped potential well, which provides an example of so-called *soft chaos* in which ordered quasi-periodic solutions coexist with chaotic ones. In Figure 5 the generating functions of scars (a) and (b) in Figure 3 are applied again for 3300 iterations in the rectangular well resulting in ordered wave functions (a, b), and a vertical generating function in the lemon well produces an ordered wave function corresponding to the quasi-periodic stable orbit superimposed below (c, d).

Significantly the cellular automaton is able to well handle oblique generating functions with periods not reducible directly to an integer number of iterations. In Figure 2 are shown both the probability superposition scars and the amplitude superposition scars, confirming that even when an oblique wave with an irrational period is modeled, the cellular automaton is able to successfully superimpose amplitudes in phase over thousands of iterations.

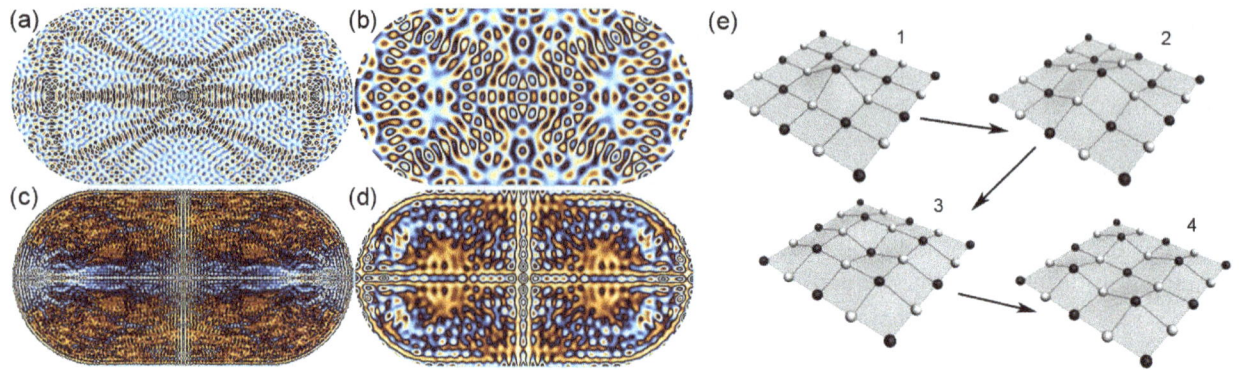

Figure 2: *Evidence of scarring: Superposition of the first 3300 iterations of a wave function initiated by two different generating functions (a and c in Figure 3) with wavelength chosen to be a fraction of the associated classical repelling orbit. Panels (a) and (b) show superposition of phase amplitudes, whereas (c) and (d) the corresponding superimposed probability distributions of the squared amplitudes, showing evident scarring. Generally the probability distributions give better resolution of the scarring. (e) The wave function cellular automaton iterates in two stages. First in the transition 1 → 2 white positions move using black neighbour values. Then in the transition 2 → 3 black positions move. Finally in the transition 3 → 4 white positions move again. This enables a discrete simulation of a Hamiltonian process. Supplement 1: Quicktime movie of the wave function cellular automaton operating.*

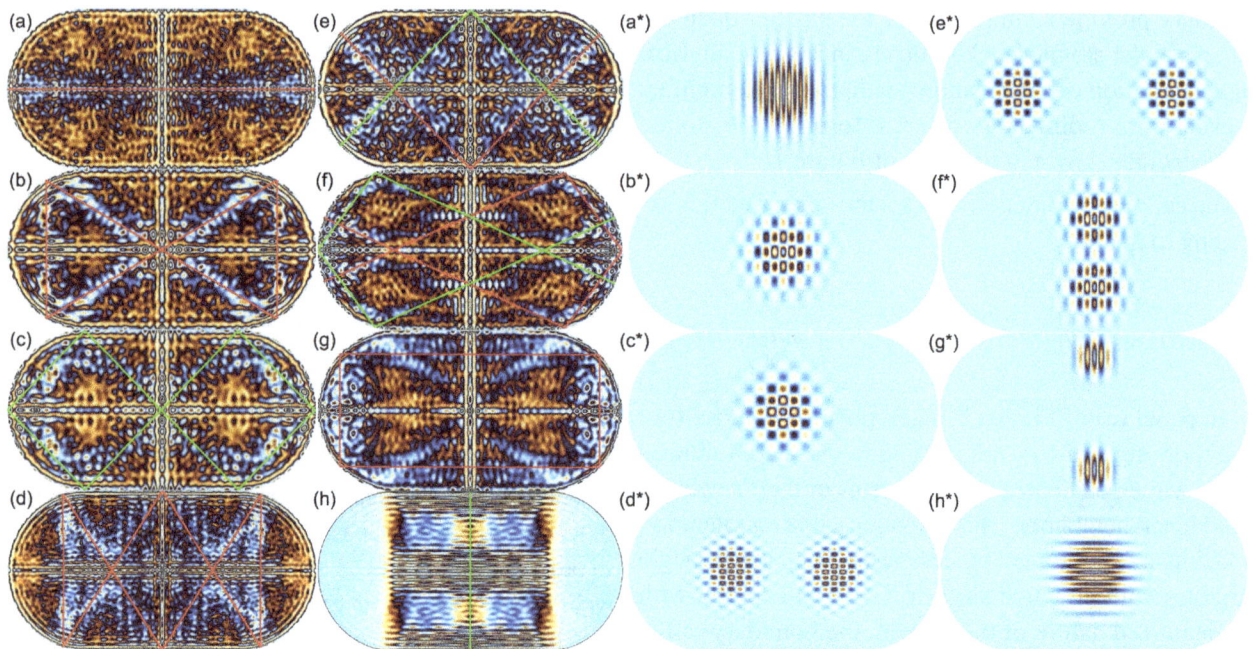

Figure 3: *Scars of the wave stadium function illustrated corresponding to 3,300 iterations of 6 distinct repelling periodic orbits in the classical system. Each of the wave functions in panels (a-h) is iterated from the corresponding generating function in panels (a*-h*), whose wavelength and direction is chosen to correspond to a fraction of the orbit length. The scars are formed by superimposing the squared wave functions at each complete revolution of the phase angle (corresponding to 11 steps of iteration because of the choice of stadium size and initial generator). The rectangular orbit scar in panel (g) also shows resonance with the bow tie in panel (b). Its orbit-length-dividing period of 1.2 relative to panel (a) is close to the bow tie's 1.29 and it closely shares the vertical section of its orbit with panel (b). (h) A vertical wave generator corresponding to the neutral vertical orbit, produces an ordered wave function. (a*-h*) The generating functions for the scars in panels (a-h) consisted of radial Gaussian wave packets of plane cosine waves aligned along the classical repelling orbits shown with wave periods a fixed fraction of the orbit length.*

The cellular automaton can also be used to model the simulations of traveling wave packets illustrated from other research. In panel (e) of Figure 5 is shown a series of images of iterations of a right-moving wave packet giving a similar evolution profile to the one generated by Heller and Tomsovic in panel (c) of Figure 1. The cellular

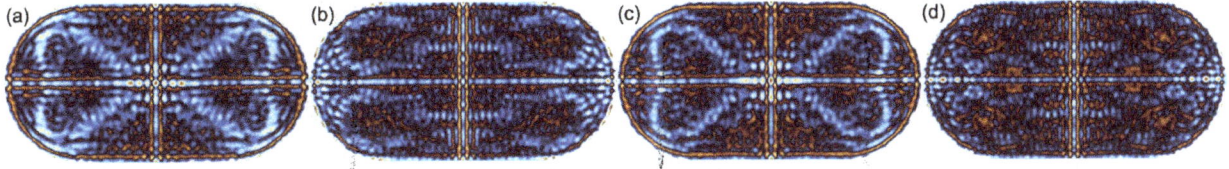

Figure 4: *The long-term stability of the scars in the cellular automaton simulation is confirmed when the cases (b) and (f) in Figure 3 are run for a further 3300 iterations and the initial transient states of the generator are thus omitted entirely. (a, b) Iterations 1-3300 superimposed. (c, d) Iterations 3300-6600 superimposed. In these figures the absolute wave amplitudes have been added rather than the squared amplitudes to illustrate various methods of scar highlighting.*

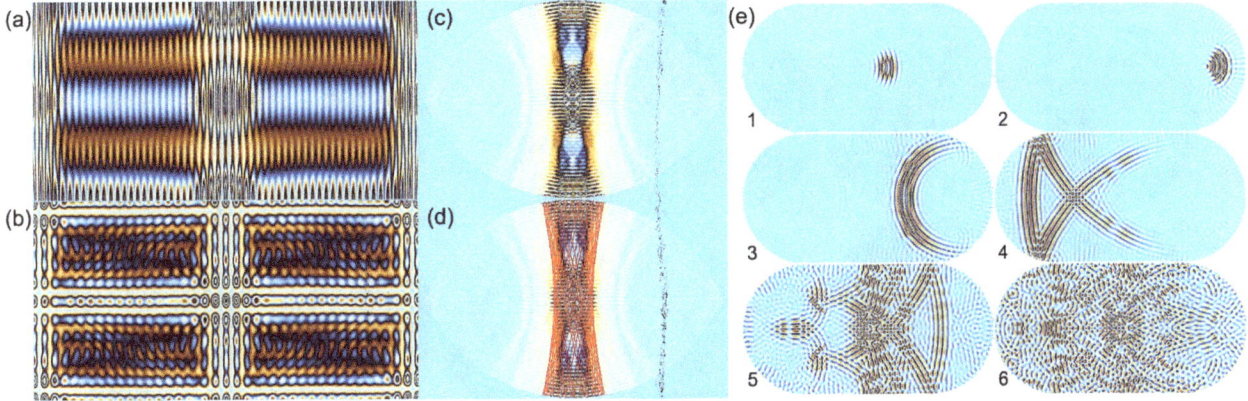

Figure 5: *(a, b) The corresponding non-chaotic wave functions for generators a*, b* in Figure 3 in a rectangular potential well do not show scarring, but form regular ordered solutions. (c, d) Non-chaotic wave function corresponding to a quasi-periodic orbit in a lemon-shaped potential well with the quasi-periodic orbit superimposed. (e) The cellular automaton also well models the semi-classical view of the spreading of a traveling wave packet. A small traveling wavelet, generated by clipping sector of a circular ripple with a Gaussian envelope, bounces back and forth, ultimately forming a periodic wave pattern, due to internal resonances generated by its characteristic frequency and wave spreading. Wave spreading compensates for the exponential spreading of the classical repelling orbit and the wave function can form standing wave constructive interference when its energy and consequently its frequency and period corresponds to a fraction of the length of the periodic orbit. Supplement 2: Avi movie of the spreading of the traveling wave packet.*

automaton can also well-model other wave experiments in physics such as the double-slit experiment illustrated in panel (a) of Figure 7.

It is also possible to compare the effects of different boundaries directly by choosing a circular wave generator of period a fraction of the overall dimensions (both dimensions for stadium and rectangle and the vertical dimension for the lemon). In panels (c-e) of Figure 7 the three boundaries are compared and the scarring characteristic of the stadium shows up clearly and quite elegantly suggestive of several superimposed orbit solutions.

The cellular automaton can also be easily interrogated to produce an autocorrelation function

$$C(\psi_0, \psi_t) = \sum_{i=1}^{myh} \sum_{j=1}^{myw} \psi_0(i, j)\psi_t(i, j) \qquad (5)$$

by multiplying the states of the generating wave function $\psi_0(x, y)$ by the propagated wave function $\psi_t(x, y)$ and integrating over all cells in the array. In panels

(f, g) of Figure 7 are shown the autocorrelation function for the bow tie generator up to 3300 iterations consisting of about 7 sweeps across the length of the stadium and the fast Fourier transform, giving back the energy/frequency/period of the generated wave function.

3 Orbits in the Classical Stadium

To complement the wave cellular automaton and draw superimposed classical orbits, a ray drawing algorithm was developed for the same matrix of elements used in the wave cellular automaton algorithm. This also enables portrayal of several features of the associated classical chaotic system.

In panels (h-k) of Figure 7 are shown how sensitive dependence on initial conditions (in this case a small angle difference in the initial orbit leads on multiple reflection in the curved sections of the stadium to exponentially increasing angular divergences between the orbits, even-

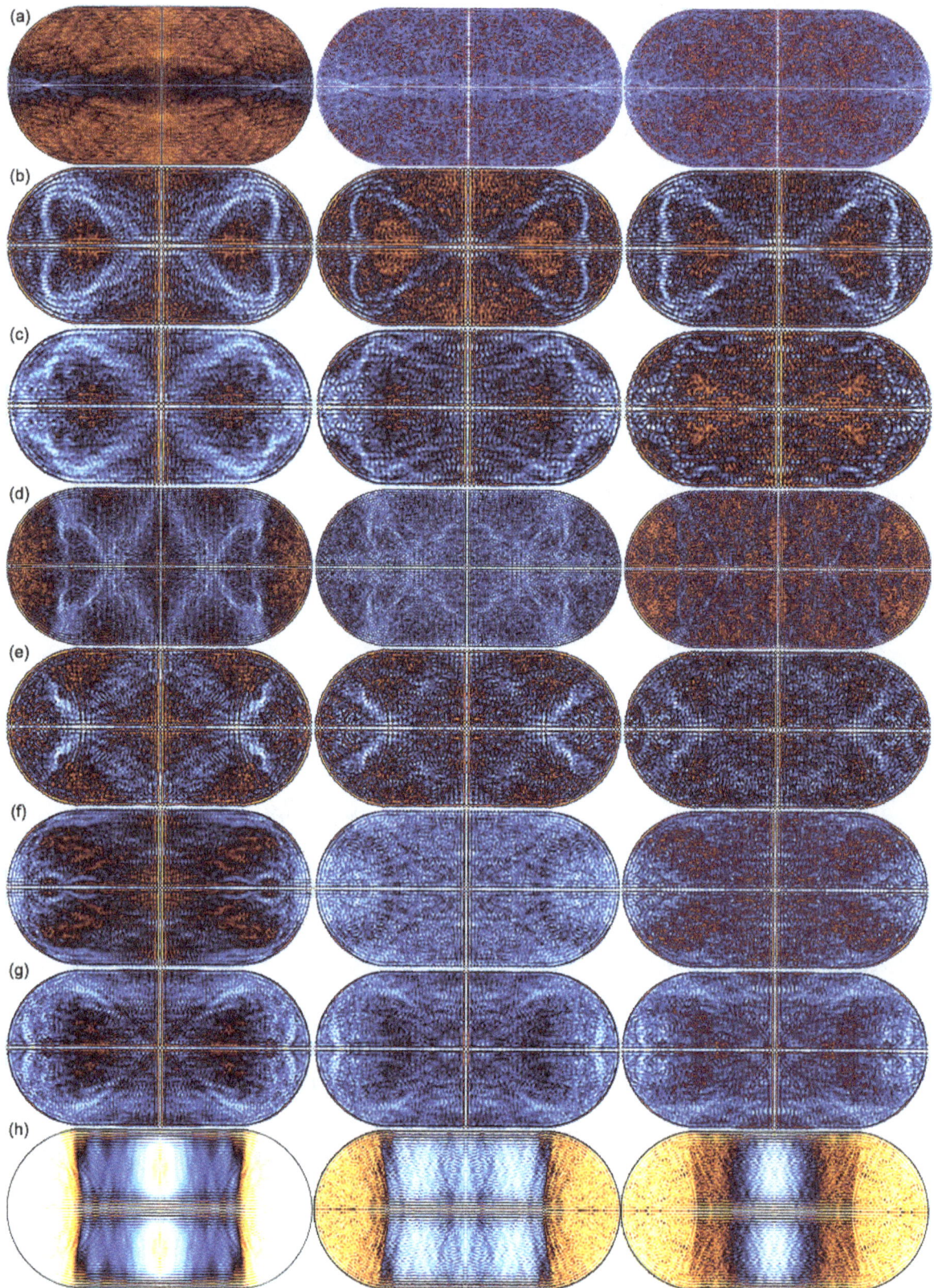

Figure 6: *A more detailed long-term evolution of scarring of a wave packet with twice the frequency and half the Gaussian variance for the same sequence of orbits as in Figure 3, superimposing stages: 0-3300, 3300-6600, 6600-9900, 9900-13200 for (f) and (g). The long-term evolution suggests that the systems are converging to a linear combination of the principal orbits due to the initial waveform exciting more than one orbital solution. Hence (b) and (e), being two of the strongest orbits, retain their excitation almost unchanged through 9900 iterations, while for example (g) already shows features of (b) in its initial excitations.*

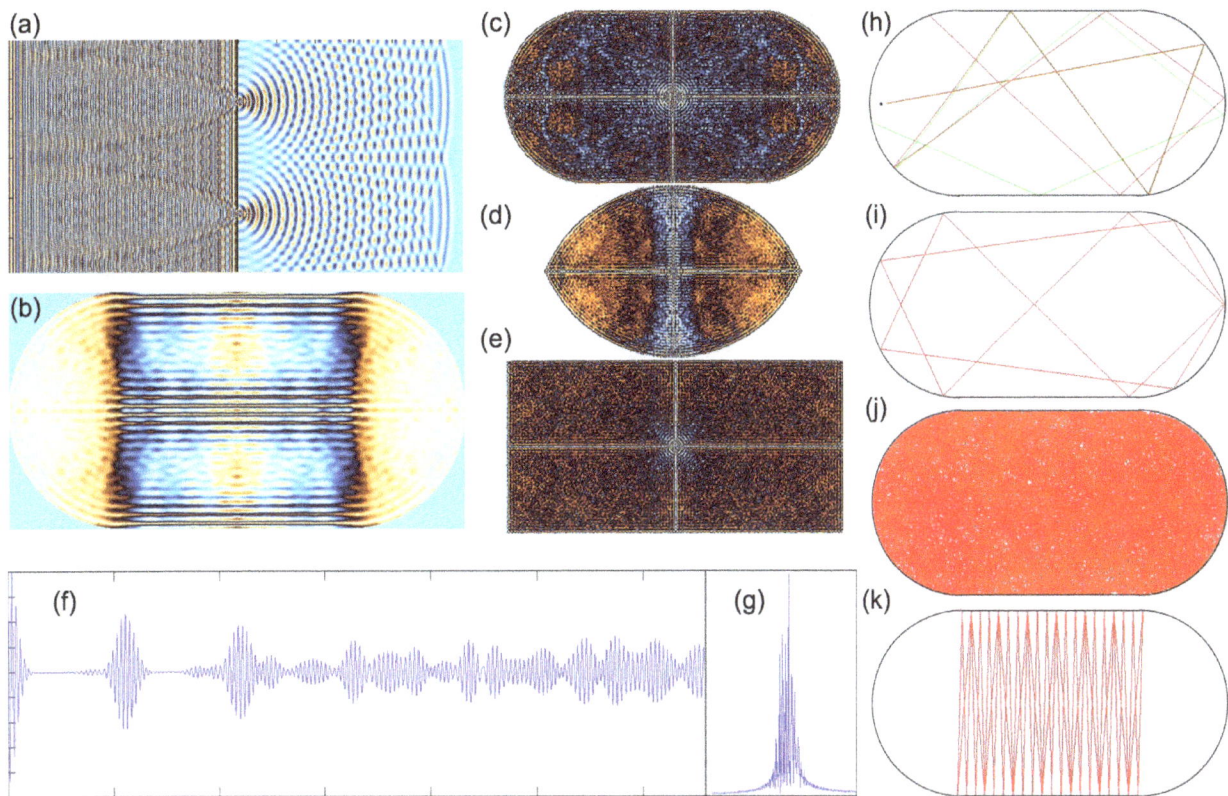

Figure 7: *(a) A modification of the rectangular potential well consists of a discrete model of a double slit interference experiment, showing the cellular automaton well models circular wave fronts emerging from a double slit and forming constructive and destructive interference fringes. (b) A vertical wave generator produces an ordered wave function on the stadium. (c-e) Probability distributions for a circular wave packet of period a fraction of the overall dimensions for 3300 iterations for the stadium (c), lemon (d) and rectangle (e) display the differences between chaotic and ordered wave forms. (f) Autocorrelation function of the generated* bow tie *scar in panel (b) of Figure 3. (g) Fast Fourier transform giving frequency domain. (h-k) Classical orbits in the stadium display the classic features of chaotic billiards. (h) Sensitive dependence on initial conditions illustrated for two closely related initial conditions (red and green) show exponentiating divergence with each reflection. Chaos also causes the space to be densely permeated with repelling periodic orbits, two of which are illustrated in (i) and (k). Because they are repelling, neighboring orbits are thrown further away, rather than being attracted into a stable periodic orbit as in an ordered system. Consequently almost all orbits are recurrent and densely fill phase space (j).*

tually ending in essentially a random relationship with one another. Letting an orbit run for a large number of reflections also shows that most orbits are not periodic, but rather densely permeate the phase space of the stadium. Although the repelling periodic points are rare, they are dense which means there are periodic orbits arbitrarily close to any trajectory. Most orbits are recurrent so that they eventually come arbitrarily close to a given trajectory, as shown in panel (a) of Figure 9.

To more closely model the classical chaotic stadium, a second version of the ray-drawing algorithm was developed which rather than operating on a matrix of pixels, solves the reflection equations at the boundary on a classical continuous stadium. In turn this algorithm was refined to search for and display the closed orbits through a given point for reflection periods up to 20. Because these are testing numerically for a closed orbit by close approx-

imation, they are really detecting close recurrence in a finite number of reflections, rather than an exact periodic solution.

As a dramatic increase is made in the number of reflections, such an orbit will eventually break stability, as the periodic solution is repelling, and it may over time recur closely to other repelling orbits. In panels (a-c) of Figure 9, the computed solution to the *bow tie* 4-period in panel (b) of Figure 3 is iterated for 200 steps in each direction, showing noticeable recurrences to another period 4 skewed bow tie and a rotationally symmetric period 8 orbit.

In panels (a-h) of Figure 8 sample orbits displaying reflective symmetry in the axes, rotational symmetry by half a revolution and asymmetric orbits are shown illustrating the variety of closed orbits and their symmetries.

The algorithm was set to compute all close recurrences

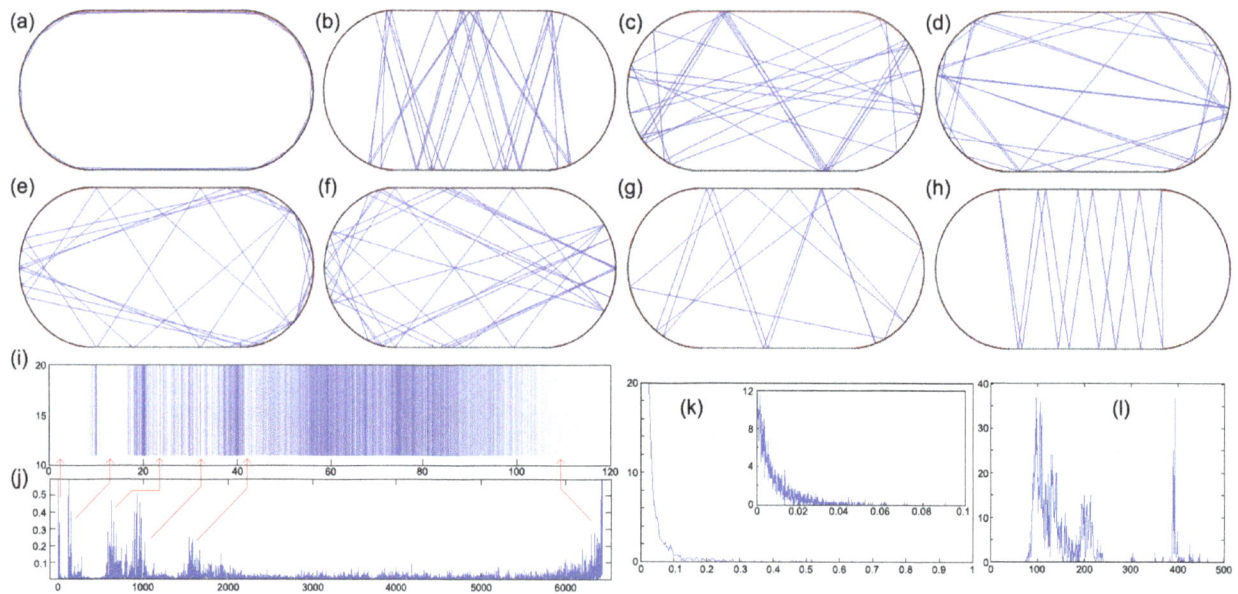

Figure 8: *A collection of classical 16-step close recurrences approximating periodic repelling orbits, some with reverse orbit symmetrization to 32 steps. Repelling orbits symmetric by horizontal reflection (a, b), symmetric by a rotation by 180 degrees (c, d), symmetric by vertical reflection (e, f), and asymmetric (g, h). Supplements 3-5: Quicktime movies of symmetric period 16 orbits passing through (0,0), (0,1) or (2,0). (i) Distribution of lengths of 6444 close orbit recurrences of up to 20 reflections generated from positions (0,0), (2,0) and (0,1), having rotational and reflective symmetry shaded to give relative densities. (j) Corresponding distribution of orbit length differences of the ranked length array. The right hand peak is an artifact caused by missing orbits with higher reflection numbers. (k) Distribution of length differences by frequency shows a descending law in which larger gaps are rarer. This pattern is repeated in the inset when the outliers are removed by choosing only the values from 2000 to 5000. This distribution differs markedly from the distribution of primes (see panel (e) of Figure 11) and is unlikely to be an artifact of the orbit selection because it is clear there are sequences of orbits with orbit lengths converging in the limit, such as those exemplified in (a), which tend asymptotically to the circumference of the stadium. According to the trace formula given by Equation 1 the phase shifts and sensitivity of the orbits also have to be included before a GUE distribution is arrived at, however, as shown in (l) a roughly GUE distribution, with prominent outliers corresponding to the shortest groups of orbits, does arise when the inverses of the orbit lengths are taken as a measure of energy.*

back to the starting trajectory for 2 up to 20 reflections. The algorithm initially uses 80000 steps of angle through a sample right angle of directions from a given point and tests for local minima of a metric combining the distance from the initial vertex and divergence of angular direction. The local minima are then scanned down a further four orders of magnitude to find as close a recurrence to a closed orbit as possible given the numerical resolution of the floating point process. These solutions are then plotted as closed orbits and their lengths and sensitivity to initial conditions are determined as an initial step towards calculations based on the trace formula given by Equation 1.

The distribution of the lengths was also investigated as a comparison with the prime numbers whose separations increase and Riemann zeta zero distributions whose separations decrease with increasing imaginary value and whose step differences have the statistics of a GUE. Panels (i-l) of Figure 8 highlight the differences between the distribution of orbits and these two.

The orbit length profile shows some regions that are sparsely filled, and others which are more densely filled. Certain small finite values of length, such as the perimeter distance of the stadium and its multiples have an infinite number of orbits of length tending to these values. The statistic of gap distances shows a descending power law trend, which is repeated internally, for example when the orbits from 3000 to 5000 are chosen, which, from panels (i, j) of Figure 8, is a relatively consistent central section of the orbit lengths not suffering from anomalies of small value divergences or missing large orbits due to cutting off the search at 20 reflections. The perimeter-angle coordinate system described in panels (d-g) of Figure 9 makes it possible to investigate the classical billiard as a discrete dynamical system. In this system orbits over a finite number of reflections become homologous in a symbolic dynamical representation in terms only of which curved and straight sections the orbits reflect off (e.g. rtblltrrb...). This makes it possible to reduce summations over all orbits, in forming classical or semiclassical correlation functions and wave plots, to summations over a single member from each family.

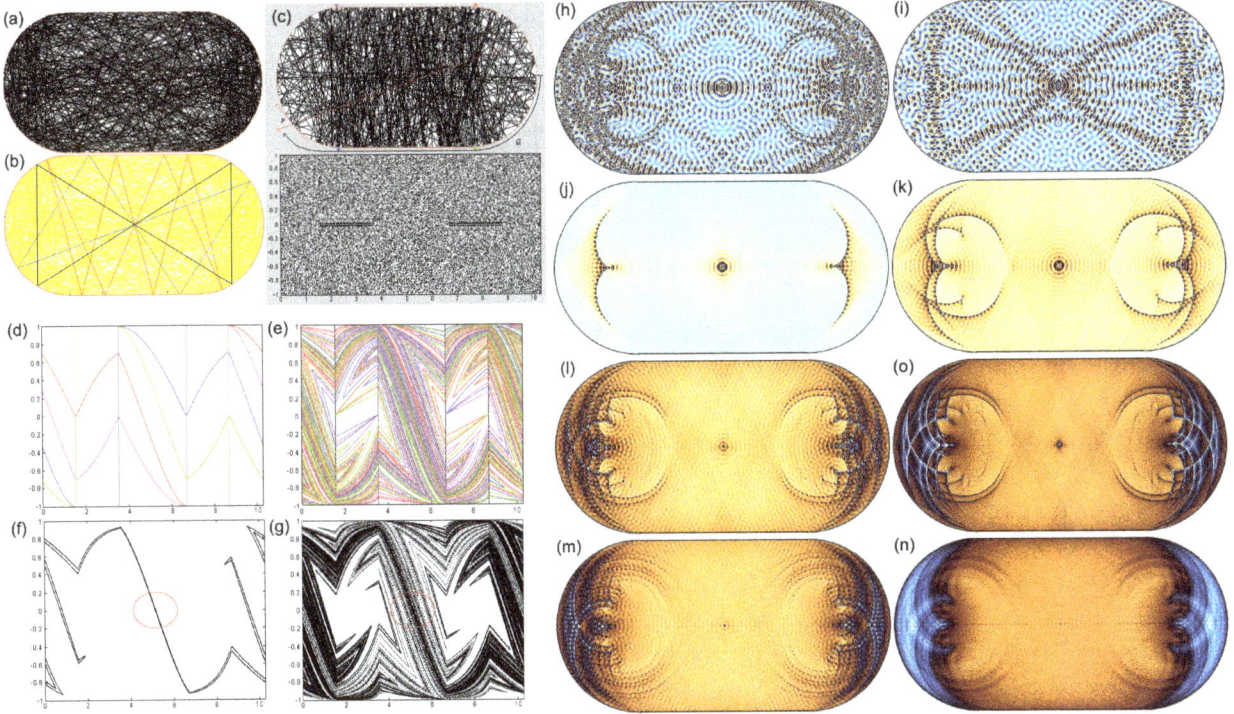

Figure 9: *(a) Extended orbit of 200 reflections in both directions from a close homoclinic recurrence generating in (b) the repelling bow tie orbit representation (black) later shows heteroclinic recurrence to other orbits, a related period 4 skewed bow tie (blue) and a period 8 orbit (red). Because the dynamics is conservative and area in phase space is conserved, the chaotic process consists of horseshoe saddles and repulsion in one eigenspace is compensated by attraction on the other. (c) Single directed orbit of 500 reflections from (0, 1) with direction (1, −3) can be transformed by the coordinate transformation shown into a momentum-position representation in which P is the cosine of the angle of the ray to the tangent and Q is the perimeter distance e.g. from the mid-point of the right circular arc. This 2D coordinate system results in the discrete dynamic below for 50000 iterations, in which phase space is ergodically (pseudo-randomly) filled except for a diminishing pair of central regions corresponding to the neighborhood of vertical stable orbits. (d) The one-step partition of phase space induced by colouring all the orbits fanning off (±1, ±1), colour-coded as in the discussion. (e) the corresponding partition for four steps. (f) The region mapped from the red ellipse in two reflection steps becomes fractally recurrent after six steps (g). (h) Semiclassical superimposed periodic ray trace of the stadium from a fan of rays at the centre with propagator having period a fraction of the stadium dimensions for 4 reflections. (i) The wave function cellular automaton, with a circular wave packet generator, having a similar fractional period, relative to the stadium width, to the semiclassical example. (j-o) Semiclassical superimposed periodic ray trace of the stadium from a circular wave generator at the centre with period a fraction of the stadium dimensions for (j) 2, (k) 3, (l) 4, (m) 5, and (n) 8 reflections. Rather than highlighting the scars around existing periodic orbits, the method highlights periodic versions of the successive caustics caused by repeated reflection also evident in (o), the classical ray trace for 4 reflections, shown for comparison.*

The dynamic can thus be described in terms of the Markov partition of phase space determined by the orbits which strike or emanate from the four junctions at (±1, ±1), labelled in red, magenta, green and blue dots in panels (d-g) of Figure 9.

4 The Semi-Classical Approach

The semi-classical approach retrieves very comparable results to a full quantum representation, panel (g) in Figure 1, by tracing a key member of each family of orbits, in the partition of the classical billiard, whose momentum divides the length by the time $p = l/t$ and applying an os-

cillating wave propagator to the member of the form [7]:

$$\left(\frac{1}{2\pi \iota \hbar}\right)^{\frac{d}{2}} \sum_j \left|\text{Det}\left(\frac{\partial^2 S_j(\mathbf{q}, \mathbf{q}'; t)}{\partial \mathbf{q} \partial \mathbf{q}'}\right)\right|^{\frac{1}{2}}$$
$$\times e^{\iota S_j(\mathbf{q},\mathbf{q}';t)/\hbar - \iota \pi v_j/2} \qquad (6)$$

where d is the number of degrees of freedom, S is the action, with j weighted according to the product of the orthogonal stability matrices:

$$M_{\parallel}(t) = \begin{bmatrix} 1 & 0 \\ t & 1 \end{bmatrix} \qquad (7)$$

$$M_{\perp}(t) = \begin{bmatrix} 1 & 0 \\ l_{n+1} & 1 \end{bmatrix} \prod_{i=1}^{n} \begin{bmatrix} -1 & \frac{2}{R\sin\theta_i} \\ 0 & -1 \end{bmatrix} \begin{bmatrix} 1 & 0 \\ l_i & 1 \end{bmatrix} \qquad (8)$$

along the orbit, giving a correlation function for a single orbital family member as follows:

$$C_{\beta\alpha}(t) = \left(\frac{2}{A_0}\right)^{\frac{1}{2}} e^{i[S_\gamma(\mathbf{q},\mathbf{q}';t)+\ldots]/\hbar - i\pi v/2 - 1/(2A_0)(\ldots)} \quad (9)$$

where ... denote further terms, and

$$A_0 = m_{11} + m_{22} + i\left(\frac{\hbar m_{21}}{\sigma^2} - \frac{\sigma^2 m_{12}}{\hbar}\right) \quad (10)$$

To explore a simplified semi-classical process, illustrating some of the principles of the above approach, we developed a Matlab program superimposing successive reflections of a 360 degree fan of $2^{12} - 2^{18}$ rays from the centre of the stadium with an oscillating propagator having period a fraction of the linear dimensions of the stadium. Cells are summed by superimposing the amplitude of each propagator at the point its trajectory crosses a given cell, weighted by the stability factor m_{11} as above. In theory this should have a similar effect to the selection of the orbit with the fractional momentum arriving at a given cell at a given time, since, as we are adding the cumulative effects over successive times, orbits with an integral number of wavelengths over a given path will interfere constructively and those not maintaining phase will effectively cancel one another out.

A more refined method would be to consider the heteroclinic orbits passing between two Gaussian distributions around the initial and final states, by selecting from the fractal intersections of the forward orbits from the initial state and the backward orbits from the final state those with a fractional momentum/period. One can then integrate this process over all the final states to give a time evolving distribution.

In panels (h, i) of Figure 9 the semi-classical method for a propagator having 1/60th of the width of a 600 pixel stadium is compared with the cellular automaton wave function method with a circular wave packet of wavelength a 1/54th fraction of the 594 cell width of the stadium. Although the profiles of the two simplified methods and their scarring patterns are not identical, both display similar features of scarring by wave coherence, illustrating how the two methods can highlight the quantum suppression of chaos. The semi-classical version also shows prominent effects of caustics caused by the curved ends as verified in panels (j-o) of Figure 9.

As shown in panels (j-o) of Figure 9, with the stability factor m_{11} inverted to highlight unstable orbit sections, because it is developed from a single fan of rays, the corresponding plots for successive reflection paths using superimposed absolute (or squared) amplitudes in panels (j-n) do not manifest closed orbit scarring, but rather display periodic versions of the caustics found when classical rays are superimposed in panel (o). We have thus made a complete transition from classical chaos to caustic order.

5 Matlab Files

Source Matlab files for the project can be downloaded as Supplement 6.

6 Summary

The central method introduced in the paper, provides a straightforward cellular automata model for a semi-classical approach to quantum chaos, more natural than the semi-classical methods of [5–8], demonstrating quantum scarring of the wave function around repelling classical orbits in fine detail. It thus both replicates the quantum and semi-classical approaches of Tomsovic and Heller and provides a transparent intuitively clear way of imaging and envisaging how quantum chaos actually comes about. The classical orbit simulations further test some of the assumptions of the semi-classical theory using rays, confirming some of the statistical results while providing a significantly different profile from other previous results.

7 Appendix: The Quantum Chaos Connection with the Riemann Zeta Function

A variety of lines of evidence [12] link phenomena of physics, and in particular quantum chaos, to the Riemann zeta function

$$\zeta(z) = \sum_{n=1}^{\infty} n^{-z} = \prod_{p\,\text{prime}} \frac{1}{1 - \frac{1}{p^z}} \quad (11)$$

expressible both as a harmonic Dirichlet sum and an Euler product over primes [13]. This has in turn led to hopes that quantum chaos might help solve the Riemann hypothesis that the non-trivial zeros of zeta, those not lying regularly on the negative real axis, all lie on the critical line $x = \frac{1}{2}$.

A major breakthrough was thought to have been made when it was discovered that pair correlations in the gaps between the zeta zeros followed the same GUE statistics as chaotic quantum systems and energy levels of large nuclei

$$\int_\alpha^\beta \left(1 - \left(\frac{\sin \pi u}{\pi u}\right)^2\right) du \quad (12)$$

The GUE statistic, and its time-reversible real variant, the Gaussian orthogonal ensemble (GOE), appear in many forms of quantum systems whose classical analogue is chaotic. These include the many body problem of nuclear energetics, highly excited atoms in a magnetic field and the quantum stadium problem. One of the defining moments in this interaction of fields was Keating's development [14], along with Berry [15] of a formula for the zeta moments, using characteristic polynomials of unitary matrices, based on the operator $H_{BK} = -i\hbar\left(x\frac{d}{dx} + \frac{1}{2}\right)$, which however has a continuous spectrum and cannot yield the hypothetical operator with eigenvalues the nontrivial zeros of the Riemann zeta function [16]. Initially this correspondence between fields caused great excitation in both the mathematics and physics communities, and a number of eminent researchers tried to prove Riemann hypothesis by discovering a system of random Hermitian matrices whose eigenvalues would be real and might correspond to the zeros of the related function

$$\xi(t) = \Gamma\left(\frac{s}{2} + 1\right)(s - 1)\pi^{-\frac{s}{2}}\zeta(s), \quad s = \frac{1}{2} + it \qquad (13)$$

thus showing they had to be real and hence those of $\zeta(s)$ would be on $x = \frac{1}{2}$. However this program has so far not borne fruit.

In attempting to create a convergence between Hermitian operators and the zeta function, researchers have constructed a variety of candidates. Berry [17] has presented one of the most straightforward of these, the semi-classical operator $H = xp$, and attempting to modify it to establish an operator having correspondence with zeta, demonstrating several putative connections between this and the zeta zeros. However the space on which this acts is not elucidated and the complex plane would need to be 'sewn up' in Berry's own words into a region, which would make the dynamics quantally bound. Secondly, there is no elucidated relationship between the primes and the periodic orbits of the Riemann dynamics. In an ingenious application of the approach, Berry [18, 19], on the basis the zeta zeros correspond to eigenvalues of a hypothetical Hamiltonian whose classical trajectories are chaotic and without time-reversal symmetry successfully predicted the longer range deviations of the zeta zeros from the GUE, which received confirmation in Odlyzko's high accuracy super-computer calculation of higher zeta zeros [20].

However, an actual candidate Hamiltonian has remained elusive and despite several further candidates such as $H = \left(x + \frac{1}{x}\right)\left(p + \frac{1}{p}\right)$, Berry and Keating as of 2011, although showing this candidate has the same asymptotic mean spectral density as the Riemann zeros, state:

> We are not claiming that our hamiltonian has
> an immediate connection with the Riemann
> zeta function. This is ruled out not only by
> the fact that the mean eigenvalue density dif-

fers from the density of Riemann zeros after the first terms, but by a more fundamental difference in the periodic orbits. For [the above formula] there is a single primitive periodic orbit for each energy E; and for the conjectured dynamics underlying the zeta function, there is a family of primitive orbits for each 'energy' t, labelled by primes p, with periods log p. This absence of connection with the primes is shared by all variants of xp, and our analysis gives no hint of a resolution of this difficulty. [21, p.13]

In a more experimental vein Berry has in 2012 developed far field radiative profiles for two zeta variants [22].

As an indication of the nature of the problem, Schumayer and colleagues have constructed potentials with energy eigenvalues equal to the prime numbers and to the zeros of the zeta function and show that these turn out to be multifractals [23].

Connes has constructed a Hermitian operator whose eigenvalues are the non-trivial Riemann zeros [24]. His operator is the transfer (Perron-Frobenius) operator of a classical transformation. Berry comments that such operators formally resemble quantum hamiltonians, but these usually have very complex non-discrete spectra with singular eigenfunctions. Connes gets a discrete spectrum by making the operator act on an abstract space where the primes acting on the Euler product are built in using a space of p-adic numbers and their units. The proof of the Riemann hypothesis is then transformed into establishing the proof of a certain classical trace formula.

Selberg has constructed a zeta function related to hyperbolic motion on constant curvature surfaces generated by discrete groups [25]. The product formula is not over primes, but over all primitive periodic orbits (ppo) for the motion of the surface considered

$$Z(s) = \prod_{ppo}\prod_{m=0}^{\infty}\left(1 - e^{-l_p(s+m)}\right) \qquad (14)$$

where l_p are the lengths of the orbits, and s is complex. This function like the Euler product is defined only for real $Z(s) > 1$, but can be analytically continued to the entire complex plane

$$Z(s) = e^{\mu\int_0^{s-\frac{1}{2}} u\tan\pi u du}Z(1 - s) \qquad (15)$$

As a result, $Z(s)$ has both trivial zeros at $1, 0, -1, -2, \ldots$ and a set of non-trivial zeros putatively on the critical line $x = \frac{1}{2}$. $Z(s)$ has a similar trace formula to the Weil explicit formula for sums over the zeros of zeta. The correspondence between primes and periodic orbits, provides a correspondence between zeros and eigen-momenta in which $\ln p$ corresponds to the orbital period T_p, resulting

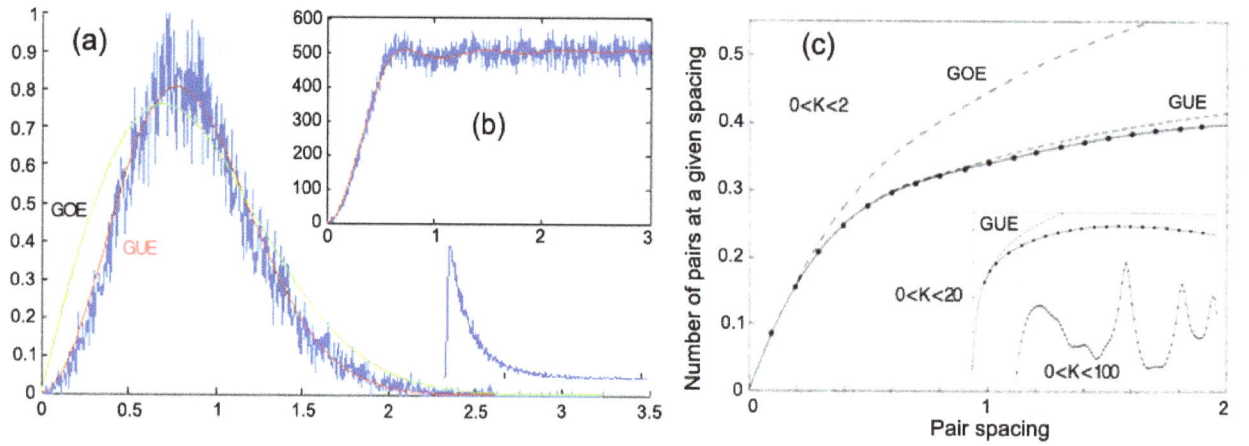

Figure 10: *(a) The spacing between neighbouring zeta zeros $10^2 - 10^4$ is compared with a GUE distribution $32\pi^{-2}s^2e^{-4s^2/\pi}$ (red), showing coincidence of the two statistics, and a GOE $\frac{1}{2}\pi se^{-\pi s^2}$ (green) characteristic of the Wigner distribution of atomic nuclear energies. The relationship is short range only and deviates for larger spacings [20] such as 100 apart (c). (b) Pair correlations for the first 10^5 neighbouring zeros compared with the theoretical GUE distribution. (c) Berry's predictions using semi-classical analysis (dashed line) of longer-range deviations of the zeta zeros from the GUE were confirmed numerically (black circles) using Odlyzko's numerical calculations.*

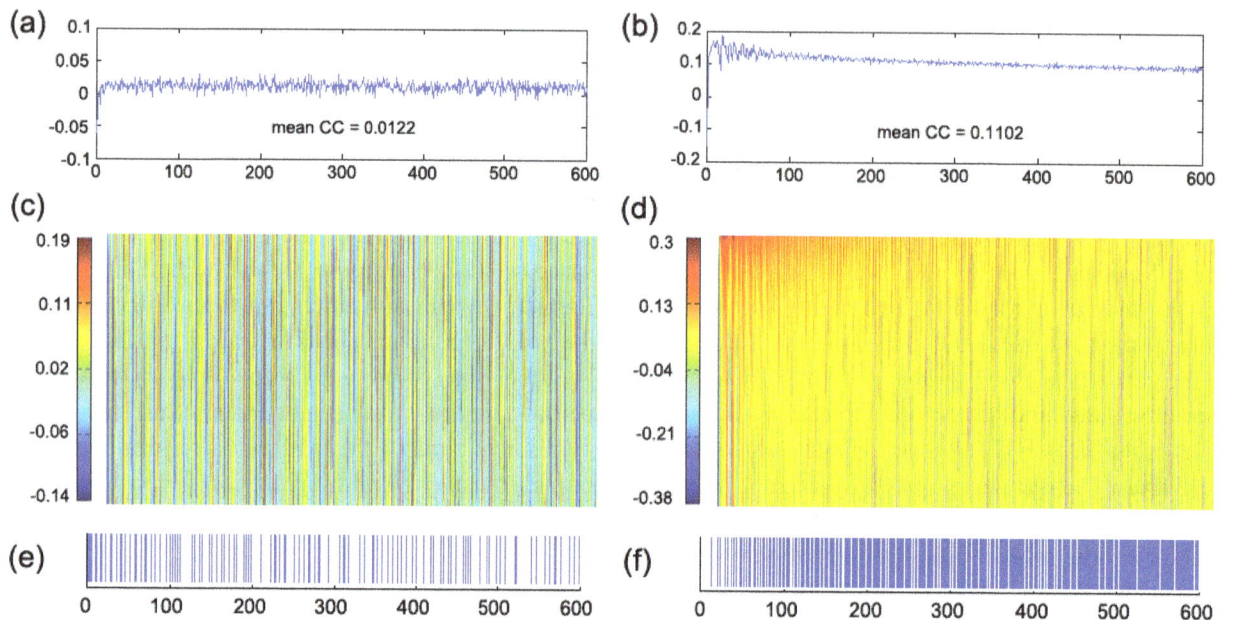

Figure 11: *Correlation coefficients of primes (a, c) and zeta zeros (b, d). While prime pairs have little correlation in their gap sizes, zeta zeros show consistent correlation, varying both as one moves up the zeros and as one examines longer range effects. (a, b) Graph of correlation coefficients using the first 20000 primes and zeros correlating differences of adjacent values with those j units further on. (c, d) Colour map where (i, j) is the correlation coefficient of the 500 pairs (X, Y) where $X(k) = (v(k), v(k + 1))$ and $Y(k) = (v(k + j), v(k + j + 1))$, $k = i, \dots, i + 500$. (e, f) Distributions of the first 600 primes (e) and zeta zeros (f).*

in an equivalent expression of the prime/periodic orbit number theorem

$$N(\ln p < T) \xrightarrow{T \to \infty} \frac{e^T}{T} \leftrightarrow N(T_p < T) \xrightarrow{T \to \infty} \frac{e^T}{T} \quad (16)$$

Fundamental to the problem are two issues. The first is that the duality already seen in the relationship between

zeta and the prime products is already the duality transform one is seeking. That is the system that decodes the zeta zeros is the distribution of numerical primes itself, so seeking an analogue from other mathematical areas cannot necessarily simplify the problem. Secondly, these GUE systems may show similarities in their statistics to zeta's zeros, because they share overall features combining structured constraints and pseudo-randomness, in common with the primes and zeta zeros, without necessarily being isomorphic to them. In a sense there is a regress occurring, in which attempting to model GUE systems to zeta's zeros results in more elaborate mathematical constructions which share zeta's characteristics but neither provide a breakthrough in proving Riemann hypothesis, nor result in a real valued quantum operator.

The quantum stadium is a direct analogue of the classical chaotic stadium billiard which displays the classical butterfly effect of chaos - sensitive dependence on initial conditions - and for almost all orbits produces a dense trajectory filling the stadium as shown in panel (j) of Figure 7. Within this classical system is a dense set of repelling periodicities, any arbitrarily small deviation from which results in a dense orbit, or transition to a differing periodicity.

The quantum versions of this system behave in a fundamentally different manner. While the initial stages of a trajectory follow the classical picture, after a limited period of time, called the quantum break time, they have a cumulatively increasing probability of entering one of the eigenvalues of the system. These eigenvalues turn out to correspond to the closed orbits of the classical system, which have now become probability maxima of the quantum system because wave spreading has effectively compensated for sensitive instability of the orbit, resulting in wave-periodicity and so-called scarring of the quantum wave function by probability maxima along these closed orbits, which also extend to fractal eigenstates of open chaotic systems [26].

In systems like the quantum stadium, the closed orbits are playing a role similar to the primes in that they are orthogonal or uncoupled to one another, are determined by constraints which result in a discrete spectrum and form an irrationally related subset of the phase space. Primes among the numbers behave similarly in that they have no common factors, form a discrete spectrum having no consistent rational formulation and act as a set of discrete generators of all the other integers. Thus the correspondence may be generic, but not homologous.

Evidence supporting differences between these two types of system comes from studies of the fractal dimension of the graph of zeta zero gaps for large zeros, which shows a Hurst exponent of 0.095 corresponding to a fractal dimension of 1.9, with anti-persistence, indicating large gaps are followed by smaller ones, self-similarity over a wide range of values and significant differences from corresponding GUE systems. When corresponding block sizes of zeros and random matrices are used, Hurst exponents for the zeros and matrices are 0.34 and 0.65, suggesting fundamental differences in fractal structure. The search for quantum systems reflecting the primes

and/or zeta zeros may thus suffer from the same problem facing researchers into L-functions that generalize zeta, associated with structures such as the elliptic curves and modular forms pivotal in solving Fermat's last theorem, the primes of other number fields, differential Maass and automorphic forms, referred to by Brian Conrey [27] in his AMS review:

There is a growing body of evidence that there is a conspiracy among L-functions – a conspiracy that is preventing us from solving Riemann hypothesis! [27, p.351]

The program of seeking to prove the Riemann hypothesis using common properties of L-functions seems to suffer from the same regress we have seen above [28]. Really the disparate L-functions, rather than shedding new light on the problem, are in a deep sense more elaborate analytic complexifications of the root prime-zeta relationship. The L-functions of elliptic curves and other number fields, for example, are more complicated representations, whose product formulae still ultimately depend on the prime distribution of the integers. Gaussian primes for example, although they come in three types, are ultimately definable in terms of the integer primes. Likewise elliptic curve L-functions are defined in terms of quadratic prime products. Even the varied Maass forms, modular differential functions satisfying the hyperbolic Laplace wave function $\Delta = -y^2 \left(\frac{\partial^2}{\partial x^2} + \frac{\partial^2}{\partial y^2} \right)$, have Euler products which are again generated in terms of the integer primes p, as illustrated by the cubic Mass form

$$L(z, \varphi \times \chi) = \prod_{p \text{ prime}} \frac{1}{1 - \frac{A(1,p)\chi(p)}{p^z} + \frac{A(p,1)\chi^2(p)}{p^{2z}} - \frac{\chi^3(p)}{p^{3z}}} \quad (17)$$

So ultimately we come back to the long-proven fact that the non-trivial zeros lie on the critical line $x = \frac{1}{2}$ if and only if the primes are distributed with fluctuations limited by order $x^{\frac{1}{2}}$, since from Riemann himself [13] and Ingham [29], the supremum of real parts of the zeros is the infimum of numbers β such that the error in the prime number theorem $\pi(x) \sim \int_2^x \frac{dt}{\ln t}$ is $O(x^\beta)$. Given the fact that the zeros also have to be symmetric around the critical line due to the analytic extension

$$\Gamma\left(\frac{s}{2}\right)\pi^{-\frac{s}{2}}\zeta(s) = \Gamma\left(1 - \frac{s}{2}\right)\pi^{-\frac{1-s}{2}}\zeta(1 - s) \quad (18)$$

the fixing of the zeros to the critical line simply reflects the fact that the primes are as evenly distributed as they can be given the fact that they cannot. Thus the tantalizing and elegant simply of Riemann hypothesis belies the fact that it is the real integer primes' root asymptotic behavior determining the entire picture.

The orbits of chaotic quantum systems provide an intriguing parallel to the primes as 'orthogonal generators',

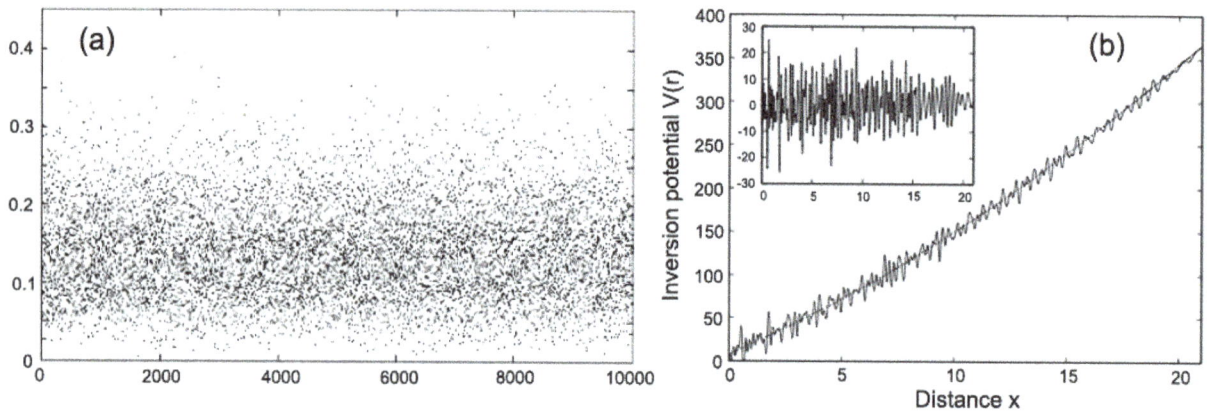

Figure 12: *(a) Pattern of differences between successive zeros of zeta in the range from 10^{21} for 10^4 successive zeros determined by Andrew Odlyzko shows a Hurst exponent corresponding to a fractal dimension of 1.9 with pronounced negative persistence, differing significantly from corresponding statistics of GUE random matrices. (b) The semi-classical potential (dashed line), and the fractal potential (solid line) supporting the first two hundred zeros of $\zeta(s)$ as energy eigenvalues [23]. Inset is the difference.*

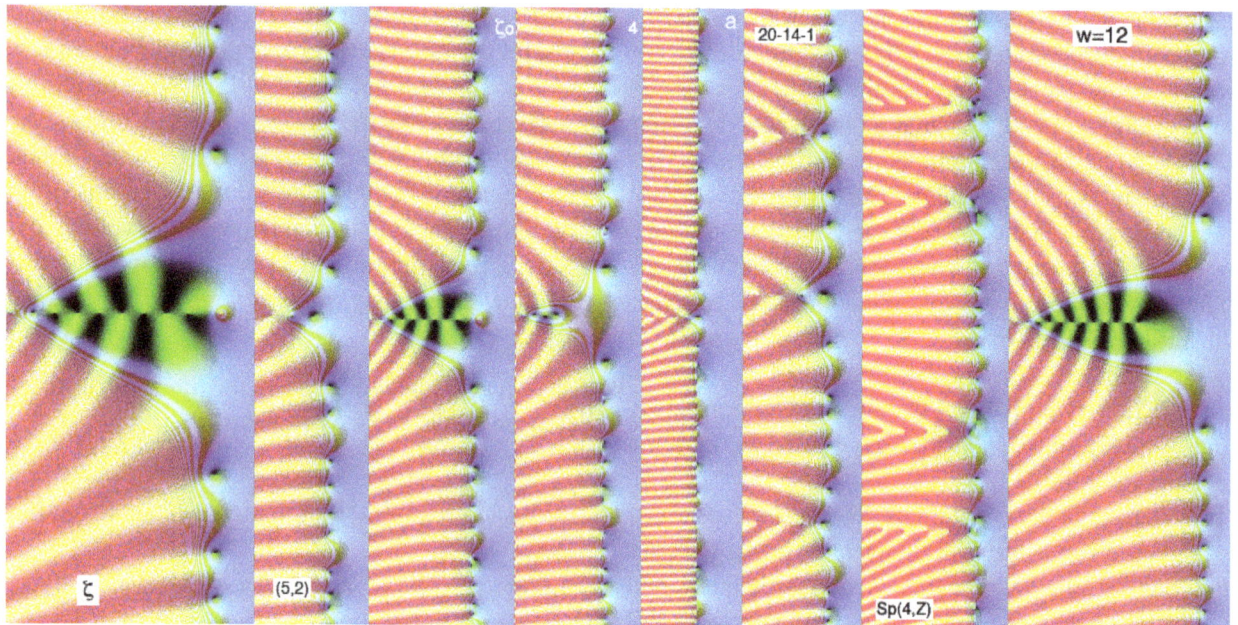

Figure 13: *A menagerie of L-functions: Riemann zeta, a Dirichlet L-function modulo 5, Dedekind zeta and a Hecke L-function of the complex primes of the Gaussian integers as an extension field, the L-function of the elliptic curve $y^2 + xy = x^3 + x^2 - 210x - 441$, third and fourth degree Maass forms and the modular discriminant all appear to have their non-trivial zeros on their weighted critical line, but so far have not shed new light on Riemann hypothesis [28].*

but to form a full analogue of Riemann hypothesis we need an equivalent of the prime sieving that generates the Euler product relation defining zeta as a complex harmonic series, because the product formula does not hold in the critical strip and the analytic continuation is possible only for the harmonic sum form. This underlying property at the root of what it is to be prime remains distinct from the notion of a quantum candidate having real eigenvalues, and may explain the difficulty of finding a suitable Hamiltonian, which would, in effect, have to sieve the dynamics in terms of the eigenvalue orbits. It

thus remains unclear whether quantum chaotic systems can generate a fully-fledged *L*-function for which a generalized Riemann hypothesis would be applicable.

This brings us full circle. Quantum theory poses a unique challenge to our notions of reality. Far from the quasi-theory of knowledge the Copenhagen interpretation asserts it to be, aspects of quantum mechanics, from bit to it, to fuzzy quantum logic and the concept of entanglement appear to be more fundamental even than our classical notions of points and sets. However primes and integers also go as close to the foundations of existence as we can envisage. From the nothingness of 0 to the unity of 1 we derive addition and the integers by simple concatenation $1 + 1 = 2$ and multiplication and hence the primes, and the prime distribution determining Riemann hypothesis, as a recursive form of concatenation $2 \times 3 = 2 + 2 + 2$. Whether quantum theory is the foundation of numbers or vice versa remains an enigma to be resolved.

References

[1] Bunimovich LA. On ergodic properties of certain billiards. Functional Analysis and Its Applications 1974; 8 (3): 254-255. http://dx.doi.org/10.1007/BF01075700

[2] Bunimovich LA. On the ergodic properties of nowhere dispersing billiards. Communications in Mathematical Physics 1979; 65 (3): 295-312. http://projecteuclid.org/euclid.cmp/1103904878

[3] Gutzwiller MC. Periodic orbits and classical quantization conditions. Journal of Mathematical Physics 1971; 12 (3): 343-358. http://dx.doi.org/10.1063/1.1665596

[4] Gutzwiller MC. Quantum chaos. Scientific American 1992; 266 (1): 78-84. http://dx.doi.org/10.1038/scientificamerican0192-78

[5] Heller EJ, Tomsovic S. Postmodern quantum mechanics. Physics Today 1993; 46 (7): 38-46. http://dx.doi.org/10.1063/1.881358

[6] Tomsovic S, Heller EJ. Semiclassical construction of chaotic eigenstates. Physical Review Letters 1993; 70 (10): 1405-1408. http://dx.doi.org/10.1103/PhysRevLett.70.1405

[7] Tomsovic S, Heller EJ. Long-time semiclassical dynamics of chaos: The stadium billiard. Physical Review E 1993; 47 (1): 282-299. http://dx.doi.org/10.1103/PhysRevE.47.282

[8] Peterson I. Back to the quantum future: evading chaos to make quantum predictions. Science News 1991; 140 (11): 282-285.

[9] Bokulich A. Can classical structures explain quantum phenomena? British Journal for the Philosophy of Science 2008; 59 (2): 217-235. http://dx.doi.org/10.1093/bjps/axn004

[10] Liu B, Zhang G-C, Dai J-H, Zhang H-J. Eigenvalues and eigenfunctions of a stadium-shaped quantum dot subjected to a perpendicular magnetic field. Chinese Physics Letters 1998; 15 (9): 628. http://dx.doi.org/10.1088/0256-307X/15/9/002

[11] Crommie MF, Lutz CP, Eigler DM, Heller EJ. Waves on a metal surface and quantum corrals. Surface Review and Letters 1995; 2 (1): 127-137. http://dx.doi.org/10.1142/S0218625X95000121

[12] Schumayer D, Hutchinson DAW. Physics of the Riemann hypothesis. Reviews of Modern Physics 2011; 83 (2): 307-330. http://dx.doi.org/10.1103/RevModPhys.83.307 http://arxiv.org/abs/1101.3116

[13] Riemann B. Ueber die Anzahl der Primzahlen unter einer gegebenen Grösse [On the number of prime numbers less than a given quantity]. In: Monatsberichte der Königlich Preußischen Akademie der Wissenschaften zu Berlin. 1859, pp.671-680. http://www.claymath.org/sites/default/files/ezeta.pdf

[14] Keating JP. Random matrices and the Riemann zeta-function. In: Highlights of Mathematical Physics. Proceedings, 13th International Congress, ICMP 2000, London, UK, July 17-22, 2000. Fokas AS, Halliwell J, Kibble T, Zegarlinski B (editors), Providence, Rhode Island: American Mathematical Society, 2002, pp.153-163.

[15] Berry M, Keating J. The Riemann zeros and eigenvalue asymptotics. SIAM Review 1999; 41 (2): 236-266. http://dx.doi.org/10.1137/S0036144598347497

[16] Endres S, Steiner F. The Berry-Keating operator on $L^2(\mathbb{R}_>, dx)$ and on compact quantum graphs with general self-adjoint realizations. Journal of Physics A: Mathematical and Theoretical 2010; 43 (9): 095204. http://dx.doi.org/10.1088/1751-8113/43/9/095204 http://arxiv.org/abs/0912.3183

[17] Berry MV, Keating JP. $H = xp$ and the Riemann zeros. In: Supersymmetry and Trace Formulae: Chaos and Disorder. Lerner IV, P.Keating J, Khmelnitskii DE (editors), New York: Plenum Press, 1999, pp.355-367.

[18] Berry MV. Semiclassical theory of spectral rigidity. Proceedings of the Royal Society of London. A. Mathematical and Physical Sciences 1985; 400 (1819): 229-251. http://dx.doi.org/10.1098/rspa.1985.0078

[19] Berry MV. Semiclassical formula for the number variance of the Riemann zeros. Nonlinearity 1988; 1 (3): 399-407. http://dx.doi.org/10.1088/0951-7715/1/3/001

[20] Odlyzko AM. On the distribution of spacings between zeros of the zeta function. Mathematics of Computation 1987; 48 (177): 273-308. http://dx.doi.org/10.1090/S0025-5718-1987-0866115-0

[21] Berry MV, Keating JP. A compact hamiltonian with the same asymptotic mean spectral density as the Riemann zeros. Journal of Physics A: Mathematical and Theoretical 2011; 44 (28): 285203. http://dx.doi.org/10.1088/1751-8113/44/28/285203

[22] Berry MV. Riemann zeros in radiation patterns. Journal of Physics A: Mathematical and Theoretical 2012; 45 (30): 302001. http://dx.doi.org/10.1088/1751-8113/45/30/302001

[23] Schumayer D, van Zyl BP, Hutchinson DAW. Quantum mechanical potentials related to the prime numbers and Riemann zeros. Physical Review E 2008; 78 (5): 056215. http://dx.doi.org/10.1103/PhysRevE.78.056215 http://arxiv.org/abs/0811.1389

[24] Connes A. Trace formula in noncommutative geometry and the zeros of the Riemann zeta function. Comptes Rendus de l'Académie des Sciences - Series I - Mathematics 1996; 323 (12): 1231-1236. http://arxiv.org/abs/math/9811068

[25] Bogomolny E. Riemann zeta function and quantum chaos. Progress of Theoretical Physics Supplement 2007; 166: 19-36. http://dx.doi.org/10.1143/PTPS.166.19 http://arxiv.org/abs/0708.4223

[26] Casati G, Maspero G, Shepelyansky DL. Quantum fractal eigenstates. Physica D: Nonlinear Phenomena 1999; 131 (1-4): 311-316. http://dx.doi.org/10.1016/S0167-2789(98)00265-6 http://arxiv.org/abs/cond-mat/9710118

[27] Conrey JB. The Riemann hypothesis. Notices of the American Mathematical Society 2003; 50 (3): 341-353. http://www.ams.org/notices/200303/fea-conrey-web.pdf

[28] King CC. A dynamical key to the Riemann hypothesis. 2011: http://arxiv.org/abs/1105.2103 http://www.dhushara.com/DarkHeart/key/key2.htm

[29] Ingham AE. The Distribution of Prime Numbers. Cambridge: Cambridge University Press, 1932.

Timeless Approach to Quantum Jumps

Ignazio Licata [1,2] *& Leonardo Chiatti* [3]

[1] *ISEM Institute for Scientific Methodology, Palermo, Italy. E-mail: ignazio.licata@ejtp.info*
[2] *School of Advanced International Studies on Applied Theoretical and Nonlinear Methodologies in Physics, Bari, Italy*
[3] *AUSL VT Medical Physics Laboratory, Via Enrico Fermi 15, Viterbo, Italy. E-mail: leonardo.chiatti@asl.vt.it*

Editors: *Eliahu Cohen* & *John Ashmead*

According to the usual quantum description, the time evolution of the quantum state is continuous and deterministic except when a discontinuous and indeterministic collapse of state vector occurs. The collapse has been a central topic since the origin of the theory, although there are remarkable theoretical proposals to understand its nature, such as the Ghirardi–Rimini–Weber. Another possibility could be the assimilation of collapse with the now experimentally well established phenomenon of quantum jump, postulated by Bohr already in 1913. The challenge of nonlocality offers an opportunity to reconsider the quantum jump as a fundamental element of the logic of the physical world, rather than a subsidiary accident. We propose here a simple preliminary model that considers quantum jumps as processes of entry to and exit from the usual temporal domain to a timeless vacuum, without contradicting the quantum relativistic formalism, and we present some potential connections with particle physics.
Quanta 2015; 4: 10–26.

1 Introduction

During the first phase of the development of quantum theory (1913–1927) three fundamental questions were posed: 1) the quantization of material motion, represented by discrete electron orbits; 2) the quantization of the field, in terms of the hypothesis of the emission and absorption of photons; and 3) the discontinuity of motion, represented by quantum jumps between stationary orbits. With regard to points 1) and 2), the subsequent definition of quantum formalism led to quantum mechanics and quantum field theory, or, in other words, respectively to first and second quantization. It has been noted that quantum mechanics and quantum field theory still do not constitute an organic structure; wave-particle duality in particular, which proved so useful for the description of Einstein–Podolsky–Rosen–Bell phenomena, has no place in descriptions based strictly on quantum field theory [1].

The point 3) did not bring any particular development; it is still the source of a periodically renewing debate, in particular about the questions related to the *wavefunction collapse* – the anomalous postulate of quantum mechanics – and its incompatibility with special relativity [1]. The concept of quantum jump is closely connected to the wavefunction collapse: the decay of an atom which transits to its ground state is both a quantum jump and the event of the collapse of the wavefunction associated to the state of that atom. The actual manifestation of this kind of events can be easily showed, for example through the detection of the photon emitted (if the transition is radiative). Various theoretical proposals contributed to get the collapse *irreducibility* into perspective, bringing it back, at least partially, to the dynamics of the system; let us remember here the results of the Milan–Pavia School, the Bohm–Bub theory, the Ghirardi–Rimini–Weber dynamical reduction program and decoherent histories [2–6]. On the experimental side, since the end of the 1980s, new technologies have allowed extraordinary realizations in revealing single quantum jumps in any kind of quantum system, so confirming and extending the original 1913 Bohr intuition. Nowadays, the direct observation of quantum jumps has been widely confirmed for trapped atoms and ions [7–9], single molecules [10], photons [11], single electrons in cyclotron [12], Josephson junctions [13], nuclear [14] and electronic [15] spin, superconducting cavities [16] thus providing an impressive demonstration of the helpful Bohr's intuition. The initial hesitancy about the real existence of quantum jumps, in particular by the community of quantum optics, is now only a distant memory of long time ago [17]. In conclusion, in the quantum jump the undulatory *ubiquity* of a quantum object (or state) and its *particle-like* localized aspect meet in a collapse event. So, we expect that a new theory about quantum jumps can tell us something on the nature of a quantum entity [18–20].

In his classical book on quantum mechanical principles [21], Heisenberg delineates with his usual clarity two possible ways to build quantum mechanics, both of which were coherent with the uncertainty principle: the first, a space-time description tending to save the classical concepts by means of a statistical connection; the second, adopting a pre-spatial mathematical scheme and a different concept of causality. In the 1930s there was no cogent reason to choose the second way. Nowadays, quantum nonlocality offers a valid reason to explore a different approach where the nonlocal features of quantum mechanics are not necessarily restricted to the entangled amplitudes, but are instead connected to a time-reversal approach [22], or some timeless background.

In this paper we propose a model for the discontinuous evolution of the quantum amplitude associated to a system: the so-called *quantum jump*. We assume here the objective nature of the wavefunction collapse that is actually identified with a quantum jump associated to a real micro-interaction. We define an explanatory model, by broadening and redefining the quantum field theory vacuum with the introduction of a complex time that regulates the structure of interaction vertices in real time as measured by an observer. We discuss a thermalized vacuum in terms of the relationship between imaginary time and temperature. The strategy of introducing imaginary time is well-known in cosmology and field theory, and has been proven effective in the removal of singularities

and the treatment of deterministic fields perturbed by an appropriate stochastic noise [23–26]. In this context, it is equivalent to a description of the vacuum as a structure of relationships between complex events characterized by a fundamental time scale, which replaces the standard quantum mechanical concept of a *particle*. The particle concept appears as still less credible after recent experiments where packets of physical quantities related to the same particle are manifested along different paths within an interferometer [27].

The theoretical choice adopted in this paper aims to avoid the ambiguities about the concept of quantum jump connected to the semi-classical nature of the wave-particle dualism. Our line of reasoning follows the transactional interpretation of quantum mechanics which sees the generic quantum system as a network of micro-events [28] and is thus closer to the spirit of quantum field theory. In this sense, we can agree with D. Zeh ("nor are there particles!") [29] yet not dismissing the idea of quantum jump, which is here reformulated within a new theoretical frame. Some potential applications to particle physics are even shortly outlined.

2 Clicks and interaction vertices

Let us consider a prepared system, at time 0, in a physical state associated with $|A_0\rangle$. We assume that in the interval $[0, t]$ the ket associated with the state of the system evolves under the effect of a unitary time evolution operator S, and that $|A_t\rangle = S(t, 0)|A_0\rangle$. At instant t, in response to an interaction, the system abruptly transits to the state represented by $|B\rangle$. We could say that the system prepared in the initial state A_0 is detected in the final state B. This is also the initial state, associated with ket $|B\rangle$, of a new preparation-detection pair. The full model can be understood by reading the following line from right to left: $|B\rangle\langle B|S(t, 0)|A_0\rangle$, in which a unitary evolution represented by the amplitude $\langle B|S(t, 0)|A_0\rangle = \langle B|A_t\rangle$ is concatenated with a quantum jump $|B\rangle\langle B|$. By way of example, let A_0 be the state of a particle beam emitted by a source, A_t the state of the same beam incident on a screen with a single slit, B the state of the beam selected by the slit and diffracted beyond the screen. The bra portion $\langle B|$ of the quantum jump is turned toward the past light cone and closes the previous unitary propagation; the ket portion $|B\rangle$ is directed toward the future light cone and opens a new segment of unitary evolution.

Naturally, the entire process is symmetrical in time and can be read in reverse; thus bra $\langle B|$ becomes the *initial* condition of the time evolution described by the evolution operator $S^{-1} = S^+$, which ends with the *final* condition represented by ket $|A_0\rangle$.

There is no substantial difference in the application of this model in quantum mechanics and in quantum field theory, except that in quantum field theory it is applied to elementary particles and combinations of their creation and annihilation operators. For example, $|B\rangle\langle B|$ could represent an interaction vertex between particles and $|B\rangle$ the creation of a set of particles exiting from it (we suppose that $|B\rangle\langle B|$ acts on its right). It should be understood, however, that this is a real interaction vertex with real particles. Virtual interactions and virtual particles are aspects of the expansion of the operator S into partial amplitudes and have no physical reality in this context. From this perspective, therefore, a *quantum jump* is synonymous with a micro-event or *click*. In this paper we aim to present a simple basic physical model for jumps, represented by the projector $|B\rangle\langle B|$. A click is considered to consist of two simultaneous (semi-) events: the destruction of the ingoing state B and the creation of the outgoing state B. We assume a jump is an interaction vertex between elementary particles and B is the state of the particles leaving the vertex.

The emphasis is therefore placed on interaction vertices and particles are considered as links between these vertices. Vertices and links form a network of relationships. This position is consistent with other theoretical proposals; for example, with the transactional interpretation, which is an interesting attempt to create a unified language to describe quantum phenomena [30–33].

In quantum formalism, the state of a many particle system is represented by a mixture of several states which are themselves entanglements of single particle states. The actualization of one of these single particle states corresponds to the selection, in the superposition that represents the entanglement, of the term wherein it appears and the consequent readjustment of the coefficients of the density matrix associated with the mixture. In an attempt to develop a model of the actualization process it is therefore necessary to focus on the single particle states, because the actualization at a higher level is a consequence of the actualization of the single particle states. On the other hand, the actualization of single particle state ensues from the interaction of the particle with material elements (measurement devices, etc.) which in turn consist of particles. We must therefore consider the single particle states entering a real interaction vertex, and those exiting from it. Both the former and the latter may be entangled.

We consider the action of the operators $|A\rangle\langle A|, |B\rangle\langle B|$ on the single particle state $|\Psi\rangle = c_A|A\rangle + c_B|B\rangle$, with $c_A c_A^* + c_B c_B^* = 1, |A\rangle = |A_1\rangle + |A_2\rangle + \ldots + |A_n\rangle$.

The action of $|B\rangle\langle B|$ will, by assumption, correspond to the localization of the particle in time (i.e. the event of its emission or absorption). Following this action, the particle will remain delocalized in accordance with the wavefunction $\langle x|B\rangle$, which will not necessarily be a Dirac δ in the x position coordinates. In other words, the event $|B\rangle\langle B|$ does not necessarily imply a maximally precise spatial localization of the particle and is therefore aspatial.

The action of $|A\rangle\langle A|$ shall not correspond to any particle localization, i.e. it will not be associated with any emission or absorption of the same. It will preserve the phase relation between the states $|A_i\rangle (i = 1, 2, \ldots)$. The event $|A\rangle\langle A|$ leaves the particle delocalized according to the wavefunction $\langle x|A\rangle$, and will therefore be aspatial. That said, we will call *quantum jump* the event $|B\rangle\langle B|$ that begins or ends a segment of unitary time evolution of the particle state. The event $|A\rangle\langle A|$ on the other hand preserves the unitary nature of the evolution. To better understand the relationship between quantum jump and unitary evolution, we consider three distinct examples.

a) *Reflection from a specular surface.* A particle incident on a perfectly reflecting surface devoid of absorption is subject to the action of the sole operator $|A\rangle\langle A|$, where $|A\rangle$ is the reflected state. If $|\Psi\rangle$ is the incident state, $\langle A|\Psi\rangle$ is the amplitude of the reflected state. The reflection will occur over the entire surface of the mirror, thus the states $|A_i\rangle$ may be the states reflected by various points of the mirror, labelled by the (continuous) index i. Because there is no localized reflection $|A_i\rangle\langle A_i|$, the phase relation between the several states $|A_i\rangle$ is preserved. No quantum jump occurs and the particle state evolution remains unitary.

b) *Impact on an opaque screen with double slit.* A particle incident on the screen has two possibilities: either it is absorbed by the screen (event $|B\rangle\langle B|$) or it goes through the double slit (event $|A\rangle\langle A|$). In the first case, the absorption occurs at a precise moment in time and has the effect of destroying the particle (for example, a photon) or of localizing it in the spatial volume corresponding to its new state (for example, atomic capture of an electron). $|B\rangle$ is in the first case the electromagnetic vacuum state of quantum electrodynamics, in the second case it is the atomic orbital of the captured electron. If the incident particle is transmuted into other particles, $|B\rangle$ will be the initial state of these particles. In the case where the event $|A\rangle\langle A|$ occurs, the states $|A_i\rangle (i = 1, 2)$ are those relative to crossing slit 1 or slit 2 respectively. These two states are in phase relation, because no process of localized crossing $|A_i\rangle\langle A_i|$ occurs and the evolution of the state $|A\rangle$ formed by their superposition remains unitary. In this case as well, the particle is delocalized according to the new wavefunction $\langle x|A\rangle$ (Fig. 1).

c) *Atomic decay.* Let us consider a two-level atom: the ground level $|B\rangle$ and the excited level $|A\rangle$. The atom is prepared in $|A\rangle$ at time 0 and the subsequent unitary evolution

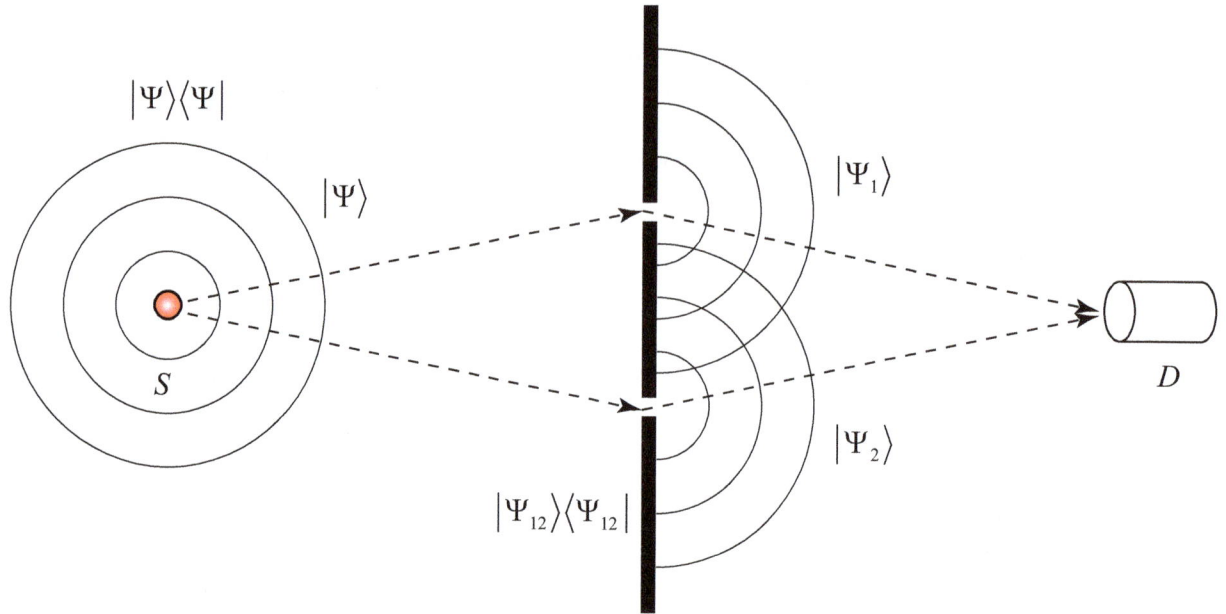

Figure 1: *The double slit experiment. The action of the double slit upon $|\Psi\rangle$ is non-unitary with respect to the absorption. However, the aspatial reduction event $|\Psi_{12}\rangle\langle\Psi_{12}|$ maintains the phase coherence of $|\Psi_1\rangle$ and $|\Psi_2\rangle$ components according to the usual path integral formalism. Legend: S = source, D = detector, $|\Psi_{12}\rangle = |\Psi_1\rangle + |\Psi_2\rangle$.*

of its state produces the superposition $|\Psi(t)\rangle = \alpha|A\rangle + \beta|B\rangle$. If we verify the state of the atom at a certain time t_0 we have two possibilities: either the atom has decayed (i.e., $|B\rangle\langle B|$ has acted at a given time t with $0 < t < t_0$), or it has not decayed (i.e. $|B\rangle\langle B|$ has not acted at any given time t with $0 < t < t_0$). In the first case, the verification action is equivalent to destroying the previous state $|B\rangle$ and recreating it, and it is therefore expressed as the projector $|B\rangle\langle B|$; in the second case, it is equivalent to acting with $|A\rangle\langle A|$ on the superposition $|\Psi(t_0)\rangle$, reinitializing its unitary evolution. In both cases, the verification action corresponds to a physical process represented by a quantum jump. It is important not to confuse this jump with that possibly occurring in the interval $0 < t < t_0$ and completed autonomously by the atom coupled with the electromagnetic field (Fig. 2).

These examples clarify the boundary between the unitary evolution of the quantum state and the quantum jump phenomenon. The latter is always associated with the precise localization of the particle over time (although not necessarily in space) and therefore its emission or absorption, or with the restart of its state. We have indeed seen cases of absorption on a screen, of emission by an atom which decays or of restart of a superposition of atomic states. The quantum jump may correspond to an observation-measurement process (as in the verification of example c), or not (as in the case of absorption on the screen in example b or of the spontaneous decay of the atom in example c). Thus, the state reduction ensuing

from a projective measurement process should be considered as a particular example of the more general concept of quantum jump. We propose to stop thinking in terms of persistent *particles* and their *states*, even though so far we conformed to this language by convention, but rather in terms of *clicks* bi-oriented in time, with reference to the projectors $|B\rangle\langle B|$ associated with the quantum jumps.

The time symmetry of a click, which appears as a kind of *two-faced Janus* along the time line, has several equivalents in quantum physics. For example, we can split the $x(t)$ coordinate of a quantum object into two coordinates, $x_+(t)$ (forward) and $x_-(t)$ (backward) using the Wigner–Feynman distribution, which incorporates nonlocal aspects of quantum mechanics [34, 35]. The evolution of the density matrix associated with this distribution leads to two copies of the Schrödinger equation, backward and forward, controlled by two Hamiltonian operators, yielding the Bohr frequency transitions. This doubling of the degrees of freedom ($x_\pm$, $p_\pm$) also occurs in dissipative quantum field theory and is therefore a fundamental structural aspect of both quantum mechanics and quantum field theory [36–40].

In the next sections we provide an in-depth description of what happens in a single quantum jump $|B\rangle\langle B|$. We will try to clarify the non-unitary aspects of the real interaction between elementary particles, not described by the unitary evolution of the state vector. For this purpose, we have to add new concepts to current quantum formalism, in order to specify the *collapse postulate*.

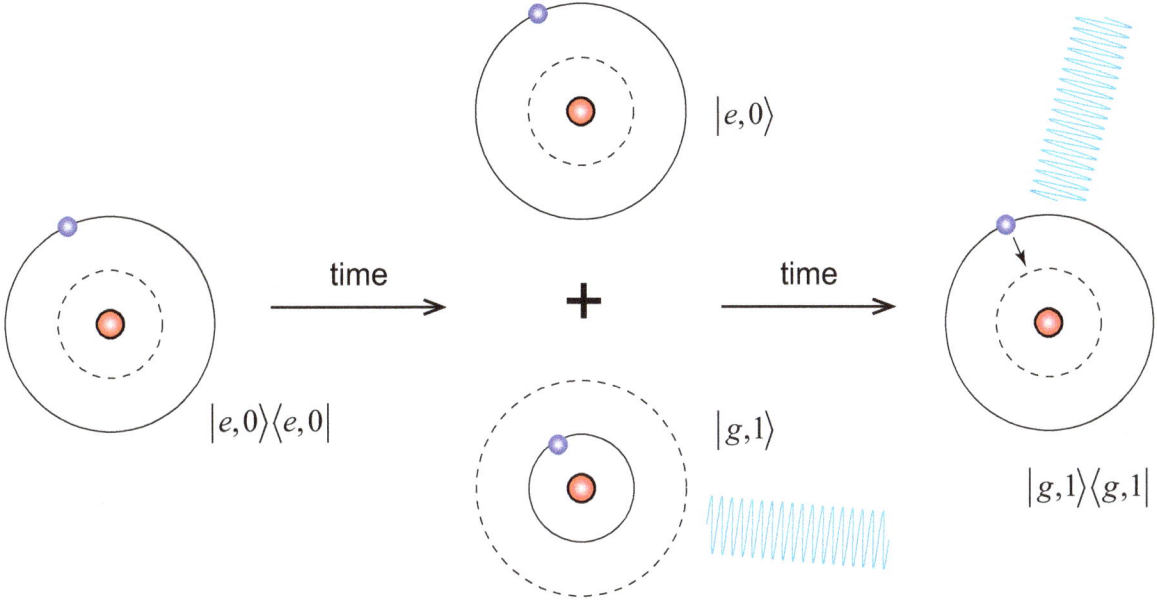

Figure 2: *The decay of an excited atom (on the right) is a non-unitary process which breaks the superposition of states $|e, 0\rangle$ (excited atom, 0 photons) and $|g, 1\rangle$ (atom in ground state, 1 photons).*

3 Complex time

We initially assume that B is a single-particle state, postponing the discussion of the general case to a later stage. The basic idea is that the destruction of B represents the stop of the time evolution of the wavefunction associated with the state B, and the creation of B represents the start of the time evolution of the wavefunction associated with the state B. *Time* here refers to laboratory time, i.e. normal *external* time measured by an experimenter. If the particles created/destroyed are provided with a rest reference frame, this time corresponds (except for a Lorentz transformation) to each particle's proper time. In this case, it would be more appropriate to speak of stopping and starting in motion in their respective proper times.

The energy required to set in motion a body at rest, or to restore a body in motion to rest, is by definition kinetic energy. The kinetic energy of a particle in its rest reference is reduced to its rest energy, which is thus the energy required to set the particle in motion in its proper time (creation) or the energy released by the stop of such motion (annihilation); the involved process must therefore also define the mass of the particles created/destroyed. Therefore, proper time and mass both appear as emerging parameters in the description we are proposing.

The second feature of the model must be a proper characterization of the *intermediate* condition of timelessness, so to speak, between the destruction of B and its successive re-creation. We follow the idea that the forerunner of time (its precursor) is a complex time $\tau = \tau' + \iota\tau''$, in

which $\tau' \in [-\theta_0, +\theta_0] \subset \mathbb{R}$ and $\tau'' \in [0, +\theta_0] \subset \mathbb{R}$. The parameter θ_0, which has the dimensions of a time interval, is assumed to be a new fundamental constant of Nature connected to the size of elementary particles, as detailed in a subsequent section.

We now focus on the creation of B. We assume the precursor of the outgoing wavefunction associated with state B is as follows

$$\Psi(y, \tau', \tau'') = Y(y)\Phi(\tau')\Lambda(\tau'') \tag{1}$$

The factor Y is only present if the particle created in state B actually has spatial extension in the internal space-time coordinates y. This is the case of hadrons, although this factor is absent in the case of particles with no spatial extension, such as leptons. The other two factors are assumed to satisfy the following equations

$$-\hbar^2 \frac{\partial^2}{[\partial(2\pi\tau')]^2}\Phi = (M_{sk}c^2)^2\Phi \tag{2}$$

with the condition $\Phi = 0$ for $\tau' \leq -\theta_0/2$, $\tau' \geq +\theta_0/2$ (the meaning of M_{sk} is discussed below);

$$-\iota\hbar\frac{\partial}{[\partial(\iota\tau'')]}\Lambda = \frac{\hbar}{2\theta_0}\Lambda \tag{3}$$

From Eq. 3, which is basically a Schrödinger equation in the imaginary component of complex time, it immediately follows that

$$\Lambda \propto \exp\left(-\frac{\tau''}{2\theta_0}\right) \tag{4}$$

From Eq. 2, which is the square of a Schrödinger equation in the real component of complex time, we get the even solutions

$$\Phi \propto \cos\left[\left(n + \frac{1}{2}\right)\frac{2\pi\tau'}{\theta_0}\right] \tag{5}$$

and the odd solutions

$$\Phi \propto \sin\left(\frac{2\pi n\tau'}{\theta_0}\right) \tag{6}$$

with integer $n \geq 0$. Given, in both cases, that

$$n' = n, \quad n + \frac{1}{2} \tag{7}$$

we have

$$M_{sk}c^2 = n'\frac{\hbar}{\theta_0} \tag{8}$$

We note that Eq. 4 can be rewritten in *thermal* form

$$|\Lambda|^2 \propto \exp\left(-\frac{E_0}{kT}\right) \tag{9}$$

where $E_0 = \hbar/\theta_0$ and the formal temperature

$$T = \frac{\hbar}{k\tau''} \tag{10}$$

has been introduced, which is infinite for $\tau'' = 0$, but assumes the minimum value for $\tau'' = \theta_0$.

Finally, with regard to $Y(y)$, we note that when it exists, i.e. in the hadronic case, each single coordinate y represents a distance in the internal space-time. If τ' is interpreted as a kind of *internal time* of the particle, the chronological distance at which an internal observer places an internal event should be limited to the interval $[-\theta_0/2, +\theta_0/2]$, whichever observer is chosen. Therefore, the coordinate transformations that lead from one internal observer to another should retain the condition

$$y \cdot y \leq \left(\frac{c\theta_0}{2}\right)^2 \tag{11}$$

which is precisely what defines a de Sitter space-time related to that single quantum jump. The space-time coordinates y are, of course, internal coordinates that distinguish internal events not accessed by an outside observer. Hence there is an *external* Minkowskian relativity and an *internal* de Sitter relativity. $Y(y)$ presumably satisfies a wave equation which also includes terms related to the interaction between the subcomponents of the hadron. Leptons have no spatial extension and only the fluctuations in internal time τ' described by Eqs. 5–8 exist for them.

4 Physical meaning of complex time

Before going further, it is worth pausing to consider the physical meaning of the real and imaginary parts of complex time τ. Note that we are considering the timeless vacuum state following the destruction of B and prior to its new creation: the system has stopped its course in external time and has not yet resumed it. Therefore, complex time τ is necessarily an *internal time* of this vacuum state, inaccessible to the external observer. From the perspective of external time, the timeless vacuum is an instant without duration; indeed, quantum jumps have no external duration.

It may be reasonably assumed that all particle states exist *in potentia* in this vacuum state as *virtual* fluctuations. However, the adjective *virtual* has to be intended in a radically different sense respect to the entirely fictitious *virtual* particles derived from the expansion of the S operator; vacuum fluctuations are instead physically real and *virtual* is here used as a mere synonimous of unobservable. If a particle is at a chronological distance τ'' from the *singularity* $\tau'' = 0$ (i.e. on the τ'' ordinate axis in the complex time plane), this means that it is associated with an energy fluctuation with amplitude $\hbar/\tau'' = kT$. Eq. 9 therefore provides the relative probability of such a fluctuation. The imaginary part τ'' of complex time is therefore a measure of the energy amplitude of the fluctuation associated with that particle.

As seen previously, the real part τ' is a kind of *internal time* of the vacuum, in which the wavefunctions of elementary particles *live* when they are *dormant* relative to external time (i.e. after their annihilation or before their creation). In this internal time, these functions are oscillatory, i.e. the *vacuum* is characterised by internal periodic phenomena; these are the different types of elementary particles that can be created or destroyed.

5 Creation and annihilation of particles in a thermalized vacuum

We now return to the discussion of the creation of state B. This is defined as a particular mapping of Eq. 1. This mapping firstly redefines the domain of the factors Φ and Λ, which now becomes the circumference, with centre $0 + i0$ and radius θ_0

$$(\tau')^2 + (\tau'')^2 = (\theta_0)^2 \tag{12}$$

The factor Φ becomes constant over the entire circumference in Eq. 12. The factor Λ becomes

$$\Lambda \propto \exp\left(-\frac{\tilde{\tau}}{2\theta_0}\right) \tag{13}$$

where

$$\tilde{\tau} = \pm 2n'\iota\omega \quad \omega = \theta_0 \arctan\left(\frac{\tau''}{\tau'}\right) \quad (14)$$

ω is the arc described on the circumference in Eq. 12, counted in $\tilde{\tau}$ as positive if it is described in an anti-clockwise direction and as negative if it is described in a clockwise direction. Eq. 13 can be written as

$$\Lambda \propto \exp\left(\pm \iota M_{sk}c^2 \frac{\omega}{\hbar}\right) \quad (15)$$

which represents the phase factor in the particle's proper *external* time, if we identify this time with ω. The appearance of the factor in Eq. 15 implies that Ψ has resumed its course in laboratory time. If the + sign applies, we are looking at the creation of a particle (positive mass); if the - sign applies, this indicates the creation of an anti-particle (*negative* mass).

Eq. 15 is just the complex factor that expresses the rotation of an arc ω on the circumference in Eq. 12, travelled with the frequency n'. Laboratory time is co-emergent with the de Broglie oscillation in Eq. 15 and with the particle's rest frame of reference. The *magic* of this triple emergence is not in the Wick rotation in Eq. 14 but in the transition from the complex plane to Eq. 12. Once the domain of the factors Φ and Λ, initially two-dimensional, is reduced to a closed line, time is reduced to an one-dimensional variable, whose domain can be traveled an infinite number of times. The infinite recurrences of a given domain point constitute the *external* time line.

Ultimately, the mapping converts a rectangular domain of complex time $\tau' + \iota\tau''$ into a circular domain ω (Fig. 3). The fluctuations of amplitude τ'' become an oscillation $\Phi\Lambda$ of fixed amplitude, represented by a vector with a free end on the circumference in Eq. 12 and the point of application in $0 + \iota 0$. The frequency of the previous oscillation Φ in τ' becomes the frequency of $\Phi\Lambda$ in ω. Thus, there is a recoding of the relevant information and the transition from a fluctuation to a state vector of constant norm *persistent* in ω.

We now come to the mapping action on the coordinates y and on the factor $Y(y)$. Firstly, the new quantum of the particle's internal time is $\hbar/M_{sk}c^2 = \theta_0/n'$ since, based on the reasoning presented above, the internal chronotope (if any) should change into a de Sitter chronotope with radius $c\theta_0/n'$. We refrain from further discussion of the implications for the structure of hadrons, leaving this for future research. We limit ourselves to observing that the coordinates y and the factor $Y(y)$ have to undergo consequent scale transformations.

However, the appearance of proper time leads to the appearance of external spatial coordinates. Indeed, a generic observer in motion respect to the particle sees the particle's proper time line as its trajectory in space.

Relativistic covariance therefore requires the appearance of a complete system of external space-time coordinates x and the simultaneous appearance of a factor $X(x)$ in the wavefunction of the particle emerging from the vertex. The outgoing wavefunction is therefore, in conclusion (and omitting internal dynamics in the case of hadrons), $\Psi = X\Lambda$. We have $X(x) = \langle x|B\rangle = \langle B|x\rangle^*$, i.e., the outgoing wavefunction is the complex conjugate of the incoming one. In general, X will be a spinor, thus the conjugation includes a transposition. This spinor represents the new initial condition for the unitary evolution described by the appropriate wave equation (Dirac, Proca, etc.) possibly with external fields. Each component of X satisfies the Klein-Gordon equation, which can be written in the usual form

$$-\hbar^2 D^2 X = (M_{sk}c)^2 X \quad (16)$$

where D^2 is the usual D'Alembert operator in Minkowski coordinates x. Note that the outgoing $X(x)$ is generally not an eigenstate of the position, and so the outgoing particle is normally delocalized.

The destruction of the state B entering the vertex is described by the exact opposite mapping. The constraint by Eq. 12 is removed and we return to the complex time plane. The factors Φ and Λ resume their original shape. External time disappears, and with it the external spatial coordinates, too. There is no longer a rest frame of reference for the particle, or a de Broglie oscillation. The *archaic* vacuum state prior to any physical manifestation is restored.

We are now able to address the general case involving a plurality of particles entering and leaving the vertex. The crucial observation is that the timeless vacuum state is described by two temporal parameters but no spatial parameter. In other words, *dormant* state precursor of B (after B annihilation and before its re-creation) does not contain external space-time coordinates x; these are created or destroyed on the occasion. The quantum jump is therefore an aspatial event, and this allows us to define B as an appropriate entanglement of the amplitudes of the individual particles, respectively entering or exiting, i.e. linear combinations of products of these amplitudes. The amplitude of each individual particle is defined on the specific configuration space of that particle, thus B will *live* in the total configuration space of all the particles involved. It is also possible to represent B through second quantization creation/annihilation operators defined on a suitable Fock space, and this leads to the quantum field theory description.

The following is an indicative example. Let A be the product of the two (non entangled) spatial wavefunctions of two identical particles of spin $\frac{1}{2}$ and their spin wave-function, which we assume to be a singlet. Let B be the

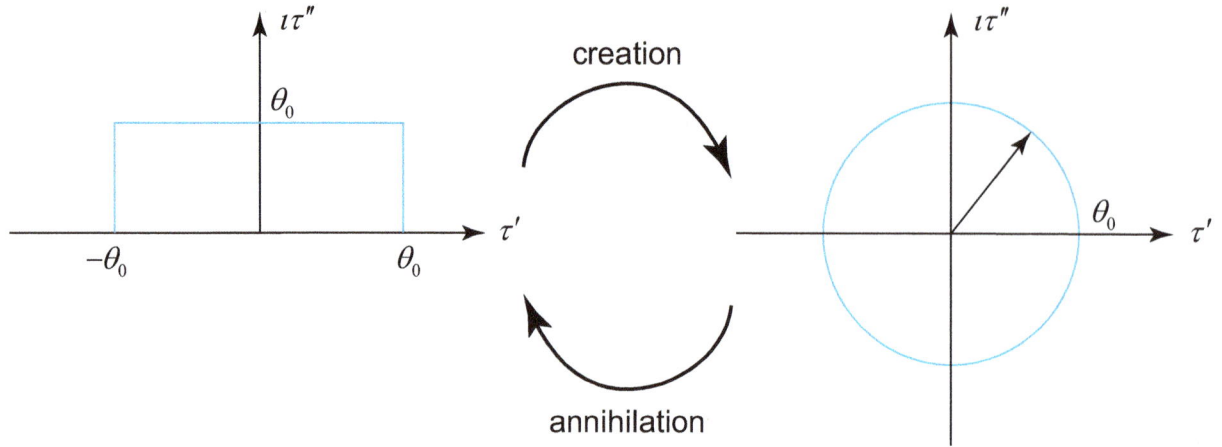

Figure 3: *Illustration of the wavefunction creation and annihilation in the proposed model.*

product of the spatial wavefunctions of the same two particles (the one peaked around the position x_0 at time t_0, the other identical to that in A) and their spin wavefunction equivalent to one of the two components of singlet A, considered along the z axis. Let us consider the preparation $|A\rangle\langle A|$ followed by detection $|B\rangle\langle B|$. This second event refers to the interaction, at time t_0, between one of the two particles and a measurement device placed in x_0, with simultaneous measurement of its spin along the z axis. In the language of present model the aspatial (and therefore in a way *ubiquitous*) background has accepted B in input and returned it as a new initial state, thus inducing the spatial localization of one of the particles and a sharper definition of spin of *both particles*. We consider here that the projectors $|A\rangle\langle A|, |B\rangle\langle B|$ ... are acting on their right. If we consider them acting on the left we have the description of the same phenomenon but reversed in time. This example illustrates the connection between background aspatiality and non-separability of an entangled state. This second form of nonlocality follows from the first, which is in this sense more radical.

6 Digression on skeleton mass and self-interaction

The interpretation of the mass M_{sk} is important. The creation (annihilation) of particles always occurs in a vertex of interaction with other particles and has a finite duration, so the mass of each particle becomes indeterminate by virtue of Heisenberg relationships. Only the *free* particle, i.e. the asymptotic particle state exiting from the vertex or entering it, has a definite mass. In terms of normal quantum field theory language, the particle is born *bare* and is then *dressed* by its own self-interaction processes, so that the nascent (bare) particle is not the particle physically

observed away from the vertex (*dressed* particle).

We can assume the physical mass m of the particle to be the sum of the *nascent* mass M_{sk} and a term ε/c^2 associated with the self-interaction. If energy mc^2 is applied to the vacuum, the creation of that particle in it becomes possible, i.e. its actual emergence from the vertex. From this perspective, the time interval θ_0/n' represents a sort of minimum chronological distance between two virtual self-interactions belonging to the same interaction vertex. The creation sequence of the mass is therefore as follows:

(1) The particle is initially massless (vacuum state);

(2) Its localization in an interaction event, for a duration of θ_0/n', requires an amount of energy equal to the ratio of $\hbar$ and this duration; this ratio is the rest energy in Eq. 8. Thus the particle *skeleton* mass M_{sk} appears.

(3) The particle self-interacts for a duration of $\hbar/M_{sk}c^2$, and therefore on a scale of lengths equal to $\hbar/M_{sk}c$. The total mass m is therefore the sum of the skeleton mass and the ε/c^2 mass derived from this self-interaction.

(4) There is no self-interaction for chronological distances from the vertex greater than $\hbar/mc^2$; the particle's rest energy is the minimum energy required to extract the particle from the vertex.

(5) The actual Klein-Gordon equation does not contain the skeleton mass M_{sk}, but the actual mass m. In other words, a term ε/c^2 must be added to the right-hand side of Eq. 16.

(6) If we interpret the skeleton mass M_{sk} as *bare* mass, we obtain the interesting result that it is finite.

If $n' = 0$, only the term of self-interaction survives in m. Only a fraction of the energy $\hbar/\theta_0$ needed to locate the particle in a temporal extension θ_0 is used, expressed by the dimensionless self-coupling constant $(g^2/\hbar c)$. This energy is therefore $g^2/(c\theta_0)$, and the particle is delocalized to a *dressing* region $(\hbar c/g^2)$ times larger than $c\theta_0$. It is possible that this is the situation of lighter particles, i.e. electrons and neutrino mass eigenstates. With electrons, the self-interaction will be essentially electrostatic and therefore the self-interaction energy will be $e^2/(c\theta_0)$, where e is the elementary electric charge. Equating this expression to the rest energy of the electron, we obtain $c\theta_0$ is the classical radius of the electron. As a result, the fundamental skeleton mass interval $\hbar/\theta_0$ seen in Eq. 8 is approximatively 70 MeV.

Although a serious formulation of the entire conjecture would require further explanation of the relationship between n' and the internal quantum numbers (and a precise calculation method for the self-interaction term), it should nevertheless be noted that the bare mass in this context is finite and that there are no divergences.

In our model, θ_0 coincides, at least by a factor of $\frac{2}{3}$, with the *chronon* introduced by Caldirola in his classical model of the electron [39, 40]. Note, however, that the fundamental interval θ_0 is a property of the *vacuum* or background that manifests itself in external time only as a minimum duration θ_0/n' associated with the localization of the particle in a quantum jump. This interval does not play any role in the next (or previous) unitary evolution of the particle state vector, which is described by current quantum formalism. However, it could be relevant in selecting base states (elementary particles) and defining their properties, such as a finite mass spectrum. A more formal illustration of the scheme can be the following. Let us consider a massive particle endowed with a proper rest frame of reference. Let t be the particle proper time and σ a scalar quantity (respect to the Poincaré group of coordinate transformations) such that $t = t(\sigma)$ and

$$\sigma_2 - \sigma_1 = |t(\sigma_2) - t(\sigma_1)| \tag{17}$$

The integrals of $d\sigma$ and dt along a given segment of a four-dimensional line respectively measure its length and the extension in the particle proper time t. They coincide if the time orientation of the line is the same in each point of the segment. From this definition, the following relations can be immediately derived

$$\frac{dt}{d\sigma} = \lim_{\sigma_2 \to \sigma_1} \frac{t(\sigma_2) - t(\sigma_1)}{\sigma_2 - \sigma_1} = \begin{cases} +1 & \text{for } t(\sigma_2) > t(\sigma_1) \\ -1 & \text{for } t(\sigma_2) < t(\sigma_1) \end{cases} \tag{18}$$

These two relations can be summarized in a single equation

$$\frac{dt}{d\sigma} = \gamma_0 = \begin{pmatrix} 1 & 0 & 0 & 0 \\ 0 & 1 & 0 & 0 \\ 0 & 0 & -1 & 0 \\ 0 & 0 & 0 & -1 \end{pmatrix} \tag{19}$$

The left hand derivative is, in this case, an operator with eigenvectors

$$\begin{pmatrix} 1 \\ 0 \\ 0 \\ 0 \end{pmatrix}, \begin{pmatrix} 0 \\ 1 \\ 0 \\ 0 \end{pmatrix}, \begin{pmatrix} 0 \\ 0 \\ 1 \\ 0 \end{pmatrix}, \begin{pmatrix} 0 \\ 0 \\ 0 \\ 1 \end{pmatrix} \tag{20}$$

having $+1$ and -1 as respective eigenvalues. By setting $x_0 = ct$ we therefore have

$$\frac{dx_0}{d\sigma} = c\gamma_0 \tag{21}$$

In a frame of reference in uniform rectilinear motion with respect to the particle rest frame, this relationship takes the following form

$$\frac{dx_\mu}{d\sigma} = c\gamma_\mu; \ \mu = 0, 1, 2, 3 \tag{22}$$

All the spacetime coordinates in this relationship have an implicit dependency on σ. If the particle wavefunction ψ does not explicitly depend on σ, but only through the coordinates, we have

$$\frac{1}{c}d_\sigma\psi = \gamma^\mu \partial_\mu \psi \tag{23}$$

In general, the gamma operators will be the Dirac matrices and the wavefunction will therefore be a spinor. Given the assumptions made in this section, the outgoing wavefunction from the interaction vertex is

$$\psi = \phi\left(x_\mu\right) \exp\left(-\imath M_{sk}c^2\sigma/\hbar - \imath c^2\sigma\delta M/\hbar\right) \tag{24}$$

This is the result of mapping, with the addition of a corrective term to the exponent containing the perturbative correction $\delta M = \varepsilon/c^2$ to the skeleton mass due to the particle self-interaction (limited to the vertex where it is created). To free the particle and remove it from the vertex it is necessary to administer a Mc^2 energy, with $M = M_{sk} + \delta M$. The ordinary Dirac equation for a free particle of mass M thus follows

$$\imath\hbar\gamma^\mu\partial_\mu\phi = Mc\phi \tag{25}$$

It is possible to treat the particle subjected to gauge fields through the typical replacement of four-momentum with the canonical four-momentum. If this procedure is applied at the interaction vertex where the particle is created, taking into account only the particle self-field, it produces expressions such as the following (where we consider only the electromagnetic self-interaction of a particle with charge e and self-field A_μ)

$$\delta M = -\frac{e}{c^2} \int \bar{\phi}\gamma^\mu A_\mu \phi \, dV \tag{26}$$

The integral is extended to a volume of diameter $\hbar/M_{sk}c$ around the vertex (we assume that $M_{sk} \neq 0$) and the minimum interaction distance is $c\theta_0/n'$, which is equal to this diameter. In the particle rest frame of reference this is a Coulomb integral whose order of magnitude is

$$\approx \frac{e^2}{\left(\frac{\hbar}{M_{sk}c}\right)} = \left(\frac{e^2}{\hbar c}\right) M_{sk}c^2 = \alpha M_{sk}c^2 \qquad (27)$$

where α is the fine structure constant.

A separate case, discussed above, is the electron for which $M_{sk} = 0$, $M = \delta M$ and the integral in Eq. 26 does not exist. In the external time of the observer the electron is localized in the interval $\hbar/Mc^2$, while in the internal time τ' it is a virtual oscillation with zero frequency and duration θ_0. The ratio between the external and internal temporal extension represents the number of times in which the interval $\hbar/Mc^2$, containing a single real electron, contains the electron as virtual oscillation. The inverse of this number is the electron adimensional constant of interaction, that is, the probability to actualize a virtual electron in response to an interaction. It is essentially the electromagnetic fine structure constant (the electron interacts substantially through the electromagnetic field). We therefore have $\hbar/Mc^2\theta_0 = \hbar c/e^2$, from which we obtain $\theta_0 = e^2/Mc^3$ namely $\hbar/\theta_0 = 70$ MeV. We could describe the situation by saying that the coupling with a real photon consists of the actualization of one (on average) of $\hbar c/e^2$ virtual electrons. In the external time period $T_e = h/Mc^2$ the phase of the electron in Eq. 24 varies by 2π. Since the electron spin is $\frac{1}{2}$, we would obtain the same phase variation by rotating the electron around any spatial axis of a 4π angle. In this sense the phase pulsation $2\pi/T_e$ is equivalent to a *spatial rotation* pulsation $4\pi/T_e = 2Mc^2/\hbar$ [41–43]. It is possible to define the electric current $I = e(2Mc^2/\hbar)$ associated with this *rotation* and the *rotation radius* $L = c(\hbar/2Mc^2)$. Therefore, the magnetic moment is defined

$$\frac{1}{c}IL^2 = \frac{e\hbar}{2Mc} \qquad (28)$$

which is the Dirac magnetic moment of the electron. On the other hand, in an interaction vertex the not yet actualized electron appears as a virtual fluctuation of minimum temporal extension θ_0, therefore associated with a transitory phase pulsation $2\pi/\theta_0$ corresponding, for the same principle, to an angular pulsation $4\pi/\theta_0$. By setting $I = e(4\pi/\theta_0)$, $L = c(\theta_0/4\pi)$, and taking into account that $c\theta_0 = \alpha\hbar/Mc$, we achieve an additional magnetic moment from this high frequency oscillation equal to

$$\frac{1}{c}IL^2 = \frac{e\hbar}{2Mc}\frac{\alpha}{2\pi} \qquad (29)$$

which is the anomalous magnetic moment at the first order. The magnetic moment is naturally a latent property which becomes effective only in presence of an external magnetic field that, with its direction, selects a spatial *rotation* axis.

However, a transitory state as that of a virtual electron in an interacion vertex cannot be described only by the monochromatic pulsation $2\pi/\theta_0$ and a wave packet with pulsations ranging from $2\pi/T_e$ to $2\pi/\theta_0$ will be actually involved. The Fourier components will exchange virtual photons so generating radiative corrections of higher order to the magnetic moment. Under this perspective α represents the probability of actualization of a virtual self-interacting electron, what justifies the perturbative origin of M.

In summary, the introduction of the chronon as a vacuum constant seems to enable, at least in principle, the derivation of a finite mass spectrum for elementary particles. The finite value of the chronon implies a finite value of the skeleton mass, which appears in the phase factor in Eq. 24 in a completely adynamic manner, together with the particle (external) proper time. The interval of proper time given by the inverse (in natural units) of the mass skeleton represents the duration of the particle self-interaction and also the minimum interval between two virtual self-interactions. The contribution of self-interaction to the effective mass of the particle is therefore finite. The finite value of the effective mass in turn justifies the finite value of the magnetic moment.

The conjugation of the adynamic mechanism for generation of the masses (such as mapping) with quantum field theory requires further elaboration and here we only highlight the effects that the chronon has on the choice of the cut-off. For example, consider the following relationship for the mass M of a fermionic $\frac{1}{2}$ spin field coupled with a gauge field of spin 1 and (non null) mass μ, with an adimensional coupling constant g [44]

$$M = M_0 + \frac{4g^2M}{2\pi^2\mu^2}\left[\Lambda^2 - M^2\log\left(\frac{\Lambda}{M}\right)^2\right] \qquad (30)$$

where M_0 is the fermion bare mass and Λ is the cut-off. Setting $M_0 = M_{sk}$ and $\Lambda = M_{sk}$ we achieve a transcendental equation in M which can be solved iteratively. For example for $2(g/\pi\mu)^2 = 0.1$, $M_{sk} = 1$ we get $M = 1.16459\ldots$ a clearly finite result. The skeleton mass is defined by the minimum time extension of the particle and, therefore, constitutes a cut-off for virtual coupling; this fact leads to a finite mass. A more consistent treatment, however, should lead to a unification of the ideas expressed in this paper with the quantum field theory formalism, a task which we must leave for a future work.

7 Suggestions for hadronic physics

The equation

$$|\Lambda|^2 = \exp\left(-\frac{\tau''}{\theta_0}\right) \tag{31}$$

represents the background as a set of thermostats with different absolute temperatures, which are included in the range between $T = \hbar/k\theta_0$ (for $\tau'' = \theta_0$) and $T = \infty$ (for $\tau'' = 0$). The thermostat corresponding to the temperature T shall contribute to the creation/annihilation of a particle with rest energy Mc^2 at that temperature (at $\tau'' = \hbar/kT$) through the heat exchange

$$dQ = Mc^2 \frac{\exp(-\frac{\tau''}{\theta_0})d\tau''}{\int_0^{\theta_0} \exp(-\frac{\tau''}{\theta_0})\,d\tau''} \tag{32}$$

equal to the product of the rest energy by the probability of its release. Now imagine an *equivalent* thermostat such that: 1) the entropy variation of the whole set of thermostats is equal in value to the entropy variation of the equivalent thermostat; 2) the sum of the thermal contributions of the different thermostats is equal to the total thermal contribution of the equivalent thermostat. Having to do with reversible processes only, these two conditions are reflected in the relation

$$\int_{\tau''=0}^{\tau''=\theta_0} dQ/T = \frac{\int_{\tau''=0}^{\tau''=\theta_0} dQ}{T_H} \tag{33}$$

where dQ is the heat exchanged by the thermostat at temperature T and T_H is the temperature of the equivalent thermostat. Now add the additional condition that the thermal exchanges dQ occur in the form of rest energy of massive particles exchanged within the same interaction vertex, i.e., in a contact interaction. Thus, we basically limit ourselves to the strong interaction between hadrons entering a vertex where a quark exchange occurs, with the possible creation/annihilation of quark pairs. The exchange takes place within that same vertex, thus generating new hadrons exiting from it. Conversely, we exclude electroweak and gravitational interactions from our consideration because the absorption and emission of their gauge quanta occur in distinct vertices; furthermore, photon and graviton are massless.

Substituting Eq. 32 into Eq. 33 we obtain

$$kT_H = \frac{\hbar}{\langle\tau''\rangle} = \frac{\hbar}{\theta_0} \frac{e-1}{e-2} \tag{34}$$

where

$$\langle\tau''\rangle = \frac{\int_0^{\theta_0} \tau'' \exp(-\frac{\tau''}{\theta_0})\,d\tau''}{\int_0^{\theta_0} \exp(-\frac{\tau''}{\theta_0})\,d\tau''} = \theta_0 \frac{e-2}{e-1} \tag{35}$$

If, according to the argument developed in the previous section, we assume $\hbar/\theta_0 = 70$ MeV we obtain $kT_H = 167.5$ MeV. This value is practically coincident with that currently accepted for the Hagedorn temperature (which is included in the range 160-190 MeV). In order to exclude a mere numerical coincidence, we now consider the case where the equivalent thermostat may be placed in thermal contact with a process consisting of the creation of a single hadron of rest energy Mc^2 at temperature T (at $\tau'' = \hbar/kT$). The related (finite) heat transfer is $Q_0 = Mc^2$. The system made of equivalent thermostat and the process undergoes a total entropy variation equal to the difference between Eq. 33 and the entropy variation associated with the process

$$\Delta S = -\frac{Q_0}{kT} + \frac{\int_{\tau''=0}^{\tau''=\theta_0} dQ}{kT_H} = -\frac{Mc^2}{kT} + \frac{Mc^2}{kT_H} \tag{36}$$

The probability of the fluctuation corresponding to the actualization of the hadron is therefore proportional to

$$\exp(\Delta S) = \exp\left(-\frac{Mc^2}{kT} + \frac{Mc^2}{kT_H}\right) \tag{37}$$

The number of fluctuations likely to occur within the localization volume $(\hbar/Mc)^3$ of the exchanged hadron can be estimated from their *thermal* volume $(\hbar c/kT)^3$ as

$$\approx \left(\frac{\hbar}{Mc}\right)^3 \left(\frac{\hbar c}{kT}\right)^{-3} = \left(\frac{kT}{Mc^2}\right)^3 \tag{38}$$

The partition of the system made of equivalent thermostat and the process will therefore be the integral in M of the following expression

$$\left(\frac{kT}{Mc^2}\right)^3 \exp\left(-\frac{Mc^2}{kT} + \frac{Mc^2}{kT_H}\right) \tag{39}$$

It involves the same density of hadron mass states

$$\rho(M) \propto M^{-3} \exp\left(\frac{Mc^2}{kT_H}\right) \tag{40}$$

as that derived from Hagedorn's [45] original *statistical bootstrap* model; this confirms the identification of T_H with the Hagedorn temperature. To conclude, the *thermostat equivalent* to the background is manifested in each vertex of strong interaction between hadrons, and it consists of the self-similar pattern of fluctuations of quark and gluon plasma, according to the current interpretation of Hagedorn's model [46, 47]. This pattern will absorb heat at $T > T_H$ (released by the annihilation of hadrons entering the vertex) and will release it at $T < T_H$ through the creation of new hadrons, thus originating a temperature T_H characteristic of the hadronization process. Hadrons entering the vertex will be annihilated at the same temperature, at which the deconfinement of quarks and gluons will occur (Fig. 4). We believe that the salient point here consists of the relation between the Hagedorn temperature, which expresses the time-scale in Eq. 35 at which the scale invariance of the fluctuations is broken, and the constant θ_0. The value of this constant, as derived from a discussion about the electron, produces the correct value of T_H. This connection between the lepton and hadron worlds, resulting from the universality of the constant θ_0, remains hidden in the conventional treatment.

The introduction of the fundamental constant of nature θ_0 also has another effect. It implies the existence of a fundamental moment of inertia $\hbar\theta_0$, which should not be interpreted in the classical terms of a mass distribution, but rather as a conversion factor between time intervals characteristic of elementary particles and their angular momentum. Before re-scaling $\theta_0 \rightarrow \theta_0/n$ of de Sitter time of the particle micro-universe, the relevant moment of inertia is $mc^2\theta_0^2 = n\hbar\theta_0$, where m is the particle mass (substantially, the skeleton mass) and n is the ratio of the skeleton mass and 70 MeV; it probably plays a role in defining the Regge trajectories [48]. After re-scaling, the relevant moment of inertia is instead

$$I = mc^2 \left(\frac{\theta_0}{n}\right)^2 = \left(n\frac{\hbar}{\theta_0}\right)\left(\frac{\theta_0}{n}\right)^2 = \frac{\hbar\theta_0}{n} \qquad (41)$$

The angular momentum J of the particle is given by the product of I for an angular frequency ω typical of the particle. It is natural to set $\omega = j/(\theta_0/n)$, where j is the eigenvalue of the spin (even in the *dormant* state, the wavefunction of a particle nevertheless has a total spin eigenvalue of j, which defines its number of components). Thus

$$J = I\omega = \left(\frac{\hbar\theta_0}{n}\right)\left(\frac{nj}{\theta_0}\right) = j\hbar \qquad (42)$$

Eq. 42 holds for all particles, yet for hadrons a word of caution seems necessary. Indeed, the quark substructure could admit a different time-scale θ' from de Sitter time θ_0/n of the hadron it belongs to. In this case, the angular momentum associated with this substructure is therefore

$$J' = I\omega' = \frac{\hbar\theta_0}{n}\frac{j}{\theta'} \qquad (43)$$

An external probe capable of selectively detecting this substructure will therefore not see the kinematic spin J, but rather the apparent spin

$$J' = J\frac{\theta_0}{n\theta'} \qquad (44)$$

If the hadron is a polarized proton, the probe can also be a polarized charged lepton (electron or muon). In a process of deep inelastic scattering of the probe on the proton, characterized by a square of the transferred four-momentum Q^2 and a fraction x of the hadron momentum carried by the interacting quark, the spin *seen* by the probe is given by the integral of the quark spin structure function $g_1(x)$

$$\frac{1}{2}\int_0^1 g_1(x)\,dx = \frac{\Delta\Sigma}{2} \qquad (45)$$

The experiments carried out by several collaborations reveal that $\Delta\Sigma \approx 0.25$ instead of 1 as expected from the quark model [49, 50]. From our point of view, however, the function $g_1(x)$ should be scaled by the factor $n\theta'/\theta_0$ independent from x. Since $n = 13.4$ for the proton, this rescaling provides $\Delta\Sigma = 1$ if $\theta' \approx \theta_0/3$. According to this point of view, the eigenvalue j of the proton spin is always derived from the contributions of individual quarks according to the rules of the quark model, but the scattering process measures J' instead of the correct *kinematic* spin J.

The reason for the result $\theta' \approx \theta_0/3$ has yet to be determined. Since each quark is in one of three possible, and statistically equivalent, colour states, we can conjecture that the interval θ' derives from the ratio between the fundamental interval θ_0 and the number of these states. If so, the introduction of the constant θ_0 would provide a different way of looking at the proton *spin crisis* and the contribution of the *sea of virtual quarks* and gluons.

8 Comparison with other models

In this paper we have repeatedly stressed that the state vector reduction postulate (von Neumann's projection postulate) is the real expression of the quantum discontinuity. This discontinuity finds its complete realisation in quantum jumps that occur in quantum systems as a result of interactions with internal degrees of freedom or external systems. Among the latter, we must include coupling with measuring devices that represent only one possibility among many. Otherwise it would be difficult to understand how gaseous oxygen and hydrogen combine giving molecules perfectly defined as water, when placed inside a container with entirely opaque walls and without any observation of the process. The scope of the von Neumann postulate then goes well beyond the measurement procedures; it is rather an essential ingredient for formulating the quantum mechanics.

Von Neumann's postulate, on the other hand, does not specify the ontology for the reduction process; this specification must, in any case, be compatible with the process of unitary evolution of the state vector between two successive collapses, described by the quantum mechanical equations of motion. Otherwise, we would be dealing

Figure 4: *The Hagedorn thermostat. F = self-similar pattern of virtual fluctuations; H = hadrons; F → H heat flow = hadronization; H → F heat flow = deconfinement.*

with a theory that is entirely different from quantum mechanics and this is what occurs with the dynamic reduction mechanisms invoked by Ghirardi–Rimini–Weber [5] or by Penrose [51]. The minimal ontology of the projection operator proposed in this paper is fully consistent with usual quantum mechanics. It does not lead to new effects on the dynamics, but possibly only to constraints on the selection of the states (for example, particle masses). The possibility has been considered that the unitary evolution process of the state vector is, under appropriate conditions, sufficient to reduce the wavefunction. Within the area of quantum measurement theory, this possibility has already been explored in early models of Milan–Pavia [2] and Rome schools [52] and has re-emerged with the principle of *decoherent histories*. Although this approach is valid in the field of quantum measurement, it seems, however, to have less effect on interaction micro-events between micro-entities. The idea of the decoherent histories, proposed in particular by Zeh [53] and Zurek [54], requires the interaction of micro-entities with an environment with many degrees of freedom. Average operations on environmental degrees of freedom lead to mixtures of states of the system composed of micro-entities and apparatus, which are effectively indistinguishable from those derived as a result of the action of projection operators defined on a base selected by dynamic itself. From our point of view, the measurement processes are expressions of quantum discontinuity like the atomic quantum jumps which can occur in fully decoupled and virtually isolated quantum systems. Consider, for example, the electromagnetic emission from neutral galactic hydrogen which can be measured by any radio-telescope operating at a wavelength of 21.1 cm. This emission is generated by the transition between the two levels of the

hyperfine structure of neutral hydrogen atoms that are virtually isolated in interstellar space (their numerical density is often lower than or approximately equal to 1 cm^{-3}). Quantum state reduction therefore occurs also in entirely decoupled systems and cannot be caused by the interaction with the surrounding environment. In this example, decoherence should in addition act on an absolutely improbable entanglement of emitting atoms with the receiving radio-telescope, two physical systems that have never been in contact. On the other hand, if the reduction process is assumed already at the atomic or molecular level in the form of quantum jumps, the measurement process can easily be interpreted as a quantum jump of the system composed of measurement apparatus and micro-entities. By doing so, there is no need to hypothesize the splitting of the state vector of this system in a multiplicity of independent *worlds*, as assumed by Everett. Moreover, it seems appropriate to mention the similarities and differences between our approach and the two-state formalism proposed by Aharanov, Vaidman and collaborators [55, 56]. In both cases, time symmetry is emphasized and the propagation of *backward* state vector is acknowledged on an equal footing with the *forward* vector. In our view, the main difference is the fact that the one, proposed by Aharanov and colleagues, is a formalism and not an ontology. This formalism assumes pre-selection and post-selection, followed by the application of the Aharanov–Bergmann–Lebowitz rule (in the original version or one of its variants) for calculating the relevant averages. It thus implicitly assumes that it is possible to perform an initial preparation of the system and a *complete* measurement of it. The real issue, however, is exactly how to ensure these conditions necessary for the applicability of the formalism which is, in itself, correct.

It is not a coincidence that the two main supporters of this idea have identified this logical gap and have responded in their own way, opting for different ontologies [57]. Vaidman opts for a time-reversal version of Everett many-worlds model; Aharanov states he is not ready for that step to which he would (reluctantly) prefer an *objective collapse*. He assumes, in each case, the physical reality of the backward vector. This option inevitably leads to the transaction concept and its importance in the context of the two-state formalism is also considered by Elitzur and Cohen [58].

9 Conclusions

We proposed an extremely preliminary model of what a *quantum jump* is, in the broad sense of a discontinuous change in the quantum state of a system eventually related with particle localization in time. Our aim was to focus on the urgent need to provide the concept proposed by Bohr a century ago in the context of atomic theory with formal maturity, as this is the only concept from his model that is still not fully formalized. The matter was reopened by Erwin Schrödinger in an article written in 1952, where he states

> It is better to consider a particle not so much as a permanent entity, but rather as an instantaneous event (...). At times these events form chains that give the illusion of something permanent [59].

At the core of the proposed model is the concept of the circularity of the time coordinate that appears in the wavefunction of a particle, a circularity expressed by the de Broglie oscillation. The annihilation of the wavefunction is described through the removal of the topological constraint of circularity. The reverse process describes its creation. The removal of the circularity constraint leads to a vacuum state represented by a complex time that is without temporality, in the ordinary sense of the term, and spatiality. Oscillations occur in the real part of complex time that represents the internal state of the various elementary particles contained in the vacuum and which can be extracted from it or lead back to it. These particles are present in the form of virtual fluctuations and the imaginary part of complex time describes the energy amplitude of these fluctuations. This vacuum structure seems to introduce new constraints to the structure of elementary particles and their internal dynamics, resulting from the introduction of a new constant of nature with the dimensions of a time interval, which we assume to be identifiable with Caldirola's *chronon*. Numerically, the chronon is the time taken by light to travel a distance equal to the classical radius of the electron. The constraints in

question are perhaps reflected in already known facts, such as the proton *spin crisis* or the Hagedorn thermostat.

We have tried to show how the hypothesis of a time precursor allows to introduce nonlocality *ab initio*, in a consistent way with both the non-separability of entangled states and the quantum equations of motion as that of Dirac. It is possible therefore, in principle, to find a connection with the more usual, stochastic *hydrodynamic* representation of these equations and the role there played by the Bohm potential [60]. It will not be specified enough that nonlocality and entanglement are two distinct phenomena, although intimately linked to the comlex background of quantum systems. This fact has been well pointed out by some recent works in quantum information [61–63]. On the other hand, we have discussed the relations with the transactional representation and the two states formalism. In our opinion, the quantum mechanics is not a closed system, as the different interpretations seem to suggest; any reading of quantum mechanics must face the challenges posed by the complexity of the actual physical world at different scales, from the cosmological to the elementary particles. That raises the question of defining the relationship between the complex time and the polarized vacuum in *ordinary* quantum field theory. It is clear that while in quantum field theory actual physical properties are re-scaled according to the equations of the renormalization semi-group, localization requires instead a defined skeleton mass. One possibility would be simply to assign the task of describing the interaction vertices with a unitary formalism to the quantum field theory, and to extend the model under consideration to the asymptotic states (obviously, a no-rescalable notion). However, the discussion of this delicate problem requires further reflections targeted to the formal definition of a relationship between the two descriptions. At the present time we have to leave this question open. At a higher speculative level, the question of the Planck scale remains, considering that between this one and the typical range of quantum mechanics there is a gap of 16 orders of magnitude and 20 between the chronon and the Planck time. Eventually, even if quantum mechanics is revealed to be emergent, this will be in favor of a pre-space and purely algebraic theory of the foundations of physics.

Acknowledgements

The authors wish to thank Eliahu Cohen and John Ashmead for valuable suggestions that have improved the quality of this article.

References

[1] Ghirardi G. Sneaking a Look at God's Cards: Unraveling the Mysteries of Quantum Mechanics. Malsbary G (translator), Princeton: Princeton University Press, 2007.

[2] Daneri A, Loinger A, Prosperi GM. Quantum theory of measurement and ergodicity conditions. Nuclear Physics 1962; 33: 297–319. doi:10.1016/0029-5582(62)90528-X

[3] Bohm D, Bub J. A proposed solution of the measurement problem in quantum mechanics by a hidden variable theory. Reviews of Modern Physics 1966; 38 (3): 453–469. doi:10.1103/RevModPhys.38.453

[4] Longtin L, Mattuck RD. Relativistically covariant Bohm-Bub hidden-variable theory for spin measurement of a single particle. Foundations of Physics 1984; 14 (8): 685–703. doi:10.1007/bf00736616

[5] Ghirardi GC, Rimini A, Weber T. Unified dynamics for microscopic and macroscopic systems. Physical Review D 1986; 34 (2): 470–491. doi:10.1103/PhysRevD.34.470

[6] Schlosshauer M. Decoherence, the measurement problem, and interpretations of quantum mechanics. Reviews of Modern Physics 2005; 76 (4): 1267–1305. arXiv:quant-ph/0312059, doi:10.1103/RevModPhys.76.1267

[7] Bergquist JC, Hulet RG, Itano WM, Wineland DJ. Observation of quantum jumps in a single atom. Physical Review Letters 1986; 57 (14): 1699–1702. doi:10.1103/PhysRevLett.57.1699

[8] Nagourney W, Sandberg J, Dehmelt H. Shelved optical electron amplifier: observation of quantum jumps. Physical Review Letters 1986; 56 (26): 2797–2799. doi:10.1103/PhysRevLett.56.2797

[9] Sauter T, Neuhauser W, Blatt R, Toschek PE. Observation of quantum jumps. Physical Review Letters 1986; 57 (14): 1696–1698. doi:10.1103/PhysRevLett.57.1696

[10] Basche T, Kummer S, Brauchle C. Direct spectroscopic observation of quantum jumps of a single molecule. Nature 1995; 373 (6510): 132–134. doi:10.1038/373132a0

[11] Gleyzes S, Kuhr S, Guerlin C, Bernu J, Deleglise S, Busk Hoff U, Brune M, Raimond J-M, Haroche S. Quantum jumps of light recording the birth and death of a photon in a cavity. Nature 2007; 446 (7133): 297–300. doi:10.1038/nature05589

[12] Peil S, Gabrielse G. Observing the quantum limit of an electron cyclotron: QND measurements of quantum jumps between Fock states. Physical Review Letters 1999; 83 (7): 1287–1290. doi:10.1103/PhysRevLett.83.1287

[13] Yu Y, Zhu S-L, Sun G, Wen X, Dong N, Chen J, Wu P, Han S. Quantum jumps between macroscopic quantum states of a superconducting qubit coupled to a microscopic two-level system. Physical Review Letters 2008; 101 (15): 157001. doi:10.1103/PhysRevLett.101.157001

[14] Neumann P, Beck J, Steiner M, Rempp F, Fedder H, Hemmer PR, Wrachtrup J, Jelezko F. Single-shot readout of a single nuclear spin. Science 2010; 329 (5991): 542–544. doi:10.1126/science.1189075

[15] Vamivakas AN, Lu CY, Matthiesen C, Zhao Y, Falt S, Badolato A, Atature M. Observation of spin-dependent quantum jumps via quantum dot resonance fluorescence. Nature 2010; 467 (7313): 297–300. doi:10.1038/nature09359

[16] Vijay R, Slichter DH, Siddiqi I. Observation of quantum jumps in a superconducting artificial atom. Physical Review Letters 2011; 106 (11): 110502. arXiv:1009.2969, doi:10.1103/PhysRevLett.106.110502

[17] Itano WM, Bergquist JC, Wineland DJ. Early observations of macroscopic quantum jumps in single atoms. International Journal of Mass Spectrometry 2015; 377: 403–409. doi:10.1016/j.ijms.2014.07.005

[18] Davies PCW. Particles do not exist. In: Quantum Theory of Gravity: Essays in Honor of the 60th Birthday of Bryce S. DeWitt. Christensen SM (editor), Bristol, England: Adam Hilger, 1984, pp.66–77.

[19] Pessa E. The concept of particle in quantum field theory. In: Vision of Oneness. Licata I, Sakaji AJ (editors), Rome: Aracne, 2011, pp.13–40.

[20] Colosi D, Rovelli C. What is a particle? Classical and Quantum Gravity 2009; 26 (2): 025002. doi:10.1088/0264-9381/26/2/025002

[21] Heisenberg W. Die physikalischen Prinzipien der Quantentheorie. Leipzig: S. Hirzel, 1941.

[22] Chiatti L, Licata I. Relativity with respect to measurement: collapse and quantum events from Fock to Cramer. Systems 2014; 2 (4): 576–589. doi: 10.3390/systems2040576

[23] Hartle JB, Hawking SW. Wave function of the universe. Physical Review D 1983; 28 (12): 2960–2975. doi:10.1103/PhysRevD.28.2960

[24] Licata I, Chiatti L. The archaic universe: Big Bang, cosmological term and the quantum origin of time in projective cosmology. International Journal of Theoretical Physics 2009; 48 (4): 1003–1018. doi: 10.1007/s10773-008-9874-z

[25] Licata I, Chiatti L. Archaic universe and cosmological model: "Big-Bang" as nucleation by vacuum. International Journal of Theoretical Physics 2010; 49 (10): 2379–2402. arXiv:0808.1339, doi:10.1007/s10773-010-0424-0

[26] Parisi G. Statistical Field Theory. Advanced Book Classics, Boulder: Westview Press, 1998.

[27] Denkmayr T, Geppert H, Sponar S, Lemmel H, Matzkin A, Tollaksen J, Hasegawa Y. Observation of a quantum Cheshire Cat in a matter-wave interferometer experiment. Nature Communications 2014; 5: 4492. doi:10.1038/ncomms5492

[28] Chiatti L. The transaction as a quantum concept. International Journal of Research and Reviews in Applied Sciences 2013; 16 (4): 28–47. arXiv: 1204.6636

[29] Zeh H-D. There are no quantum jumps, nor are there particles! Physics Letters A 1993; 172 (4): 189–192. doi:10.1016/0375-9601(93)91005-P

[30] Chiatti L. Wave function structure and transactional interpretation. In: Waves and Particles in Light and Matter. van der Merwe A, Garuccio A (editors), Berlin: Springer, 1994, pp.181–187. doi: 10.1007/978-1-4615-2550-9_15

[31] Chiatti L. Path integral and transactional interpretation. Foundations of Physics 1995; 25 (3): 481–490. doi:10.1007/bf02059232

[32] Kastner RE. The Transactional Interpretation of Quantum Mechanics: The Reality of Possibility. Cambridge: Cambridge University Press, 2013.

[33] Licata I. Transaction and non locality in quantum field theory. EPJ Web of Conferences 2014; 70: 00039. doi:10.1051/epjconf/20147000039

[34] Cini M. Field quantization and wave particle duality. Annals of Physics 2003; 305 (2): 83–95. doi:10.1016/S0003-4916(03)00042-3

[35] Feynman RP, Vernon Jr FL. The theory of a general quantum system interacting with a linear dissipative system. Annals of Physics 1963; 24: 118–173. doi: 10.1016/0003-4916(63)90068-X

[36] Schwinger J. Brownian motion of a quantum oscillator. Journal of Mathematical Physics 1961; 2 (3): 407–432. doi:10.1063/1.1703727

[37] Blasone M, Srivastava YN, Vitiello G, Widom A. Phase coherence in quantum Brownian motion. Annals of Physics 1998; 267 (1): 61–74. arXiv: quant-ph/9707048, doi:10.1006/aphy.1998. 5811

[38] Vitiello G. Classical trajectories and quantum field theory. Brazilian Journal of Physics 2005; 35 (2A): 351–358. doi:10.1590/S0103-97332005000200021

[39] Caldirola P. The introduction of the chronon in the electron theory and a charged-lepton mass formula. Lettere al Nuovo Cimento 1980; 27 (8): 225–228. doi:10.1007/bf02750348

[40] Recami E, Olkhovsky VS, Maydanyuk SP. On non-self-adjoint operators for observables in quantum mechanics and quantum field theory. International Journal of Modern Physics A 2010; 25 (9): 1785–1818. arXiv:0903.3187, doi:10.1142/S0217751X10048007

[41] Battey-Pratt EP, Racey TJ. Geometric model for fundamental particles. International Journal of Theoretical Physics 1980; 19 (6): 437–475. doi:10.1007/BF00671608

[42] Rietdijk CW. The world is realistically four-dimensional, waves contain information embodied by particles codedly, and microphysics allows understandable models I. Annales de la Fondation Louis de Broglie 1988; 13 (2): 141–182.

[43] Rietdijk CW. The world is realistically four-dimensional, waves contain information embodied by particles codedly, and microphysics allows understandable models II. Annales de la Fondation Louis de Broglie 1988; 13 (3): 299–336.

[44] Kleinert H. Hadronization of quark theories. In: Understanding the Fundamental Constituents of Matter, vol.14. The Subnuclear Series, Zichichi A (editor), New York: Plenum Publishing, 1978, pp.289–389. doi:10.1007/978-1-4684-0931-4_7

[45] Hagedorn R. Statistical thermodynamics of strong interactions at high energies. Supplemento al Nuovo Cimento 1965; 3 (2): 147–186. CERN: 938671

[46] Hagedorn R, Rafelski J. From hadron gas to quark matter I. In: Statistical Mechanics of Quarks and Hadrons. Satz H (editor), Amsterdam: North Holland, 1981, pp.237–251. CERN: 125482

[47] Rafelski J, Hagedorn R. From hadron gas to quark matter II. In: Statistical Mechanics of Quarks and Hadrons. Satz H (editor), Amsterdam: North Holland, 1981, pp.253–272. CERN: 126179

[48] Chiatti L, Licata I. Quark gluon plasma as a critical state of de Sitter geometries. 2015, doi:10.13140/2.1.1181.2167

[49] Ashman J, et al. A measurement of the spin asymmetry and determination of the structure function g_1 in deep inelastic muon-proton scattering. Physics Letters B 1988; 206 (2): 364–370. doi:10.1016/0370-2693(88)91523-7

[50] Prok Y, et al. Precision measurements of g_1 of the proton and of the deuteron with 6 GeV electrons. Physical Review C 2014; 90 (2): 025212. doi:10.1103/PhysRevC.90.025212

[51] Penrose R. The Road to Reality: A Complete Guide to the Laws of the Universe. London: Jonathan Cape, 2004.

[52] Cini M, De Maria M, Mattioli G, Nicolò F. Wave packet reduction in quantum mechanics: a model of a measuring apparatus. Foundations of Physics 1979; 9 (7-8): 479–500. doi:10.1007/bf00708364

[53] Zeh H-D. On the interpretation of measurements in quantum theory. Foundations of Physics 1970; 1 (1): 69–76. doi:10.1007/bf00708656

[54] Zurek WH. Decoherence and the transition from quantum to classical. Physics Today 1991; 44 (10): 36–44. doi:10.1063/1.881293

[55] Aharonov Y, Bergmann PG, Lebowitz JL. Time symmetry in the quantum process of measurement. Physical Review B 1964; 134 (6): 1410–1416. doi:10.1103/PhysRev.134.B1410

[56] Aharonov Y, Albert DZ, Casher A, Vaidman L. Surprising quantum effects. Physics Letters A 1987; 124 (4): 199–203. doi:10.1016/0375-9601(87)90619-0

[57] Aharonov Y, Vaidman L. The two-state vector formalism: an updated review. In: Time in Quantum Mechanics. Lecture Notes in Physics, vol.734, Muga JG, Mayato RS, Egusquiza ÍL (editors), Springer, 2008, pp.399–447. doi:10.1007/978-3-540-73473-4_13

[58] Elitzur AC, Cohen E. The retrocausal nature of quantum measurement revealed by partial and weak measurements. AIP Conference Proceedings 2011; 1408 (1): 120–131. doi:10.1063/1.3663720

[59] Bitbol M. Schrödinger's Philosophy of Quantum Mechanics. Boston Studies in the Philosophy and History of Science, vol.188, Berlin: Springer, 1996. doi:10.1007/978-94-009-1772-9

[60] Bohm D, Hiley BJ. The Undivided Universe: An Ontological Interpretation of Quantum Theory. London: Routledge, 1993.

[61] Vértesi T, Brunner N. Quantum nonlocality does not imply entanglement distillability. Physical Review Letters 2012; 108 (3): 030403. arXiv:1106.4850, doi:10.1103/PhysRevLett.108.030403

[62] Buscemi F. All entangled quantum states are nonlocal. Physical Review Letters 2012; 108 (20): 200401. arXiv:1106.6095, doi:10.1103/PhysRevLett.108.200401

[63] Fiscaletti D, Licata I. Bell length in the entanglement geometry. International Journal of Theoretical Physics 2015; 54 (7): 2362–2381. doi:10.1007/s10773-014-2461-6

4

Realism and Antirealism in Informational Foundations of Quantum Theory

Tina Bilban

Institute Nova revija, Cankarjeva cesta 10b, SI-1000 Ljubljana, Slovenia. E-mail: tinabilban@gmail.com

Editors: *George Svetlichny, Stig Stenholm & Avshalom C. Elitzur*

Zeilinger-Brukner's informational foundations of quantum theory, a theory based on Zeilinger's foundational principle for quantum mechanics that an elementary system carried one bit of information, explains seemingly unintuitive quantum behavior with simple theoretical framework. It is based on the notion that distinction between reality and information cannot be made, therefore they are the same. As the critics of informational foundations of quantum theory show, this antirealistic move captures the theory in tautology, where information only refers to itself, while the relationships outside the information with the help of which the nature of information would be defined are lost and the questions "Whose information? Information about what?" cannot be answered. The critic's solution is a return to realism, where the observer's effects on the information are neglected. We show that radical antirealism of informational foundations of quantum theory is not necessary and that the return to realism is not the only way forward. A comprehensive approach that exceeds mere realism and antirealism is also possible: we can consider both sources of the constraints on the information, those coming from the observer and those coming from the observed system/nature/reality. The information is always the observer's information about the observed. Such a comprehensive philosophical approach can still support the theoretical framework of informational foundations of quantum theory: If we take that one bit is the smallest amount of information in the form of which the observed reality can be grasped by the observer, we can say that an elementary system (grasped and defined as such by the observer) correlates to one bit of information. Our approach thus explains all the features of the quantum behavior explained by informational foundations of quantum theory: the wave function and its collapse, entanglement, complementarity and quantum randomness. However, it does so in a more comprehensive and intuitive way. The presented approach is close to Husserl's explanation of the relationship between reality and the knowledge we have about it, and to Bohr's personal explanation of quantum mechanics, the complexity of which has often been missed and simplified to mere antirealism. Our approach thus reconnects phenomenology with contemporary philosophy of science and introduces the comprehensive approach that exceeds mere realism and antirealism to the field of quantum theories with informational foundations, where such an approach has not been taken before.
Quanta 2014; 3: 32–42.

1 Introduction

A century after the establishment of fundamentals of quantum mechanics, there is still no widely accepted interpretation of the results of quantum experiments and of the mathematical description of the quantum world. However, this does not mean that different interpretations of quantum mechanics have not contributed to the understanding of the world around us. As Avshalom C. Elitzur stated: "To be sure, physics would be very dull had these interpretations not been proposed in the first place. They teased researchers' minds and stimulated experimentation and theorizing" [1, p. 4].

In the last decades, quantum information theories have been one of the most important mind teasers. They offered different theoretical frameworks explaining the characteristics of the quantum world and stimulated new experiments. As they based the explanation of the quantum world on the concept of information and considered the relationship between information, knowledge and reality, they opened some fundamental philosophical questions, previously considered as too theory-laden to be included in the formulation of fundamental theory: "Information? Whose information? Information about what?" [2, p. 34].

This also holds for Brukner's and Zeilinger's *informational foundations of quantum theory*, a theory based on Zeilinger's foundational principle for quantum mechanics that the most elementary system has the information carrying capacity of at most one bit [3]. Brukner and Zeilinger manage to explain some seemingly problematic and unintuitive characteristics of the quantum world (e.g. entanglement, collapse of the wave function) by using simple theoretical framework based on a radical philosophical proposition that distinction "between reality and information, cannot be made" [4]. However, critics of informational foundations of quantum theory emphasize that this antirealistic move captures informational foundations of quantum theory in tautology [5] where (the system of) information is explained by (the characteristics of) information [5, 6]. Furthermore, if information is all we have, questions considering the nature of information: "Information? Whose information? Information about what?" [2], stay open and the philosophical basis of informational foundations of quantum theory is undefined. Consequently, the critics of the theory suggest return to realism [5, 6].

In the present paper we will analyze the philosophical standpoint of informational foundations of quantum theory, its problems and standpoints of its critics. We will consider the need for Zeilinger's and Brukner's philosophical radicalism, the justification of critic's appeal to realism and propose the third option, which exceeds mere realism and antirealism.

2 Zeilinger-Brukner's informational foundations of quantum theory

Zeilinger-Brukner's theoretical framework is based on the concept of information. However, information is not understood in a technical sense as in classical information theory. Zeilinger and Brukner describe information as the result of the observation, as the answer about the property of the observed system. One bit of information represents one possible answer to the question about the property of the object of investigation. For example, to the question "Spin up?" there are two possible answers, "yes" (spin up) or "no" (spin down) [7]. Regarding information as the answer to the question about the measured property, Zeilinger equates the role of knowledge and information in several papers, when he describes knowledge or information about an object [3, 8] or of reality [4].

Exceeding the point of view of realism of ontic approaches, where the scientific knowledge about reality is taken as a direct manifestation of reality, while the influence of the one observing/measuring and the observation process/measurement are neglected, informational foundations of quantum theory follows the epistemic approach and closely considers the way we refer to reality, the form in which we grasp reality; as Zeilinger writes: "there is no way to refer to reality without using the information we have about it" [4]. Based on this observation, Zeilinger presupposes that: "it is important not to make distinctions that have no basis" and concludes: "the distinction between reality and our knowledge of reality, between reality and information, cannot be made" [4], therefore: "Wirklichkeit und Information sind dasselbe" ("Reality and information are the same") [7, p. 317].

This equation between reality and information is the basis of the foundational principle of informational foundations of quantum theory. If we decompose "a system which may be represented by numerous propositions into constituent systems", each "constitutent system will be represented by fewer propositions" and

> the limit is reached when an individual system finally represents the truth value to one single proposition only. Such a system we can call an elementary system. We thus suggest a principle of quantization of information as follows. An elementary system represents the truth value of one proposition. [...] We now note that the truth value of a proposition can be represented by one bit of information [...] Thus our principle becomes simply: An elementary system carries one bit of information. [3, p. 635]

However, regarding the antirealistic character of their theory, Zeilinger and Brukner emphasize:

the notions such as that a system "represents" the truth value of a proposition or that it "carries" one bit of information only implies a statement concerning what can be said about possible measurement results. For us a system is no more than a representative of a proposition. [9, p. 326]

Considering this, Zeilinger and Kofler describe the foundational principle as: "An elementary system is the manifestation of one bit of information" [8, p. 476].

On the basis of this simple foundational principle, Zeilinger and Brukner can explain the seemingly unintuitive fundamental quantum phenomena revealed by quantum experiments. The principle explains quantum randomness and complementarity: since an elementary system carries the answer to one question only, all other answers must contain an element of randomness.

> The extreme case is when the measurement direction is orthogonal to the eigenstate direction. Then for the new measurement situation the system does not carry any information whatsoever, and the result is completely random. [...] The information carried now by the system is not in any way determined by the information it carried before the measurement. Thus we conclude that the new information the system now represents has been spontaneously created in the measurement itself. We finally remark that the viewpoint just presented lends natural support to Bohr's notion of complementarity. This notion is well known, for example, for position and momentum or for the interference pattern and the path taken in a two-slit experiment; precise knowledge of one quantity excludes any knowledge of the other complementary quantity. [3, p. 636]

Furthermore, Zeilinger argues that the entanglement as another fundamental feature of quantum mechanics follows from a slight generalization of the foundational principle. In quantum mechanics, states are said to be entangled if for any composite system of two or more particles there exist pure states of the system (states that are as completely specified as the theory allows) in which parts of the system do not have pure states of their own [10].

> *N elementary systems represent the truth values of N propositions. N elementary systems carry N bits.* [...] After the interaction the N bits might still be represented by the N systems individually or, alternatively, they might all be represented by the N systems in a joint way, in

the extreme with no individual system carrying any information on its own. In the latter case we have complete entanglement. [3, p. 637]

In the case of complete entanglement of two elementary systems, two bits of information are used to describe joint properties: e.g. should the spins of the two systems be measured along the z axis, they would be found to be identical and should they be measured along the x axis, they would be also found to be identical. These two propositions now uniquely determine the entangled quantum state, which does not contain any information about the individual systems. Therefore, any measurement performed on individual systems gives completely random results [3, 11].

The framework of informational foundations of quantum theory can also be used to explain the seemingly paradoxical collapse of the wave function:

> There is never a paradox if we realize that the wave function is just an encoded mathematical representation of our knowledge of the system. When the state of a quantum system has a non-zero value at some position in space at some particular time, it does not mean that the system is physically present at that point, but only that our knowledge (or lack of knowledge) of the system allows the particle the possibility of being present at that point at that instant. What can be more natural than to change the representation of our knowledge if we gain new knowledge from a measurement performed on the system? When a measurement is performed, our knowledge of the system changes, and therefore its representation, the quantum state, also changes. In agreement with the new knowledge, it instantaneously changes all its components, even those which describe our knowledge in the regions of space quite distant from the site of the measurement. [9]

Based on equation between reality and information, the mathematical formulation describing the quantum world can be taken as a mere representation of our knowledge: with the measurement we gain, new knowledge and consequently the presentation of knowledge changes. The wave function and the measured property are just two different representations of different information. In informational foundations of quantum theory information is not information about reality, it is our only reality. Information is understood as not causally connected to anything it would be about. Consequently, the objectivity of information cannot be taken as self-evident on the basis of the common, from us independently existing outer world.

Any concept of an existing reality is then a mental construction based on these answers [("yes" or "no") answers to the questions posed to Nature]. Of course this does not imply that reality is no more than a pure subjective human construct. From our observations we are able to build up objects with a set of properties that do not change under variations of modes of observation or description. These are "invariants" with respect to these variations. Predictions based on any such specific invariants may then be checked by anyone, and as a result we may arrive at an intersubjective agreement about the model, thus lending a sense of independent reality to the mentally constructed objects. [9, p. 351]

Objectivity of the quantum world can be taken into account only on the basis of certain invariants and of the inter-subjective agreement about the gained information and its meaning. On this basis it is possible to exceed the solipsism and to conclude that a system of information, independent from us, forms, what we can call objective reality, so that the outer world (in that sense) exists [7].

In informational foundations of quantum theory we can speak about an inter-subjective world of information, however, we cannot speak about the outer world that this information is about and which would be a basis for scientific objectivity. As critics of informational foundations of quantum theory emphasize, if information and reality are the same [7], we can end up in tautology, where all the information describes is information [5], it only explains (a system of) information by (the characteristics of) information [6].

Zeilinger's argument is based on sensible notion about the relationship between reality and information that we always refer to reality with the information we have about it [4]. However, by equating reality and information the sensible observation changes in tautology. The statement that the argument should support, destroys the sensibility of the argument itself. Now information only refers to itself.

By equating information and reality, the informational foundations of quantum theory loses the relationships outside the information with the help of which the mere nature of information would be defined. If information is all we have the answer to the question "Information about what?" cannot be provided. Despite the epistemic character of informational foundations of quantum theory, even the answer to the question "Whose information?" is not clear–if we can speak about a system of information independent from us, whose information it is then? It seems that the connection between information and the one getting/possessing this information has been lost within the attempts to assure the objectivity of information. As the critics of informational foundations of quantum theory put it:

> The very concepts of knowledge and information imply a special kind of relationship between different things, appropriate correlations between a knower and what is known. Thus "the distinction between reality and our knowledge of reality" not only can be made; *it must be made if the notions of knowledge and information are to have any meaning in the first place.* [5, p. 131]

3 Critic's standpoint

Critic's response to the problems in the epistemic approach of informational foundations of quantum theory and the radical antirealism of its authors is a return to the realism of ontic approaches. In the work by Daumer and colleagues [5] this can be seen in repeated praise of "Bohm's simple deterministic" explanation of quantum mechanics and simultaneous criticism of "the convoluted indeterministic one of the Copenhagen view." Timpson's criticism [6], however, offers more direct insight into the problems faced by informational foundations of quantum theory. Timpson labels informational foundations of quantum theory's epistemic position as immaterialist metaphysics, where results of measurement do not pertain to an externally existing mind-independent world and the object is just a useful construct connecting observations. He consents to Zeilinger's point of view that the immaterialistic or antirealistic position in informational foundations of quantum theory is based on Copenhagen tradition and that it can be found in similar form in Bohr's description of his own point of view. Commenting on supposedly similar Bohr's and Zeilinger's understanding of the relationship between physics, Nature and reality, Timpson's own standpoint becomes clear:

> The last sentence [of the statement famously attributed to Bohr by Petersen] is particularly pertinent: "Physics concerns what we can say about nature." Compare again, another statement of Zeilinger's, "...what can be said about Nature has a constitutive contribution on what can be *real.*" I think we find in these sentiments a crucial strand contributing to the thought that the rise of quantum information theory supports an informational immaterialism. If quantum mechanics reveals that the true subject matter of physics is what can be said, rather than how

things are, then this seems very close to saying that what is fundamental is the play of information across our psyche. [...] However, it is important to recognize that there is a very obvious difficulty with the thought that what can be said provides a constitutive contribution to what can be real and that physics correspondingly concerns what we can say about nature. Simply reflect that some explanation needs to be given of where the relevant constraints on what can be said come from. Surely there could be no other source for these constraints than the way the world actually is, it cannot *merely* be a matter of language. [6, p. 225]

Timpson criticizes the merely epistemic approach of informational foundations of quantum theory, which lacks consideration of the constraints coming from Nature. However, he replaces it with a merely ontic approach: for Timpson "there could be no other source for [the relevant constraints on what can be said] than the world actually is" [6, p. 225]. The informational foundations of quantum theory's epistemic position–the consideration of the way we refer to reality, of the constraints coming from the observational/descriptional ability of the one observing/describing Nature–is now completely left out. For Zeilinger, reality and information are the same, we could say that reality is merely the manifestation of information, for Timpson reality is all there is to affect what can be said in physics, we could say that information is merely the manifestation of reality.

Antirealism of the epistemic approach taken in informational foundations of quantum theory leads to some serious philosophical problems: the argumentation on which the equation between reality and information and consequently the foundational principle for quantum mechanics are based is lost in tautology, while the mere nature of information cannot be comprehensively defined. But is the return to the realism of ontic approaches really the only way forward?

4 Exceeding mere realism and antirealism

On the one hand, informational foundations of quantum theory is based on radical antirealism, on the other hand, the critique of informational foundations of quantum theory is based on radical realism. However, despite very rarely used in the philosophy of quantum mechanics, a comprehensive view that would consider both constraints: those coming from the way nature actually is and those coming from the way we (can) observe, describe Nature, is also possible.

As Zeilinger pointed out, we always refer to reality with the information, we have about it. Or to put it otherwise, reality is always given to us in the form of information. We cannot grasp the reality us such, we observe the reality and get the information about it on the basis of this observation. This is always the information about the reality, or to be more precise, about *the observed system*, thus defining *the observed* as the part of reality we (the one observing) are focused on, as the object of the observation process. However, as information is always information about reality and not reality itself, the constraints defining the information cannot come just from the way *the observed system* is. The information is always information for (or to put it otherwise, according to) someone that receives this information, thus defining this someone as *the observer*: in the present article *the observer* is defined as the one receiving the information about *the observed*; based on our experiences we have an insight only into how human being's receive and process the information, so, when not stated otherwise we will speak about a *human being observer*, though the term *observer* is a broader term. Information as *the observer's* information is always co-defined by the way *the observer* (can) observe(s), measure(s), describe(s) and understand(s) *the observed*.

Following a comprehensive view that exceeds the radicalism of mere realism or mere antirealism and considers the complex relationships between information and other agents of the observation process, the fundamental philosophical questions about the nature of information: "Information? Whose information? Information about what?" [2], can be easily answered:

Information about what? It is information about *the observed*. In the case of the description of a particular measurement, we can say that information is the value of the position or of the polarization along a particular direction of the (observed) photon. However, this information only describes *the observed* in the context of that particular measurement. This knowledge cannot be generalized to *the observed* in all contexts, since *the quantum observed* is changed by the measurement. Information is causally connected with *the observed*; what we know is causally connected with what there is. However, information does not present *the observed* in itself, information is *the observed* as perceived by *the observer* in the context of the particular observation process.

Whose information? It is *the observer's* information. It is always information of the one who observes *the observed*. Information is thus causally connected to *the observer's* way of observation, context of observation and his ability to observe. We should not attribute *the observed* any a priori properties independent of *the observer* and context of description.

The question is, is such a comprehensive approach possible in quantum theory with informational foundations, or is the Zeilinger's and Brukner's radicalism, which makes informational foundations of quantum theory vulnerable to philosophical critique, necessary for the theoretical framework based on the concept of information.

Considering information as *the observer's* information about *the observed*, we can still base the foundational principle for quantum mechanics on relation (but not on equation) between reality and information about it: given that one bit is the smallest amount of information in the form of which the observed reality can be grasped by the observer, we can say that the elementary system (grasped and defined as such by the observer, on the basis of his observation of reality) correlates to one bit of information.

All the fundamental features explained by informational foundations of quantum theory can be then explained by quantum theory with informational foundations based on philosophical basis exceeding mere realism or antirealism. Such a comprehensive approach enables the insight into complex connections between information, *the observer* and *the observed* and thus a complex philosophical insight into the quantum world as described by quantum theory with informational foundations.

4.1 Connections between information, the observer and the observed

Information and *the observed* are both in two ways connected to *the observer*:

4.1.1 Ontic connection

Information and *the observed* are connected to *the observer* as to the one who, by trying to get information, already (necessary) has an influence on *the observed*, because the inclusion of *the observed* in the observation process already influences how *the observed* is. This connection is a precondition to get information about *the observed*.

When a quantum system is measured, it entangles with *(the observer's)* environment. This is described as decoherence, "the practically irreversible dislocalization (in Hilbert space) of superpositions due to ubiquitous entanglement with the environment" [12, p. 7]. However, to describe something, it is necessary to be outside the described set. If *the observer* is to describe this entanglement, he has to put the cut between the entangled system he is describing and himself. Usually the cut is put between the measurement apparatus and *the observer*, who thus describes the quantum system and the measurement apparatus as the entangled system and thus as *the quantum observed*.

The postulate that to describe something, it is necessary to be outside the described set, operationalistically explains the cut between quantum and classical in the process of measurement and is thus identical to Heisenberg's consideration of this problem known as "Heisenberg cut" [13]. This cut is a necessary condition for the possibility of empirical knowledge and is as such operationalistic, but not arbitrary; the choice depends on the nature of the experiment/approach and co-defines the way *the observer* describes *the observed*. However, since quantum description is universal, while classical physics can describe only complex classical systems, the cut cannot be shifted arbitrary in the direction of the quantum system, but it can "be shifted arbitrarily far in the direction of *the observer* in the region that can otherwise be described according to the laws of classical physics" [14, p. 12].

When considering the connection between *the observer*, information and *the observed* from the ontic point of view, the answer to the question "what is changed at measurement?" would be–*the observed*. Of course, we cannot approach *the observed* directly, *the observed* by itself, we can only claim that what is changed is our information about *the (changed) observed* in the context of a particular measurement (that caused the change).

4.1.2 Epistemic connection

Information and *the observed* are connected with *the observer* as observer *per se*, as to the one for whom the information has a meaning. On the basis of the ontic connection the information becomes available, on the basis of the epistemic connection the information is grasped by *the observer*. Information has a meaning only as long as it is information for someone. Most probably the preconditions of our comprehension are those that determine information as the form, in which everything we comprehend is given. This epistemic connection can be described with the concept of projection postulate, describing the "collapse" of the state vector.

The cut between *the quantum observed* and *the classical observer* is now emplaced between the quantum system and the measuring apparatus; between both component systems of the total entangled system as defined within the description of the ontic connection.

When considering the connection between *the observer*, *the observed* and information from the epistemic point of view, the answer to the question "what is changed at measurement?" would be–our knowledge/our information. However, this knowledge is still causally connected to *the observed*. As we can see, the ontic and the epistemic connections between *the observer*, *the observed* and information are mutually dependent.

4.1.3 Fundamental features of the quantum world

Based on the insight into complex connections between information, *the observer* and *the observed*, some of the fundamental features of the quantum world, explained by informational foundations of quantum theory, can be explained in a more comprehensive and intuitive way. The wave function is still understood as *the observer's* knowledge, however, as *the observer's* knowledge about *the observed* that describes *the observed* before or after the measurement (or more correctly about *the potential future/past observed*). Thus it can describe *the observer's* knowledge only according to the potential (future) information about *the observed*, according to potential results of potential measurements. Therefore, the wave function can be understood by *the observer* only as a probability function.

Describing the quantum world as described by quantum theory with informational foundations, we can say that from the point of view of *the observer*, physical systems carry information and this informational content is behind quantum behavior. We always get particular information about *the observed* in the context of a particular observation. From the point of view of *the observer* (which is the only point of view we can have), *the observed* has the potential to give information even when not in the observation process. However, *the observed* is only defined by the particular information when in a relationship with another (observing) system and only for that observing system. Considering this, the quantum behavior, and entanglement and randomness as its main features, can be explained in a more comprehensive and clear way:

In entangled system, one of the component systems (one of the entangled particles) is defined only as a part of the total entangled system. It is completely specified from the point of view of the system it is entangled with, but unspecified for an "outer" observer:

> one could prepare a pair of particles, A and B, in a superposition of the state "particle A is at position x_1 and particle B is at position x_3" and the state "particle A is at position x_2 and particle B is at position x_4", formally written as $(|x_1\rangle_A|x_3\rangle_B + |x_2\rangle_A|x_4\rangle_B)/\sqrt{2}$. In such an entangled state, the composite system is completely specified in the sense that the correlations between the individuals are well defined. Whenever particle A is found at position x_1 (or x_2), particle B is certainly found at x_3 (or x_4 respectively). However, there is no information at all about whether particle A is at x_1 or x_2 and whether B is at x_3 or x_4. [8, p. 472]

When particles A and B are entangled, particle A is completely determined from the point of view of particle B and vice versa, each of them is completely determined from the point of view of *the "inner" observer*, where *the observer* is not understood as a human being *observer*, but simply as a system "possessing" the information about the other (observed) system. What is fully defined, are not the particular properties as such, in the sense of objective reality for all *the observers*, but the relationship between both component systems. Therefore from the point of view of one of the component systems, with respect to himself as a reference system, the other component system is fully determined. A classical measurement apparatus can get only a random answer to the question about the position of the component system A or B. For *an "outer" observer* the relationship between component systems is determined, but not the component systems themselves (because a reference system of *the "outer" observer* cannot depend on *the observed component system*). In the same way, all our properties (and properties of other classical systems) are completely determined (from the point of view of an *"inner" observer*, e.g., *an observer* from our environment that considers himself as *the observer*, myself as *the observed* and places the cut somewhere between us), since in our classical everyday world we are completely entangled with our environment (different measurements are constantly performed on us).

Considering this, it seems that randomness is the basic "characteristic" of the world, while determinism is a consequence of decoherence. In the case of description of a measurement of a quantum system, the ontic connection between *the observer*, *the observed* and information about it, is a description from the point of view of *the "outer" observer*: the cut between *the observer* and *the observed* is placed between the measurement apparatus and the rest of our environment to describe the entanglement of the quantum system. The epistemic connection is a description from the point of view of the *"inner" observer*: according to us (as *the observer's*), *the observed quantum system* is now fully defined by the measured property.

5 Discussion

In informational foundations of quantum theory, seemingly unintuitive quantum phenomena can be explained using a simple foundational principle based on equation between reality and information. However, this radical antirealism makes informational foundations of quantum theory vulnerable to philosophical critique, which emphasizes that informational foundations of quantum theory ends up in tautology, where all the information describes is information, while the relationships outside the information are lost and the answers to the questions "Information? Whose information? Information about what?" cannot be provided. Critic's solution to these philosophical problems of informational foundations of quantum theory is a turn to mere realism, where informational foundations of quantum theory's lack of consideration of the constraints coming from Nature is replaced with the lack of consideration of the constraints coming from the way the observer observes/describes Nature.

However, as we show in section 4, mere antirealism or realism are not the only available explanations of the relationship between reality and information, a comprehensive approach that considers all the constraints–those coming from the way Nature actually is and those coming from the way *the observer* observes/describes Nature–is also possible. The information can be understood as *the observer's* information about *the observed*. There exist ontic and epistemic connections between information, *the observed* and *the observer*. When a quantum system is measured (and thus becomes *the observed*), it entangles with the measurement apparatus. Consequently, its ontic status is essentially changed and only now the information about *the observed* is available to *the observer*, thus leading to the epistemic connection. Now *the observer* can describe *the observed* by one of his concepts. However, all of *the observer's* concepts are classical concepts, based on his experiences from his classical environment and applicable only to *the observed*, which is part of this environment (decoherence). The ontic and the epistemic connection are mutually dependent and both, *the observed* and *the observer* define the information.

Zeilinger's and Brukner's radical antirealism is not a necessary philosophical standpoint for the explanation of quantum behavior with the concept of information. In a context of the presented comprehensive philosophical approach, the foundational principle of informational foundations of quantum theory can still be formed, all the fundamental features explained by informational foundations of quantum theory still explained, while some of the fundamental features of the quantum world, like the wave function, entanglement and quantum randomness, can now be explained in a more comprehensive and intuitive way.

Considering the intuitiveness of such a comprehensive approach, the question arises whether, after a century after the establishment of fundamentals of quantum mechanics, such an approach really is something new. In continental philosophy a similar explanation of the relationship between reality and the knowledge one can have about reality has been offered by Husserl's phenomenology, based on his understanding of the concept of phenomenon. Husserl's phenomenon can be described as the thing as has been given/shown to me by itself, but

essentially to me, in my horizon, with the meaning it has to me [15]. As such, phenomenon is always essentially related to both, *the observer* (me) and *the observed* (the thing). Phenomenon is always intentional phenomenon, is phenomenon of something. The core of what we observe is *the observed* itself, but always within the horizon of the observation process and according to our own orientation. It is either an orientation towards thing itself or a specific interest, e.g. admiration, esthetical contemplation, practical interest, and this difference is essential for the observation. If things smell good or bad, these are not properties of things themselves. This is how they are given to *the observer*, because of his specific physical (bodily) interest, but these are always things given to *the observer* by themselves. Belief in the outer world, in reality, is Glaubengewissheit; it is belief in itself, because the connection between phenomenon and thing is based on certainty of reason, which is the foundation of any rational action in the world [16]. Husserl applied both the constraints, those coming from the thing (*the observed*/reality) and those coming from *the observer's* way of observing, to the phenomenon and we could say that he considered phenomenon as *the observer's* information about *the observed reality*.

Despite great relevance of Husserl's work to quantum mechanics, Husserl's phenomenology and quantum mechanics have only rarely been considered together [18–22], mostly due to the general separation between physics and philosophy during the so called "shut up and calculate era" [26] in the middle of 20th century and to the more personal "philosophical and political parting of the ways [between phenomenologists and philosophers of science] in pre-war Germany" [17]. However, according to [18, 20], Bohr's personal interpretation of quantum mechanics, though not directly influenced by Husserl, also considered both constraints: those that come from Nature/reality and those that come from the way *the observer* observes/measures Nature. In the Introduction to the forth volume of his *Philosophical writings*, his position is described as ontological realism and epistemological anti-realism:

> Bohr's insistence that the description of nature involves the description of interactions between measuring instruments and the objects whose properties they are designed to measure [...] commits him to an ontological realism. [...] Not only did Bohr deny that atomic objects were purely constructions, but also he [... distinguished] his view from those philosophers who regarded the measurement interaction as in some sense 'creating' the object of measurement. [...] At the same time, however, Bohr [...]

argues strongly against those forms of realism which would attempt to describe an objectively existing, independent reality in terms of concepts which are well-defined only in relation to 'phenomena', as he uses that term. Bohr's ontological realism extends beyond the macro-realm to the atomic domain, nevertheless his epistemological anti-realism prohibits any attempt to carry the descriptive concepts of classical physics necessary for the description of phenomena beyond the phenomenal sphere to a world of things-in-themselves. [23, pp. 12–13]

Bohr's position exceeds mere realism and antirealism: the information we get in the process of the measurement is information about *the observed*, about the measured object, however, it is *the observer's* information and not a direct manifestation of *the observed*: *the observer* can describe *the measured observed* by one of his classical concepts, but this concept is not applicable to *the observed* in all contexts and cannot be taken as the property defining *the observed* independently from the context of particular observation.

Though comprehensive and influential, Bohr's interpretation has been often misunderstood and simplified, especially after the so called "shut up and calculate" era, when a direct flow of knowledge especially between physicists and contemporary philosophers outside mere philosophy of science was broken. Partly integrated into the Copenhagen interpretation, where it was combined with common views of different quantum physicists mostly gathered around Bohr's Institute of Theoretical Physics in Copenhagen, Bohr's interpretation lost its complexity and sharpness. As we have shown in section 3, both, Zeilinger from his antirealistic point of view and Timpson from his realistic point of view, understand Bohr as immaterialist and antirealist and interpret the statement famously attributed to Bohr by Petersen accordingly:

> It is wrong to think that the task of physics is to find out how nature is. Physics concerns what we can say about nature. [24, p. 8]

Simplifying Bohr's point of view, Zeilinger and Timpson both limit themselves on "Physics concerns what we can say". For Timpson this reveals the problematic immaterialism:

> If quantum mechanics reveals that the true subject matter of physics is what can be said, rather than how things are, then this seems very close to saying that what is fundamental is the play of information across our psyches. [6, p. 225]

In contrast, for Zeilinger it directly supports his understanding of quantum mechanics as only indirectly a science of reality and predominately a science of knowledge and thus of information [25]. However, such understandings of Bohr's view are only possible, because the second part of his statement is disregarded: "Physics concerns what we can say *about nature*" [24, p.8] (our emphasis). Exceeding mere realism or antirealism, Bohr's statement considers both the constraints, those coming from *the observer*–"what we can say"–and those coming from *the observed*–"about nature". Thus, Bohr's statement is not implying that "the true subject matter of physics is what can be said" [6] and neither that quantum mechanics is predominately a science of knowledge [25]. It implies that the task of physics cannot be to describe nature as such, nature as it is, but nature as given to *the observer* in the form of phenomena (the expression used by Bohr, see the comparison between Husserl's and Bohr's usage of the term phenomenon in [18]).

As a dialog within the philosophical consideration of quantum theories with informational foundations is limited on opposition between realism and antirealism, the comprehensiveness of Bohr's position is lost. The approach taken in the present paper is not completely new in philosophy or even in philosophy of quantum mechanics, but it is completely new within quantum information theory: it connects a comprehensive philosophical approach exceeding mere realism and antirealism with understanding of the relationship between reality and information, solves philosophical problems of antirealistic quantum theories with informational foundations, helps to better define the nature of information and shows the limitness of the realistic critique of antirealistic approach within informational foundations of quantum theory. Furthermore, it reconnects philosophical thought outside mere philosophy of science with contemporary quantum mechanics and shows that such a connection can definitely benefit our understanding of the world around us. Half a century after the so called "shut up and calculate" era their dialogue should finally be renewed. We believe that our study can serve as a model of a successful connection between philosophical aspects and new knowledge emerging in the field of quantum mechanics.

Acknowledgements

This work was supported by a grant from the John Templeton Foundation. I would like to thank Markus Aspelmeyer, Časlav Brukner, Johannes Kofler and Anton Zeilinger for inspiring discussions and helpful comments.

References

[1] Elitzur AC. What is the measurement problem anyway? Introductory reflections on quantum puzzles. In: Quo Vadis Quantum Mechanics? The Frontiers Collection, Elitzur AC, Dolev S, Kolenda N (editors), Berlin: Springer, 2005, pp. 1–5. http://dx.doi.org/10.1007/3-540-26669-0_1

[2] Bell JS. Against 'measurement'. Physics World 1990; 3 (8): 33–40.

[3] Zeilinger A. A foundational principle for quantum mechanics. Foundations of Physics 1999; 29 (4): 631–643. http://dx.doi.org/10.1023/A:1018820410908

[4] Zeilinger A. The message of the quantum. Nature 2005; 438 (7069): 743. http://dx.doi.org/10.1038/438743a

[5] Daumer M, Dürr D, Goldstein S, Maudlin T, Tumulka R, Zanghì N. The message of the quantum? AIP Conference Proceedings 2006; 844 (1): 129–132. http://dx.doi.org/10.1063/1.2219357 http://arxiv.org/abs/quant-ph/0604173

[6] Timpson CG. Information, immaterialism, instrumentalism: old and new in quantum information. In: Philosophy of Quantum Information and Entanglement. Bokulich A, Jaeger G (editors), Cambridge University Press, 2010, pp. 208–227.

[7] Zeilinger A. Einsteins Schleier. Die neue Welt der Quantenphysik. München: Goldmann, 2005.

[8] Kofler J, Zeilinger A. Quantum information and randomness. European Review 2010; 18 (4): 469–480. http://dx.doi.org/10.1017/S1062798710000268 http://arxiv.org/abs/1301.2515

[9] Brukner Č, Zeilinger A. Information and fundamental elements of the structure of quantum theory. In: Time, Quantum and Information. Castell L, Ischebeck O (editors), Berlin: Springer, 2003, pp. 323–354. http://dx.doi.org/10.1007/978-3-662-10557-3_21 http://arxiv.org/abs/quant-ph/0212084

[10] Wootters WK. Entanglement of formation and concurrence. Quantum Information & Computation 2001; 1 (1): 27–44.

[11] Brukner Č, Zukowski M, Zeilinger A. The essence of entanglement, 2001. http://arxiv.org/abs/quant-ph/0106119

[12] Schlosshauer M, Camilleri K. The quantum-to-classical transition: Bohr's doctrine of classical concepts, emergent classicality, and decoherence, 2008. http://arxiv.org/abs/0804.1609

[13] Schlosshauer M. Decoherence and the Quantum-To-Classical Transition. The Frontiers Collection, Berlin: Springer, 2007. http://dx.doi.org/10.1007/978-3-540-35775-9

[14] Heisenberg W. Ist eine deterministische Ergänzung der Quantenmechanik möglich? Crull E, Bacciagaluppi G (translators), 2011. http://philsci-archive.pitt.edu/8590/

[15] Hribar T. Fenomen uma. In: Ideje za čisto fenomenologijo in fenomenološko filozofijo. E. Husserl, Slovenska Matica, 1997, pp. 498–522.

[16] Husserl E. Ideas Pertaining to a Pure Phenomenology and to a Phenomenological Philosophy, Kersten F (translator). Dordrecht: Kluwer, 1983.

[17] Heelan PA. Phenomenology and the philosophy of the natural sciences. In: Phenomenology World Wide: Foundations–Expanding Dynamics–Life-Engagements, A Guide for Research and Study. Tymieniecka A-T (editor), Dordrecht: Kluwer, 2003, pp. 631–641.

[18] Bilban T. Husserl's reconsideration of the observation process and its possible connections with quantum mechanics: Supplementation of informational foundations of quantum theory. Prolegomena 2013; 12 (2): 431–458.

[19] French S. A phenomenological solution to the measurement problem? Husserl and the foundations of quantum mechanics. Studies in History and Philosophy of Science Part B 2002; 33 (3): 467–491. http://dx.doi.org/10.1016/S1355-2198(02)00019-9

[20] Lurçat F. Understanding quantum nechanics with Bohr and Husserl. In: Rediscovering Phenomenology, vol.182. Phaenomenologica, Boi L, Kerszberg P, Patras F (editors), Springer Netherlands, 2007, pp. 229–258. http://dx.doi.org/10.1007/978-1-4020-5881-3_8

[21] London F, Bauer E. The theory of observation in quantum mechanics. In: Quantum Theory and Measurement. Wheeler JA, Zurek WH (editors), Princeton: Princeton University Press, 1983, pp. 217–259.

[22] Heelan PA. The phenomenological role of consciousness in measurement. Mind and Matter 2004; 2(1): 61–84.

[23] Faye J, Folse HJ. Introduction. In: The Philosophical Writings of Niels Bohr, vol. IV. Woodbridge: Ox Bow Press, 1998.

[24] Petersen A. The philosophy of Niels Bohr. Bulletin of the Atomic Scientists 1962; 19 (7): 8–14.

[25] Brukner Č, Zeilinger A. Quantum physics as a science of information. In: Quo Vadis Quantum Mechanics? The Frontiers Collection, Elitzur AC, Dolev S, Kolenda N (editors), Berlin: Springer, 2005, pp. 47–61. http://dx.doi.org/10.1007/3-540-26669-0_3

[26] Stenholm S. The Quest for Reality: Bohr and Wittgenstein–Two Complementary Views. Oxford: Oxford University Press, 2011.

5

Is Bohm's Interpretation Consistent with Quantum Mechanics?

Michael Nauenberg

Department of Physics, University of California, Santa Cruz, United States. E-mail: michael@physics.ucsc.edu

Editors: **Joseph B. Keller & Danko Georgiev**

The supposed equivalence of the conventional interpretation of quantum mechanics with Bohm's interpretation is generally demonstrated only in the coordinate representation. It is shown, however, that in the momentum representation this equivalence is not valid.
Quanta 2014; 3: 43–46.

1 Introduction

Recently, there has been a renewed interest in David Bohm's interpretation of *non-relativistic* quantum mechanics [1–4] and many pedagogical papers on this topic have appeared [5–17], while online, arXiv.org lists over 200 submissions on this topic during the past ten years. Bohm claimed that "as long as the present general form of Schrödinger's equation is retained the physical results obtained with [this] suggested alternative interpretation are precisely the same as those obtained with the usual interpretation", and that his interpretation "leads to precisely the same results for all physical processes as does the usual interpretation" [1, p.166]. Similar assertions also have been made in references [5–17], but this equivalence is usually demonstrated only in the coordinate representa-

tion, while the implications of Bohm's interpretation in the momentum representation are usually ignored. While there have been some criticisms in the past of Bohm's interpretation of quantum mechanics [18, 19], we give here an elementary proof that the momentum distribution in this interpretation differs from that in standard quantum mechanics. We show that the definition of particle velocity in this interpretation, implies that the product of mass times velocity is not equal to momentum, which is inconsistent with both classical and quantum mechanics. The word "consistent" is used here in accordance to its definition in the World English Dictionary: "A set of statements capable of all being true at the same time or under the same interpretation".

2 Bohmian mechanics differs from conventional quantum mechanics in the momentum representation

In Bohm's interpretation of quantum mechanics, the velocity of a particle with mass m is given by

$$\vec{v}_B = \vec{\nabla}S/m \qquad (1)$$

where $S/\hbar$ is the phase of the wave function ψ obtained by solving the time dependent Schrödinger equation. According to Bohm,

$$\vec{v}_B = \frac{d\vec{q}}{dt}, \qquad (2)$$

where $\vec{q}$ is the time dependent coordinate for the position of the particle, and Equation 1 becomes a first order differential equation that determines $\vec{q}$ as a function of time t, given its initial value. (In his original papers, [1,2], Bohm introduced as fundamental, the equation of motion for the acceleration $d\vec{v}/dt$. This equation can be obtained by taking the time derivative of Equation 1, but it is misleading to regard it as *fundamental*, because it implies that the initial velocity of the particle can be assigned arbitrarily. But given the initial position $\vec{q}$, this velocity is determined uniquely by Equation 1. Bohm's equation of motion leads to the appearance of a non-local "quantum potential" that accounts for the origin of an acceleration even when the classical potential vanishes.). But it turns out that the product $m\vec{v}_B$ is not equal to the canonical momentum $\vec{p}$, because $\vec{v}_B$ does not correspond to the velocity $\vec{v}$, that is determined in quantum mechanics by the operator

$$\vec{v} = -\frac{\iota\hbar}{m}\vec{\nabla}_q = \frac{\vec{p}}{m}. \tag{3}$$

A proof of this relation is given in section 5. Setting

$$\psi = R \, exp(\iota S/\hbar), \tag{4}$$

where R is the amplitude of ψ, we obtain

$$\vec{v}\,\psi = (\vec{\nabla}_q S/m - \iota\hbar\vec{\nabla}_q R/mR)\psi. \tag{5}$$

But in Bohm's definition of the particle velocity, Equation 1, only the first term on the right hand side of this equation appears. The relevance of the second term can be illustrated by considering the mean values $\langle\vec{v}\rangle$ and $\langle\vec{v}^2\rangle$ in this representation for ψ. We have

$$\langle\vec{v}\rangle = \int d^3q\,\psi^\dagger\vec{v}\psi = \int d^3q\bar{R}^2\vec{\nabla}S/m = \langle\vec{v}_B\rangle, \tag{6}$$

and

$$\langle\vec{v}^2\rangle = \int d^3q\,\psi^\dagger(\vec{v})^2\psi = \langle(\vec{v}_B)^2\rangle + (\hbar/m)^2\int d^3q\,(\vec{\nabla}R)^2. \tag{7}$$

Hence, Equation 7 implies that the second moment of the velocity distribution in conventional quantum mechanics differs from that obtained in Bohm's interpretation of the particle velocity, Equation 1, by the appearance of the additional term $(\hbar/m)^2\langle(\vec{\nabla}R)^2/R^2\rangle$ on the right hand side of this equation. Remarkably, this discrepancy is not even mentioned in any of the recent articles on Bohm's interpretation of wave mechanics [5–15]. Similar discrepancies also appear in all the higher moments of this distribution.

3 Bohmian "osmotic velocity", "fluid flow pathlines" and particle trajectories

To get agreement with the mean value $\langle\vec{v}^2\rangle$ in quantum mechanics, Equation 7, Bohm's interpretation requires, in addition to the Bohmian particle velocity $\vec{v}_B$ given by Equation 1, the existence of an *ad hoc* random velocity

$$\vec{v}_o = \frac{\hbar}{mR}\vec{\nabla}R, \tag{8}$$

with vanishing mean value, Originally, such a contribution was introduced with an undetermined coefficient as a *random* velocity by D. Bohm and J. P. Vigier [20], who named it an "osmotic velocity", after a term introduced by Einstein to describe the chaotic Brownian motion. But now such a term has been abandoned in discussions of Bohmian mechanics.

In particular, for stationary solutions of the Schrödinger, the phase $S = 0$, and Bohm's interpretation leads to the conclusion that the particle velocity vanishes in such a state. This conclusion is explained by invoking a *quantum force* due to a non-local *quantum potential* that supposedly balances the force due to the conventional potential that gives rise to the stationary solution. This non-classical force appears when the acceleration $d^2\vec{q}/dt^2$ is calculated by taking the time derivative of Equation 1 and Equation 2. But this result contradicts the fact that in quantum mechanics the velocity or momentum distribution for stationary solutions, given by the absolute square of the Fourier transform of ψ in coordinate space, is not a delta function at $\vec{v} = 0$, as is implied by Bohm's interpretation.

The trajectories obtained by integrating Bohm's first order differential equation for the particle coordinate $\vec{q}$, Equation 2, correspond to *pathlines* associated with the probability distribution $\rho = |\psi|^2$ which satisfies, like a normal fluid of density ρ, the continuity equation,

$$\frac{\partial\rho}{\partial t} + \vec{\nabla}_q.\vec{j} = 0 \tag{9}$$

where $\vec{j} = \vec{v}_B\rho$ is the associated current. While pathlines provide a visualization of a fluid flow, these lines do not correspond to the actual motion of the particles composing the fluid that also can have a random component. Likewise, Bohmian pathlines serve to visualize the evolution of the probability distribution in quantum mechanics, but do not correspond to actual trajectories of elementary particles.

Recently, experiments have been made with water droplets surfing on the waves produced by the Faraday instability on the surface of an oscillating tank filled with a fluid [21]. The motion of these droplets mimics the suggestion of de Broigle and of Bohm that elementary particles are likewise "piloted" by the ψ function of wave mechanics. In particular, it is claimed that when the waves propagate through two slits, or are confined in a "corral", the droplets satisfy statistics that are similar to those observed for particles in quantum mechanics. But such experiments only demonstrate the universality of wave propagation, and the associated pathlines, whether governed by the equations of fluid mechanics, quantum mechanics, or of other sources of waves in physics.

4 Discussion

In his original articles [1,2], Bohm proposed an extension of de Broigle's pilot wave theory of quantum mechancs which he asserted to be equivalent to Schrödinger's formulation of wave mechanics (together with Born's statistical interpretation). In Bohm's theory, particles move along classical trajectories with a velocity determined by the phase of Schrödinger's wave function ψ, satisfying a second order Newtonian-like equation of motion, but with an additional force due to a so-called "quantum potential", that is obtained from a solution of Schrödinger's equation. In his theory, the statistical character emerges from the unknown initial velocity and position of the particle that is given by the probability distribution $|\psi|^2$. Actually, Bohm's velocity satisfies a *first* order equation, and therefore only the initial position of the particle, but not its velocity, can be imposed arbitrarily.

We have shown that in Bohm's interpretation of quantum mechanics, the product of mass m times the velocity v of a particle does not correspond to the momentum. Hence, this interpretation is not only inconsistent with the standard formulation of quantum mechanics, but also with classical mechanics, where momentum is *defined* by the relation $p = mv$. But such inconsistencies were not mentioned in Bohm's original articles, and are now generally ignored in the vast literature on this subject.

5 Appendix. The relation between velocity and momentum in non-relativistic quantum mechanics

In quantum mechanics, the velocity $\vec{v}$, like the position $\vec{q}$ and the momentum $\vec{p}$, is an operator. It is *defined* by the relation

$$\vec{v} = \frac{\imath}{\hbar}[H, \vec{q}\,], \tag{10}$$

where H is the hamiltonian operator, and $[a, b] = ab - ba$ is the commutator of the operators a and b. In non-relativistic quantum mechanics,

$$H = -\frac{\hbar^2}{2m}\nabla_q^2 + V(\vec{q}), \tag{11}$$

corresponding to the time dependent Schrödinger equation

$$\imath\hbar\frac{\partial\psi}{dt} = H\psi \tag{12}$$

Hence, substituting this expression for H in Equation 10, one finds that the velocity operator is given by

$$\vec{v} = \frac{\vec{p}}{m} \tag{13}$$

where

$$\vec{p} = -\imath\hbar\vec{\nabla}_q \tag{14}$$

is the momentum operator.

For an alternative derivation of the connection between the velocity and momentum operators, Equation 13, that does not presuppose the Schrödinger equation, Equation 11 and Equation 12, consider the commutation relation Equation 10 for the Hamiltonian of a free particle $H_0 = \vec{p}^2/2m$. Then, according to the definition of velocity, Equation 10,

$$v_i = \frac{\imath}{2\hbar m}(p_j[p_j, q_i] + [p_j, q_i]p_j), \tag{15}$$

and substituting the Heisenberg-Born commutation relation

$$[p_j, q_i] = -\imath\hbar\delta_{i,j} \tag{16}$$

leads again to Equation 13.

References

[1] Bohm D. A suggested interpretation of the quantum theory in terms of "hidden" variables I. Physical Review 1952; 85 (2): 166-179. http://dx.doi.org/10.1103/PhysRev.85.166

[2] Bohm D. A suggested interpretation of the quantum theory in terms of "hidden" variables. II. Physical Review 1952; 85 (2): 180-193. http://dx.doi.org/10.1103/PhysRev.85.180

[3] Bohm D, Hiley BJ. The Undivided Universe: An Ontological Interpretation of Quantum Theory. New York: Routledge, 1993.

[4] Bacciagaluppi G, Valentini A. Quantum Theory at the Crossroads: Reconsidering the 1927 Solvay Conference. Cambridge: Cambridge University Press, 2009. http://arxiv.org/abs/quant-ph/0609184

[5] Matzkin A. Realism and the wavefunction. European Journal of Physics 2002; 23 (3): 285-294. http://dx.doi.org/ 10.1088/0143-0807/23/3/307 http://arxiv.org/abs/quant-ph/0208018

[6] Boozer AD. Hidden variable theories and quantum nonlocality. European Journal of Physics 2009; 30 (2): 355-365. http://dx.doi.org/10.1088/ 0143-0807/30/2/015

[7] Passon O. How to teach quantum mechanics. European Journal of Physics 2004; 25 (6): 765-769. http://dx. doi.org/10.1088/0143-0807/25/6/008 http://arxiv.org/abs/quant-ph/0404128

[8] Tumulka R. Understanding Bohmian mechanics: A dialogue. American Journal of Physics 2004; 72 (9): 1220-1226. http://dx.doi.org/ 10.1119/1.1748054 http://arxiv.org/abs/ quant-ph/0408113

[9] Tumulka R. Feynman's path integrals and Bohm's particle paths. European Journal of Physics 2005; 26 (3): L11-L13. http://dx.doi.org/ 10.1088/0143-0807/26/3/L01 http://arxiv. org/abs/quant-ph/0501167

[10] Nikolić H. Would Bohr be born if Bohm were born before Born? American Journal of Physics 2008; 76 (2): 143-146. http://dx.doi.org/10.1119/ 1.2805241 http://arxiv.org/abs/physics/ 0702069

[11] Bernstein J. More about Bohm's quantum. American Journal of Physics 2011; 79 (6): 601-606. http://dx.doi.org/10.1119/1.3556713

The author asserted that "when unambiguous the predictions of the two theories [Bohmian mechanics and standard quantum mechanics] are identical."

[12] Sanz AS, Miret-Artés S. Setting up tunneling conditions by means of Bohmian mechanics. Journal of Physics A: Mathematical and Theoretical 2011; 44 (48): 485301. http://dx. doi.org/10.1088/1751-8113/44/48/485301 http://arxiv.org/abs/1104.1298

[13] Sanz AS, Miret-Artés S. Quantum phase analysis with quantum trajectories: A step towards the creation of a Bohmian thinking. American Journal of Physics 2012; 80 (6): 525-533. http://dx.doi. org/10.1119/1.3698324 http://arxiv.org/ abs/1104.1296

In a departure from the usual Bohmian interpretation of quantum mechanics, the authors suggest that "if the Bohmian equations are understood as hydrodynamic equations, the trajectories obtained from the equation of motion (see Equation 1)

should not be regarded necessarily as the trajectories pursued by real particles, but rather as the streamlines associated with the quantum fluid". This article also contains a large number of references to the literature on Bohm's interpretation of quantum mechanics.

[14] Norsen T. The pilot-wave perspective on quantum scattering and tunneling. American Journal of Physics 2013; 81 (4): 258-266. http://dx.doi. org/10.1119/1.4792375 http://arxiv.org/ abs/1210.7265

[15] Beneduci R, Schroeck Jr FE. On the unavoidability of the interpretations of quantum mechanics. American Journal of Physics 2014; 82 (1): 80-82. http://dx.doi.org/10.1119/1.4824797 http://arxiv.org/abs/1211.3883

[16] Goldstein S. Quantum theory without observers–part two. Physics Today 1998; 51 (4): 38-52. http://dx.doi.org/10.1063/1.882241

The author concluded that Bohmian mechanics "agrees completely with orthodox quantum theory in its predictions. Precise and simple it involves an almost obvious incorporation of Schrödiger's equation into an entirely deterministic reformulation of quantum theory".

[17] Goldstein S. Bohmian mechanics. In: Stanford Encyclopedia of Philosophy, Zalta EN, Nodelman U, Allen C (editors), Stanford, California: Stanford University, 2012; http://plato.stanford. edu/entries/qm-bohm/

[18] Keller JB. Bohm's interpretation of the quantum theory in terms of "hidden" variables. Physical Review 1953; 89 (5): 1040-1041. http://dx.doi.org/ 10.1103/PhysRev.89.1040

A related demonstration of the inconsistency of Bohm's interpretation of quantum mechanics was pointed out by Keller more than 60 years ago, but it has either been ignored or forgotten.

[19] van Kampen NG. The scandal of quantum mechanics. American Journal of Physics 2008; 76 (11): 989-990. http://dx.doi.org/10.1119/1.2967702

[20] Bohm D, Vigier JP. Model of the causal interpretation of quantum theory in terms of a fluid with irregular fluctuations. Physical Review 1954; 96 (1): 208-216. http://dx.doi.org/10.1103/ PhysRev.96.208

[21] Harris DM, Moukhtar J, Fort E, Couder Y, Bush JWM. Wavelike statistics from pilot-wave dynamics in a circular corral. Physical Review E 2013; 88 (1): 011001. http://dx.doi.org/10.1103/ PhysRevE.88.011001

6

On Unitary Evolution and Collapse in Quantum Mechanics

Francesco Giacosa

Institute of Physics, Jan Kochanowski University, Kielce, Poland
Institute for Theoretical Physics, Johann Wolfgang Goethe University, Frankfurt am Main, Germany
E-mail: giacosa@th.physik.uni-frankfurt.de

Editors: *Chariton A. Dreismann & Danko Georgiev*

In the framework of an interference setup in which only two outcomes are possible (such as in the case of a Mach–Zehnder interferometer), we discuss in a simple and pedagogical way the difference between a standard, unitary quantum mechanical evolution and the existence of a real collapse of the wavefunction. This is a central and not-yet resolved question of quantum mechanics and indeed of quantum field theory as well. Moreover, we also present the Elitzur–Vaidman bomb, the delayed choice experiment, and the effect of decoherence. In the end, we propose two simple experiments to visualize decoherence and to test the role of an entangled particle. Quanta 2014; 3: 156–170.

1 Introduction

Quantum mechanics is a well-established theoretical construct, which passed countless and ingenious experimental tests [1]. Still, it is renowned that quantum mechanics has some puzzling features [2–8]: are macroscopic distinguishable superpositions (Schrödinger-cat states) possible or there is a limit of validity of quantum mechanics? Do measurements imply a non-unitary (collapse-like) time evolution or are they also part of a unitary evolution? In the latter case, should we simply accept that the wavefunction splits in many branches (i.e., parallel worlds), which decohere very fast and are thus independent from

each other? It is important to stress that these issues are not only central in nonrelativistic quantum mechanics but apply also in relativistic quantum field theory. Namely, the generalization to quantized fields does not modify the role of measurements. In this work we discuss in a introductory way some of the questions mentioned above. We study the quantum interference in an idealized two-slit experiment and we analyze the effect that a detector measuring "which path has been taken" has on the system. In particular, we shall concentrate on the collapse of the wavefunction, such as the one advocated by collapse models [7–14] and show which are the implications of it.

Variants of our setup also lead us to the presentation of the famous Elitzur–Vaidman bomb [15] and to delayed choice experiments [16, 17]. Thus, we can describe in a unified framework and with simple mathematical steps (typical of a quantum mechanical course) concepts related to modern issues and experiments of quantum mechanics.

Besides the pedagogical purposes of this work, we also aim to propose two experiments (i) to see decoherence at work in an interference setup with only two possible outcomes and (ii) to test the dependence of the interference on an idler entangled particle.

2 Collapse vs no-collapse: no difference?

2.1 Interference setup

We consider an interference setup as the one depicted in Fig. 1. A particle P flies toward a barrier which contains two 'slits' and then flies further to a screen S. Usually in such a situation there is a superposition of waves which generates on the screen S many maxima and minima. We would like to avoid this unnecessary complication here but still use the language of a double-slit experiment in which a sum over paths is present. To this end, we assume that the particle can hit the screen in two points only, denoted as A and B. All the issues of quantum mechanics can be studied in this simplified framework. We assume also to 'sit on' the screen S: when the particle hits A or B we 'see' it.

First, we consider the case in which only the left slit is open, see Fig. 1a. In order to achieve our goal, the slit is actually not a simple hole in the barrier (out of which a spherical wave would emerge) but a more complicated filter which projects the particle either to a straight trajectory ending in A or to a straight trajectory ending in B. In the language of quantum mechanics, this situation amounts to a wavefunction $|L\rangle$ associated to the particle which has gone through the left slit, which is *assumed* to be

$$|L\rangle = \frac{1}{\sqrt{2}}\left(|A\rangle - |B\rangle\right). \qquad (1)$$

Then, by simply using the Born rule (i.e., by squaring the coefficient multiplying $|A\rangle$ or $|B\rangle$), we predict that the particle ends up either in the endpoint A with probability 50% or in the endpoint B with probability 50%. This is indeed what we measure by repeating the experiment many times. As we see, the probability is a fundamental ingredient of quantum mechanics, which however enters only in the very last step, i.e. when the measurement comes into the game. The state $|L\rangle$ is an equal (antisymmetric) superposition of $|A\rangle$ and $|B\rangle$, but in a single experiment we do *not* find a pale spot on A and a pale spot on B: we always find the particle either fully in A or in B. It is only after many repetitions of the experiment that we realize that the outcome A and the outcome B are equally probable.

If only the right slit is open, see Fig. 1b, we have a similar situation in which only two trajectories ending in A and in B are present. The wavefunction of the particle after having gone through the right slit is denoted by $|R\rangle$ and is described by the orthogonal combination to $|L\rangle$:

$$|R\rangle = \frac{1}{\sqrt{2}}\left(|A\rangle + |B\rangle\right). \qquad (2)$$

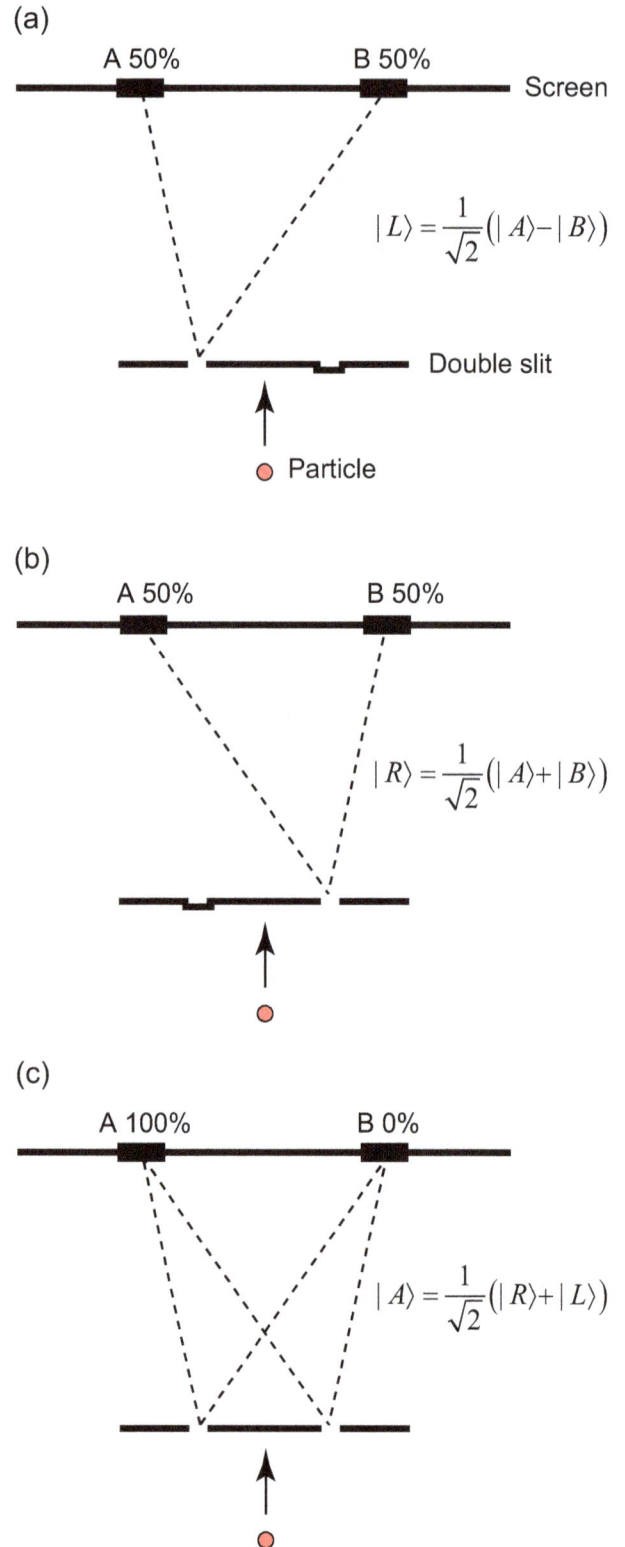

Figure 1: Hypothetical experiment with only two possible outcomes (A and B). Panel (a) only the left slit is open. Panel (b) only the right slit is open. Note, each slit is not a simple hole but acts as a filter which projects the particle either to a trajectory with endpoint A or to a trajectory with endpoint B. Panel (c) shows the experiment with both slits open: interference takes place and all particles hit the screen in A.

In this case one also finds the particle in 50% of cases in A and 50% in B.

We now turn to the case in which both slits are open, see Fig. 1c. The wave function of the particle is assumed to be the sum of the contributions of the two slits:

$$|\Psi\rangle = \frac{1}{\sqrt{2}}(|L\rangle + |R\rangle), \qquad (3)$$

i.e. the contributions of both slits add coherently. A simple calculation shows that

$$|\Psi\rangle = |A\rangle, \qquad (4)$$

which means that the particle P *always* hits the screen in A and *never* in B. Namely, in A we have a *constructive* interference, while in B we have a *destructive* interference. (Notice that the points A and B are not equidistant from the two slits. However, we take the two slits as being close to each other and the points A and B as being far from each other: the difference between the segments LA and RA (and so between LB and RB) is assumed to be negligible such that the two contributions of the wave packet of the particle P from the left and right slit arrive almost simultaneously and the depicted interference effect takes place).

In conclusion, we have chosen the language of a two-slit experiment because it is the most intuitive. The price to pay is a slit acting as a filter and not as a simple hole. However, one can easily build analogous setups as the one here described by using photon polarizations, electron spins or equivalent quantum objects, or by using a Mach–Zehnder interferometer, see details in § 2.3.3.

2.2 Detector measuring the path

As a next step we put a detector D right after the two open slits. D measures through which hole the particle has passed, without destroying it. We analyze the situation in two ways: first, by assuming the collapse of the wavefunction as induced by D and, second, by studying the entanglement of the particle with the detector. Note, we still assume that we sit on (or watch) the screen S only, but we are not directly connected to the detector D.

2.2.1 Collapse

In this case we assume that the detector D generates a collapse of the wavefunction. Suddenly after the interaction with D, the state of the particle P collapses into $|L\rangle$ with a probability of 50% or into $|R\rangle$ with a probability of 50%. Then, the state is described by either $|L\rangle$ or $|R\rangle$, but not any longer by the superposition of them. As a consequence, we have in half of the cases a situation analogous

to having only the left slit open and in the other half to having only the right slit open.

What we will then see on the screen S? The probability to find the particle in A is given by

$$P[A] = P[L,A] + P[R,A] = \frac{1}{2}\cdot\frac{1}{2} + \frac{1}{2}\cdot\frac{1}{2} = \frac{1}{2} \qquad (5)$$

where $P[L,A] = 1/4$ is the probability that the detector D has measured the particle going through the left slit and then the particle has hit the screen in A. Similarly, $P[R,A] = 1/4$ is the probability that the detector D has measured the particle going through the right slit before the latter hits A. For P[B] holds a similar description

$$P[B] = P[L,B] + P[R,B] = \frac{1}{2}\cdot\frac{1}{2} + \frac{1}{2}\cdot\frac{1}{2} = \frac{1}{2}. \qquad (6)$$

The collapse is obviously part of the standard interpretation of quantum mechanics, in which a detector is treated as a classical object which induces the collapse of the quantum state. As a result, there is no interference on the screen S. As renowned, the standard interpretation does not put any border between what is a classical system and what is a quantum system. Nevertheless, one can interpret the collapse postulate as an effective description of a physical process. Namely, in theories with the collapse of the wavefunction, the collapse is a real physical phenomenon which takes place when one has a macroscopic displacement of the position wavefunction of the detector (or, more generally, of the environment). In this framework, somewhere in between the quantum world and the classical macroscopic world, a *new* physical process takes place which realizes the collapse: this could be, for instance, the stochastic hit in the Ghirardi-Rimini-Weber model [7,9,10] or the instability due to gravitation in the Penrose-Diosi approach [8,12,13]. Neglecting details, the main point is that such collapse theories realize physically the collapse which is postulated in the standard interpretation and liberates it from inconsistencies. Still, it is an open and well posed physical question if (at least one of) such collapse theories are (is) correct.

2.2.2 No-collapse

In this case we do not assume that the detector D generates a collapse of the wavefunction, but we enlarge the whole wave function of the system by including also the wavefunction of the detector. We assume that, prior to measurement, the detector is in the state $|D_0\rangle$ (we can, for definiteness, think of a old-fashion indicator which points to 0). Then, when both slits are open, the state of the whole system just after having passed through them but not yet in contact with the detector D, is given by

$$|\Psi\rangle = \frac{1}{\sqrt{2}}(|L\rangle + |R\rangle)|D_0\rangle. \qquad (7)$$

Then, the particle-detector interaction induces a (we assume very fast) time evolution which generates the following state:

$$|\Psi\rangle = \frac{1}{\sqrt{2}} \left(|L\rangle |D_L\rangle + |R\rangle |D_R\rangle \right), \tag{8}$$

where $|D_L\rangle$ ($|D_R\rangle$) describes the pointer of the detector pointing to the left (right). Thus, no collapse is here taken into account, because the whole wavefunction still includes a superposition of $|L\rangle$ and $|R\rangle$, which, however, are now entangled with the detector states $|D_L\rangle$ and $|D_R\rangle$, respectively.

An important point is that the overlap of $|D_L\rangle$ and $|D_R\rangle$ is small

$$\langle D_L | D_R \rangle \simeq 0, \tag{9}$$

to a very good degree of accuracy. To show it, let us ignore the rest of the detector and the environment and concentrate on the pointer only, which is assumed to be made of N atoms, where N is of the order of the Avogadro constant. The atom α of the pointer is in a superposition of the type $\left(\psi_L^\alpha(\vec{x}) + \psi_R^\alpha(\vec{x}) \right) / \sqrt{2}$, where $\psi_L^\alpha(\vec{x})$ ($\psi_R^\alpha(\vec{x})$) is the wavefunction of the atom when the pointer points to the left (right). We have

$$\langle D_L | D_R \rangle = \prod_{\alpha=1}^{N} \int d^3 x \left(\psi_L^\alpha(\vec{x}) \right)^* \psi_R^\alpha(\vec{x}). \tag{10}$$

The quantity $\int d^3 x \left(\psi_L^\alpha(\vec{x}) \right)^* \psi_R^\alpha(\vec{x}) = \lambda_\alpha$ is such that $|\lambda_\alpha| < 1$. For a large displacement, λ_α is itself a very small number (small overlap), but the crucial point is to observe that $\langle D_L | D_R \rangle$ is the product of many numbers with modulus smaller then 1. Assuming that $\lambda_\alpha = \lambda$ for each α (each atom gets a similar displacement: this assumption is crude but surely sufficient for an estimate), we get

$$\langle D_L | D_R \rangle \simeq \lambda^N, \tag{11}$$

which is extremely small for large N. Even if we take $\lambda = 0.99$ (which is indeed quite large and actually overestimates the overlap of the wave functions of an atom belonging to macroscopic distinguishable configuration), we obtain

$$\langle D_L | D_R \rangle \simeq 0.99^{N_A} \sim 10^{-10^{21}} \tag{12}$$

which is tremendously small.

After having clarified the *de facto* orthogonality of $|D_L\rangle$ and $|D_R\rangle$, we rewrite the full wavefunction of the system $|S\rangle$ as

$$|\Psi\rangle = \frac{1}{2} \left[|A\rangle \left(|D_R\rangle + |D_L\rangle \right) + |B\rangle \left(|D_R\rangle - |D_L\rangle \right) \right]. \tag{13}$$

Then, the probability to find the particle P in A is obtained (now by using the Born rule, because we are observing the screen S):

$$P[A] = P[L, A] + P[R, A] = \frac{1}{2} \cdot \frac{1}{2} + \frac{1}{2} \cdot \frac{1}{2} = \frac{1}{2} \tag{14}$$

where $P[L, A] = 1/4$ is the probability that the system is described by $|A\rangle |D_L\rangle$ and $P[R, A] = 1/4$ the probability that it is described by $|A\rangle |D_R\rangle$. A similar situation holds for $P[B] = 1/2$. Thus, also in this case the presence of D causes the disappearance of interference.

The same result is obtained if we use the formalism of the statistical operator, which is defined by $\hat{\rho} = |\Psi\rangle \langle\Psi|$ (see, for instance, Refs. [1, 7]). Upon tracing over the detector states (environment states) the reduced statistical operator reads (we use here $\langle D_L | D_R \rangle = 0$):

$$\hat{\rho}_{red} = \langle D_L | \hat{\rho} | D_L \rangle + \langle D_R | \hat{\rho} | D_R \rangle$$

$$= \left(\begin{array}{cc} |A\rangle & |B\rangle \end{array} \right) \left(\begin{array}{cc} \frac{1}{2} & 0 \\ 0 & \frac{1}{2} \end{array} \right) \left(\begin{array}{c} \langle A| \\ \langle B| \end{array} \right), \tag{15}$$

where the diagonal elements represent $p[A] = p[B] = 1/2$ respectively, while the off-diagonal elements vanish in virtue of the (for all practical purposes) orthogonality of $|D_L\rangle$ and $|D_R\rangle$.

2.2.3 Summary

We find that, for us sitting on the screen S, the *very same outcome*, i.e. the absence of interference, is obtained by applying the collapse postulate as an intermediate step due to the detector D or by considering the whole quantum state (including the detector D) and by applying the Born rule *only* in the very end. This equivalence holds as long as the (anyhow very small) overlap of the detector states of Eq. 12 is neglected (see also the related discussion in § 3). The question is then: do we need the collapse? The second calculation (no-collapse) seems to answer us: 'no, we don't'. In this respect, one has a superposition of macroscopic distinct states, which coexist and are nothing else but the branches of the Everett's many worlds interpretation of quantum mechanics [18]. Thus, assuming that no collapse takes place brings us quite naturally to the many worlds interpretation [3, 19–21]. (Originally, Everett [18] introduced the concept of 'relative state formulation', which was reinterpreted as the many worlds interpretation by Wheeler and Dewitt [19,20]. The many worlds interpretation is the most natural interpretation when no collapse is present, but the definition of what is a 'world' is not trivial. Intuitively, it is a piece of the wavefunction which is a pointer-state, i.e. it does not contain spacial superpositions of macroscopic objects. Other points of view, such as 'many histories' and 'many minds' were also considered.)

However, care is needed: in fact, the 'no collapse' assumption is a general statement and means also that there is no collapse when the particle P hits the screen S (where *our own* wavefunction is part of the game). Let us clarify better this point by going back to the very first case we have studied, in which only the left slit was open and no detector D was present (Fig. 1a). The wavefunction of the particle just before hitting the screen is given by Eq. 1. But then, after the hit and assuming no collapse, the whole wavefunction (including us, who are the observers) reads:

$$|\Psi\rangle = \frac{1}{\sqrt{2}} |A\rangle \left|\text{Screen recording } A \text{ and we observing } A\right\rangle$$
$$- \frac{1}{\sqrt{2}} |B\rangle \left|\text{Screen recording } B \text{ and we observing } B\right\rangle .$$
$$(16)$$

The question is why the coefficient in front of the vector

$$|A\rangle \left|\text{Screen recording } A \text{ and we observing } A\right\rangle$$

tells us which is the *subjective* probability of observing A for the observer (us) sitting on the screen. In other words, how does the many worlds interpretation explain the probabilities according to the Born rule? The Born rule seems to be an additional postulate, which has to be put *ad hoc* into it. This situation is however not satisfactory, because the main idea of the many worlds interpretation is to eliminate the collapse from the description of the quantum mechanics and consequently to *derive* the standard Born probabilities. Although there are attempts to show that there is no need of postulating the Born rule in this context [22–24] (see also Ref. [25]), no agreement has been reached up to now [7, 26, 27]. (Notice that in the case of Eq. 16 one could understand the many worlds interpretation by noticing that there are two worlds, ergo the subjective probability to be in one of those is 50% in agreement with the Born rule. However, this is a particular case with equal coefficients. When the coefficients in front of the kets are *not* $1/\sqrt{2}$ (but say a and b with $|a|^2 + |b|^2 = 1$) one still has two worlds but the subjective probability to be in one of those is not $1/2$, but the one given by the Born rule ($|a|^2$ and $|b|^2$ respectively). This is exactly the point discussed in Refs. [22–24, 26, 27] with, however, different conclusions.) This is indeed an argumentation in favor of the possibility that a collapse really takes place. Surely, 'real collapse' scenarios deserve to be studied theoretically and experimentally [7–9].

Note, up to now we did not mention the decoherence, see e.g. Refs. [2, 28–32] and references therein. This is possible because we have put a detector that makes a measurement by evolving from the state $|D_0\rangle$ into two (almost) orthogonal states $|D_L\rangle$ and $|D_R\rangle$, but actually one

can interpret this fast change of the detector state as the result of a decoherence phenomenon. This is however a rather peculiar decoherence, because we have prepared the detector in a particular (low entropic) $|D_0\rangle$ state, which is 'ready to' evolve into $|D_L\rangle$ and $|D_R\rangle$ as soon as it interacts with the particle P. In § 3 we will describe what changes when the environment, instead of the detector, is taken into account.

2.3 Variants of the setup

2.3.1 The bomb

A simple change of the setup allows us to present the famous Elitzur–Vaidman bomb, first described in Ref. [15] and then experimentally verified in Ref. [33]. We substitute the detector with a 'bomb', which can be activated by the particle P. We place the bomb only in front of the left slit, see Fig. 2. This means that, if only the left slit is open, the bomb explodes soon after the particle has gone through the slit. If, instead, only the right slit is open, it doesn't explode. For definiteness and simplicity we assume that the particle is not destroyed nor absorbed by the bomb.

Just as previously, we can interpret the experiment applying either the collapse or by studying the whole wavefunction. In the collapse approach, the bomb simply makes a measurement. When both slits are open the wavefunction, before the interaction with the bomb, is given by Eq. 3: we will have an explosion in 50% of cases and no explosion in the remaining 50%. Notice that in the second case the bomb is doing a *null* measurement. The very fact that the bomb does *not* explode means that the particle went to the right slit (we assume 100% efficiency in our ideal experiment). When the bomb explodes there is a collapse into $|\Psi\rangle = |L\rangle$, when it does not explode there is a collapse into $|\Psi\rangle = |R\rangle$. Then, we have a situation which is very similar to the case of the detector D which we have studied previously: no interference on the screen S is observed, but we observe the particle in the endpoint A and B with probability $1/2$ each.

If we do not assume the collapse of the wavefunction, the whole wavefunction is given by (after interaction with the bomb)

$$|\Psi\rangle = \frac{1}{\sqrt{2}} (|L\rangle |B_E\rangle + |R\rangle |B_0\rangle) \qquad (17)$$
$$= \frac{1}{2} [|A\rangle (|B_0\rangle + |B_E\rangle) + |B\rangle (|B_0\rangle - |B_E\rangle)]$$

where $|B_0\rangle$ is the state describing the unexploded bomb and $|B_E\rangle$ the exploded one. Obviously, as in Eq. 12, we have $\langle B_E|B_0\rangle \simeq 0$. Again and just before no interference is seen on S but the two outcomes A and B are

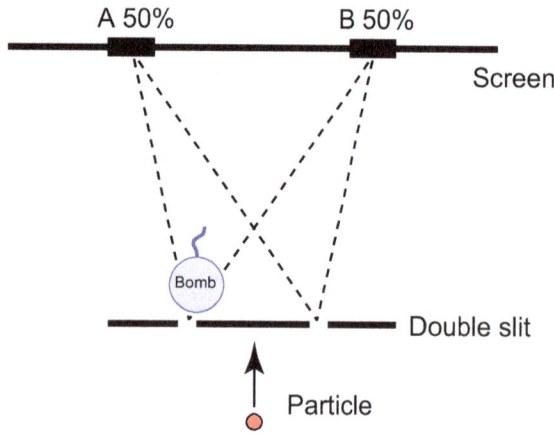

Figure 2: *Variant of the Elitzur–Vaidman experiment: a bomb is placed just after the left slit.*

equiprobable. Clearly, no difference between assuming the collapse or not is found, but the interesting fact is that the non-explosion of the bomb is enough to destroy interference. If, instead of the bomb we put a fake bomb (referred to as the dud bomb, which has the very same aspect of the real functioning bomb but does not interact *at all* with the particle P), the wavefunction of the system is given by

$$|\Psi\rangle = \frac{1}{\sqrt{2}}(|L\rangle + |R\rangle)\left|B_0^{\mathrm{dud}}\right\rangle = |A\rangle\left|B_0^{\mathrm{dud}}\right\rangle \quad (18)$$

where $\left|B_0^{\mathrm{dud}}\right\rangle$ describes the wavefunction of the dud bomb. In this case, there is interference and the particle P *always* ends up in A.

Then, the amusing part comes: if we do not know if the bomb is a dud or not, we can (in some but not all cases) find out by placing it in front of the left slit. If there is no explosion and the particle ends up in B, we deduce *for sure* that the bomb is real. Namely, this outcome is not possible for a dud, see Eq. 18. Note, we have deduced that the bomb is 'good' without making it explode (that would be easy: just send the particle P toward the bomb, if it goes 'boom' it *was* real). This situation occurs in 25% of cases in which a functioning bomb is placed behind the slit, see Eq. 17: we can immediately 'save' 25% of the good bombs. Conversely, in 50% of cases the good bomb simply explodes and we lose it (then, the particle P goes to either A (25%) or to B (25%)). In the remaining 25% the good bomb does not explode, but the particle P hits A. Then, we simply do not know if the bomb is good or fake: this situation is compatible with both hypotheses. We can, however, repeat the experiment: in the end, we will be able to save $1/4 + 1/4 \cdot 1/4 + \ldots = 1/3$ of the functioning bombs.

2.3.2 The idler particle and the delayed choice experiment

Another interesting configuration is obtained by assuming that a second entangled particle, denoted as I (for idler), is emitted when P goes through the slit(s). The system is built in the following way: if the particle P goes through the left slit, the particle I is described by the state $|I_L\rangle$. Similarly, when the particle P goes to the right slit, the particle I is described by the state $|I_R\rangle$. We assume that the two idler states are orthogonal: $\langle I_L|I_R\rangle = 0$. This situation resembles closely that of delayed choice experiments [16, 17].

When both slits are open the whole wavefunction of the system is given by

$$|\Psi\rangle = \frac{1}{\sqrt{2}}(|L\rangle|I_L\rangle + |R\rangle|I_R\rangle)$$
$$= \frac{1}{\sqrt{2}}(|A\rangle|I_+\rangle + |B\rangle|I_-\rangle), \quad (19)$$

where

$$|I_+\rangle = \frac{1}{\sqrt{2}}(|I_R\rangle + |I_L\rangle) \quad (20)$$

$$|I_-\rangle = \frac{1}{\sqrt{2}}(|I_R\rangle - |I_L\rangle). \quad (21)$$

The idler particle I is entangled with the particle P, but being a microscopic object, we surely cannot apply the collapse hypothesis because the particle I is *not* a measuring apparatus.

Do we have interference on the screen S in this case? The answer is clear: no. The states $|A\rangle|I_L\rangle$, $|A\rangle|I_R\rangle$, $|B\rangle|I_L\rangle$, $|B\rangle|I_R\rangle$ represent a basis of this system, thus the probability to obtain $|A\rangle$ (that is, the probability of P hitting S in A) is $1/4 + 1/4 = 1/2$. So for B. The presence of the entangled idler state destroys the interference on S.

It is sometime stated that this result is a consequence of the fact that the state of the idler particle I carries the information of which way P has followed. For this reason, the interference has disappeared (this is a modern reformulation of the complementarity principle). However, such expressions, although appealing, are often too vague and need to be taken with care.

As a next step we study what happens if we perform a measurement on the idler particle I. We study separately two distinct types of measurements.

Measuring the idler particle in the $|I_L\rangle,|I_R\rangle$ basis

First, we perform a measurement which tells us if the state of the idler particle is $|I_L\rangle$ or $|I_R\rangle$. For simplicity, we apply the collapse hypothesis (as usual, the results would *not* change by keeping track of the whole unitary quantum evolution). But first, we have to clarify the following issue: when do we perform the measurement on I? We have two possibilities:

- If we measure the state of I *before* the particle S hits the screen, the wavefunction reduces to $|L\rangle |I_L\rangle$ or to $|R\rangle |I_R\rangle$ with 50% probability, respectively. Then, the screen S performs a second measurement: we find as usual 50% of times A (25% $|A\rangle |I_L\rangle$ and 25% $|A\rangle |I_R\rangle$) and 50% of times B (25% $|B\rangle |I_L\rangle$ and 25% $|B\rangle |I_R\rangle$).

- If, instead, the particle P arrives *first* on the screen S, the quantum state collapses into $|A\rangle |I_+\rangle$ in 50% of cases (A has clicked), or into $|B\rangle |I_-\rangle$ in the other 50% of cases (B has clicked). The subsequent measurement of the I particle will then give $|I_L\rangle$ or $|I_R\rangle$ (50% each).

In conclusion, we realize that it is absolutely irrelevant which experiment is done before the other. In particular, for us sitting on the screen S, it does not matter at all when and if the measurement of the idler state is performed. We simply see no interference.

Measuring the idler particle in the $|I_+\rangle, |I_-\rangle$ basis

Being the particle P entangled with another particle and not with a macroscopic state, we can also decide to perform a different kind of measurement on I. For instance, we can put a detector measuring I by projecting onto the basis $|I_+\rangle$ and $|I_-\rangle$. If we do this measurement before the particle P has hit the screen S, we have the following outcome as a consequence of the collapse induced by the I-detector:

$$|\Psi\rangle = |A\rangle |I_+\rangle \text{ with probability 50\%;} \qquad (22)$$

$$|\Psi\rangle = |B\rangle |I_-\rangle \text{ with probability 50\%.} \qquad (23)$$

In the former case, the particle P will surely hit S in A, in the latter case the particle P will surely hit S in B.

One sometimes interpret the experiment in the following way: the detector measuring the state of I as being either $|I_+\rangle$ or $|I_-\rangle$ 'erases the which-way information'. When the detector measures $|I_+\rangle$ we still have interference and we see the particle P in the position A, just as the case with two open slits (Fig. 1). In the other case, when the detector measures $|I_-\rangle$, we also have a kind of interference in which the final position B is the only outcome. In the language of Ref. [16], one speaks of 'fringes' in the former case, and of 'anti-fringes' in the latter.

However, care is needed: for us sitting on S, if we do not know which measurement is performed on I, we simply see that *no* interference occurs (50%-A and 50%-B). But, if we could then speak with a colleague working

with the I-detector, we would realize that, each time we have measured A he has found the state $|I_+\rangle$, while each time we have measured B he has found $|I_-\rangle$. Thus, we have a *correlation* of our results (measurement of the screen S) with those of the I-detector. This is actually no surprise if we look at the quantum state of Eq. 19. This statement is indeed more precise than the statement of having interference because we have erased the which-way information. Namely, we do *not* have interference.

Indeed, we can perform the measurement of I even after (in principle long time after) the screen S has measured P in either A or B. Here the name 'delayed choice' comes from: we choose if we retain the which-way information or not. Still, the result is the same because there is no influence on the time-ordering of the measurements. If the measurement of the screen S occurs first, we have a collapse onto the very same Eqs. 22 and 23. Then, a measurement of the idler particle I would simply find either $|I_+\rangle$ correlated with A or $|I_-\rangle$ correlated with B. For sure, there is no change of the past by a measurement of the idler state, but simply a correlation of states. Still, such a very interesting setup visualizes many of the peculiarities of quantum mechanics and can be used for quantum cryptography.

2.3.3 Realizations of the setup

In a two-slit experiment all the peculiarities of quantum mechanics are evident due to the fact that the particle P follows (at least) two paths at the same time. This is extremely fascinating as well as counterintuitive for our imagination based on a childhood with rolling 'classical' marbles. However, as already mentioned in § 2.1, a simple implementation of the two-slit experiment does not produce only two possible outcomes, but gives rise to a superposition of waves with many maxima and minima. In the following we present two possible realizations of our Gedankenexperiment which do not make use of slits.

An interference experiment in which only two outcomes are possible can be realized by using particles with spin 1/2 (such as electrons in a Stern-Gerlach-type experiment) or photons (spin 1, but due to gauge invariance only two polarizations are realized). Clearly, all the quantum mechanics features do not depend on which particle or on which quantum number are implemented, but solely on the presence of superpositions and on the effect of measurements. In the case of photon polarizations we can use the fact that a photon can be horizontally or vertically polarized (corresponding to the kets $|h\rangle$ and $|v\rangle$ respectively). In our analogy, the state $|h\rangle$ corresponds to the state of our particle P coming out from the left slit, $|h\rangle \equiv |L\rangle$, and similarly $|v\rangle$ from the right slit, $|v\rangle \equiv |R\rangle$. Then, we place a detector which acts as the screen S by

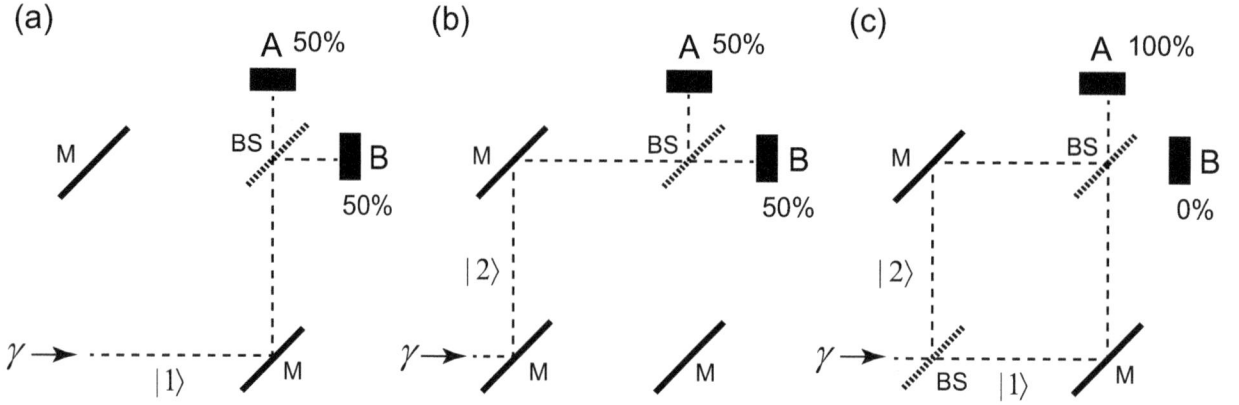

Figure 3: *The Mach–Zehnder interferometer. In analogy with the setup shown in Fig. 1, the case (a) corresponds to only the left slit open, (b) to only the right slit open, (c) to both slits open. The interferometer paths are adjusted so that constructive interference occurs at detector A and destructive interference occurs at detector B. M, mirror; BS, beam splitter; γ, photon.*

making a measurement in the basis $|A\rangle = (|v\rangle + |h\rangle)/\sqrt{2}$ and $|B\rangle = (|v\rangle - |h\rangle)/\sqrt{2}$. In addition, we can place a second detector which plays the role of the detector D by measuring the polarization in the $|h\rangle, |v\rangle$ basis. Indeed, in this case we do not need to send the photons along two different paths, because the polarization degree of freedom is enough for our purposes.

Another possible realization of our setup is the Mach–Zehnder interferometer [34, 35], see Fig. 3, which makes use of beam splitters. When a photon is sent to path $|1\rangle$ of Fig. 3a, both photon counters A and B can detect the photon with a probability of 50%. For our analogy, we have $|1\rangle \equiv |L\rangle$. Similarly, when the photon is sent to path $|2\rangle$ of Fig. 3b, we hear a click in A or in B with 50% probability. For the analogy: $|2\rangle \equiv |R\rangle$. When a beam splitter is put in the beginning of the setup as shown in Fig. 3c, after the photon passes through, we get a superposition $(|1\rangle + |2\rangle)/\sqrt{2}$. The inclusion of the detector D, the bomb, entangled particle(s) as well as the environment can be easily carried out.

In the end, notice that Mach–Zehnder interferometers can be constructed by using neutrons instead of photons. The so-called neutron interferometers (see the recent review paper [36] and references therein) can be very well controlled and allow to experimentally study quantum systems to a great level of accuracy.

3 Collapse vs no-collapse: there is a difference

In this section we show that there is a difference between the collapse and no-collapse scenarios. To this end, instead of having a detector, a bomb, or an idler entangled state, we assume that the space between the slits and the screen is not the vacuum. Then, we study the time evolu-

tion of the *environment* which interacts with the particle P. This interaction is assumed to be soft enough not to absorb or kick away the particle in such a way that the final outcomes on the screen S are still the endpoints A or B.

Before the particle P goes through the slit(s), the environment is described by the state $|E_0\rangle$. First, we study the case in which only the left slit is open. Denoting as $t = 0$ the time at which P passes through the left slit, the wavefunction of the environment evolves as function of time t as

$$|\Psi(t)\rangle = |L\rangle |E_L(t)\rangle, \qquad (24)$$

where by construction $|E_L(0)\rangle = |E_0\rangle$. Similarly, if only the right slit is open, at the time t the system is described by $|\Psi(t)\rangle = |R\rangle |E_R(t)\rangle$ with $|E_R(0)\rangle = |E_0\rangle$.

We now turn to the case in which both slits are open. It is important to stress that, by assuming a weak interaction of the particle P with the environment, we surely do not have (at first) a collapse of the wavefunction, but an evolution of the whole quantum state given by

$$|\Psi(t)\rangle = \frac{1}{\sqrt{2}} (|L\rangle |E_L(t)\rangle + |R\rangle |E_R(t)\rangle)$$
$$= \frac{1}{2} [|A\rangle (|E_R(t)\rangle + |E_L(t)\rangle) + |B\rangle (|E_R(t)\rangle - |E_L(t)\rangle)].$$
$$(25)$$

This is indeed very similar to the detector case, but there is a crucial aspect that we now take into consideration. The state $|E_L(t)\rangle$ and $|E_R(t)\rangle$ coincide at $t = 0$ and then *smoothly* depart from each other. At the time t we assume to have

$$c(t) = \langle E_L(t)|E_R(t)\rangle = e^{-\lambda t}. \qquad (26)$$

(where $c(t)$ is taken to be real for simplicity). This is nothing else than a gradual decoherence process. The states of the environment entangled with $|L\rangle$ and $|R\rangle$ overlap less and less by the time passing. The constant λ describes the speed of the decoherence and depends on the number of particles involved and the intensity of the interaction. Note, strictly speaking, this non-orthogonality is also present in the case of the detector (if no collapse is assumed), but the overlap is amazingly small, see the estimate in Eq. 12. (In the case of the detector D of § 2.2, λ is very large and consequently λ^{-1} is a very short time scale, shorter than any other time scale in that setup. For that reason we assumed that the detector state evolved for all practical purposes instantaneously from the ready-state (pointer at 0) to pointing either to the left or to the right.)

Now we ask the following question: what is the probability that the particle P hits the screen in A? We assume that the particle P hits the screen at the time τ. At this instant, the state is given by $|\Psi(\tau)\rangle$ with $\langle E_L(\tau)|E_R(\tau)\rangle = c(\tau)$.

We now present the mathematical steps leading to $p[A, \tau]$, which, although still simple, are a bit more difficult than the previous ones. The reader who is only interested in the result can go directly to Eq. 31.

At the time τ we express the state $|E_L(\tau)\rangle$ as

$$|E_L(\tau)\rangle = c(\tau)|E_R(\tau)\rangle + \sum_\alpha b_\alpha(\tau)\left|E_{R,\perp}^\alpha(\tau)\right\rangle \quad (27)$$

where the summation over α includes all states of the environment which are orthogonal to $|E_R(\tau)\rangle$: $\left\langle E_{R,\perp}^\alpha(\tau)|E_R(\tau)\right\rangle = 0$. This expression is possible because the set $\{E_R(\tau), E_{R,\perp}^\alpha(\tau)\}$ represents an orthonormal basis for the environment state. Its explicit expression will be extremely complicated, but we do not need to specify it. The normalization of the state $|E_L(\tau)\rangle$ implies that

$$|c(\tau)|^2 + \sum_\alpha |b_\alpha(\tau)|^2 = 1. \quad (28)$$

Then, the state of the system at the instant τ is given by the superposition

$$|\Psi(\tau)\rangle = \frac{1}{2}[1 + c(\tau)]|A\rangle|E_R(\tau)\rangle$$
$$+ \frac{1}{2}|A\rangle \sum_\alpha b_\alpha(\tau)\left|E_{R,\perp}^\alpha(\tau)\right\rangle$$
$$+ \frac{1}{2}[1 - c(\tau)]|B\rangle|E_R(\tau)\rangle$$
$$+ \frac{1}{2}|B\rangle \sum_\alpha b_\alpha(\tau)\left|E_{R,\perp}^\alpha(\tau)\right\rangle. \quad (29)$$

At the time τ the probability of the particle P hitting A is given by

$$p[A, \tau] = \frac{1}{4}|1 + c(\tau)|^2 + \frac{1}{4}\sum_\alpha |b_\alpha(\tau)|^2$$
$$= \frac{1}{4}|1 + c(\tau)|^2 + \frac{1}{4}\left(1 - |c(\tau)|^2\right), \quad (30)$$

where in the last step we have used Eq. 28. A simple calculation leads to

$$p[A, \tau] = \frac{1}{2} + \frac{1}{2}c(\tau) = \frac{1}{2} + \frac{1}{2}e^{-\lambda\tau}. \quad (31)$$

A similar calculation leads to the probability of the particle P hitting S in B as

$$p[B, \tau] = \frac{1}{2} - \frac{1}{2}c(\tau) = \frac{1}{2} - \frac{1}{2}e^{-\lambda\tau}. \quad (32)$$

We see that 'a bit' of interference is left (no matter how large the time interval τ is):

$$p[A, \tau] - p[B, \tau] = e^{-\lambda\tau}, \quad (33)$$

showing that there is always an (eventually very slightly) enhanced probability to see the particle in A rather than in B.

Notice that the very same result is found by using the reduced statistical operator

$$\hat\rho_{red}(\tau) = \langle E_R(\tau)|\hat\rho(\tau)|E_R(\tau)\rangle$$
$$+ \sum_\alpha \left\langle E_{R,\perp}^\alpha(\tau)|\hat\rho(\tau)|E_{R,\perp}^\alpha(\tau)\right\rangle$$
$$= \left(\begin{array}{cc} |A\rangle & |B\rangle \end{array}\right)\left(\begin{array}{cc} p[A, \tau] & c(\tau) \\ c(\tau) & p[B, \tau] \end{array}\right)\left(\begin{array}{c} \langle A| \\ \langle B| \end{array}\right)$$
(34)

where $\hat\rho(\tau) = |\Psi(\tau)\rangle\langle\Psi(\tau)|$. The diagonal elements are the usual Born probabilities, while the non-diagonal elements quantify the overlap of the two branches and become very small for increasing time. (A related subject to the quantum evolution described here is that of the weak measurement, in which the 'measurement' is performed by a weak interaction and thus a unitary evolution of the whole system is taken into account, see the recent review [37] and references therein.)

All these considerations do not require any collapse of the wavefunction due to the environment (see also Ref. [38]). Indeed, if we replace the environment with the detector D of § 2.2 (which was nothing else than a particular environment), the whole discussion is still valid (but see the comments on time scale after Eq. 26). The only point when the Born rule enters is when we see the particle being either in A or in B, but as we commented previously in this no-collapse many worlds scenario, we *do not know why* the Born rule applies [26, 27]. In this sense, decoherence alone is not a solution of the measurement problem [39]. The wavefunction is still a superposition of different and distinguishable macroscopic states. Still, because of decoherence, these states (branches) become almost orthogonal, thus decoherence is an important element of the many worlds interpretation although it does

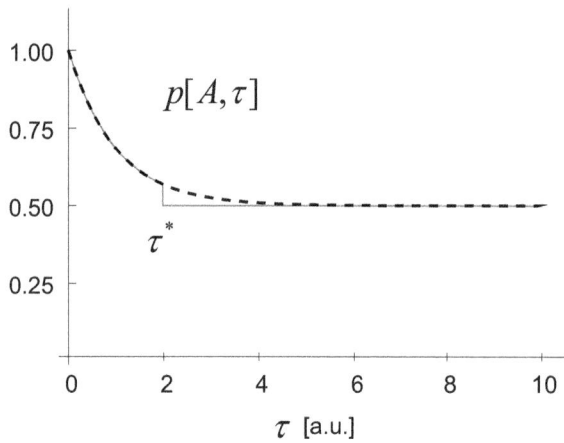

Figure 4: *The quantity $p[A, \tau]$ is plotted as function of τ. The dashed line represents the prediction of the unitary evolution of Eq. 25. The solid line represents the prediction of the collapse hypothesis of Eq. 35: if the detection of the screen takes place for τ larger than the critical value τ^*, the state has collapsed to either A or B, therefore $p[A, \tau > \tau^*] = 1/2$. Note, we use arbitrary units. The choice of τ^* is also arbitrary and serves to visualize the effect (it is expected to be much larger in reality).*

not explain the emergence of probabilities. What do theories with the collapse of the wavefunction predict? As long as few particles of the environment are involved (i.e., at small times), for sure we do not have any collapse and the entanglement in Eq. 25 is the correct description of the system. Namely, we know that interference effects occur for systems which contains about 1000 (and even more) particles [40]. But, if we wait long enough we can reach a critical number of particles at which the collapse takes place. Thus, simplifying the discussion as much as possible, according to collapse models there should be a critical time-interval τ^* at which the probability $p[A, \tau]$ *suddenly* jumps to $1/2$:

$$p[A, \tau] = \begin{cases} \frac{1}{2} + \frac{1}{2}e^{-\lambda\tau} & \text{for } \tau < \tau^*; \\ \frac{1}{2} & \text{for } \tau \geq \tau^*. \end{cases} \quad (35)$$

(In the presented example we vary the time of flight τ by keeping all the rest unchanged, but the crucial point is the number of particles involved. Alternatively, one could change the density of the particles of the environment, which induces a change of the parameter λ. In that case, one would have a critical λ_*.) Indeed, such a sudden jump is an oversimplification, but is enough for our purposes: it shows that a new phenomenon, the collapse, takes place. In Fig. 4 we show schematically the difference between the 'no-collapse' and the 'collapse' cases. Obviously, if τ^* is very large, it becomes experimentally very difficult to distinguish the two curves, but the qualitative difference between them is clear.

In Ref. [41] the gradual appearance of decoherence due to interaction of electrons with image charges has been

experimentally observed. This is analogous to our Eq. 31. (For other decoherence experiment see Ref. [32] and references therein.) Indeed, it would be very interesting to study decoherence in a setup with only two outcomes, for instance with the help of a Mach–Zehnder interferometer or by using neutron interferometers. Namely, even if the distinction between collapse/non-collapse is not yet reachable [9], a clear demonstration of decoherence and the experimental verification of Eq. 31 would be useful on its own.

As a last step, we show that the behavior $p[A, t] = 1/2$ for all $t \geq \tau^*$ is a peculiarity of the collapse approach which is *impossible* if only a unitary evolution is taken into account. The proof makes use of the Hamiltonian H of the whole system (particle+slits+environment), for which we assume that $\langle R|H|L\rangle = \langle L|H|R\rangle = 0$, i.e. the full Hamiltonian does *not* mix the states $|L\rangle$ and $|R\rangle$. (This is indeed a quite general assumption for the type of problems that we study: once the particle has gone through the left slit, its wavefunction is $|L\rangle$ and stays such (and vice versa for $|R\rangle$). Similarly, in the example of a (photon or neutron) Mach–Zehnder interferometer, after the first beam-splitter the path is either the lower or the upper and the whole Hamiltonian does not mix them.) It then follows that:

$$|\Psi(t)\rangle = e^{-iHt}\frac{1}{\sqrt{2}}(|L\rangle|E_0\rangle + |R\rangle|E_0\rangle)$$
$$= \frac{1}{\sqrt{2}}\left(|L\rangle e^{-iH_Lt}|E_0\rangle + |R\rangle e^{-iH_Rt}|E_0\rangle\right) \quad (36)$$

where we have expressed $|E_L(t)\rangle = e^{-iH_Lt}|E_0\rangle$ and $|E_R(t)\rangle = e^{-iH_Rt}|E_0\rangle$ by introducing the Hamiltonians $H_L = \langle L|H|L\rangle$ and $H_R = \langle R|H|R\rangle$ which act in the subspace of the environment. (These expressions hold because $H^n|L\rangle|E_0\rangle = |L\rangle H_L^n|E_0\rangle$ for each n). The overlap $c(t)$ defined in Eq. 26 can be formally expressed as

$$c(t) = \langle E_L(t)|E_R(t)\rangle = \left\langle E_0\left|e^{-i(H_R-H_L)t}\right|E_0\right\rangle. \quad (37)$$

The Hamiltonians H_L and H_R, as well as their difference $H_R - H_L$, are Hermitian. For a finite number of degrees of freedom of the system, the quantity $c(t)$ shows a (almost) periodic behavior and returns (very close) to the initial value 1 in the so-called Poincaré duration time (which can be very large for large systems). It is then excluded that $c(t)$ vanishes for $t > \tau^*$. (At most, it can vanish for certain discrete times, see § 4, but not continuously). Even in the limit of an infinite number of states, the quantity $c(t)$ does not vanish but approaches smoothly zero for $t \to \infty$.

4 Entanglement with a non-orthogonal idler state

As a last example, we design an ideal setup in which the environment is represented again by a single particle, the idler state (see § 2.3.2). However, we assume now that a time-evolution of the idler state takes place

$$|\Psi(t)\rangle = \frac{1}{\sqrt{2}}\left(|L\rangle\,|E_L(t)\rangle + |R\rangle\,|E_R(t)\rangle\right), \quad (38)$$

with the 'environment' states now expressed in terms of the orthonormal idler-basis $\{|I_1\rangle, |I_2\rangle\}$.

$$|E_L(t)\rangle = |I_1\rangle, \quad (39)$$

$$|E_R(t)\rangle = \cos(\omega t)\,|I_1\rangle + \sin(\omega t)\,|I_2\rangle. \quad (40)$$

Thus, while $|E_L(t)\rangle = |I_1\rangle$ is a constant over time, we assume that $|E_R(t)\rangle$ rotates in the space spanned by $|I_1\rangle$ and $|I_2\rangle$. Then, we can rewrite $|\Psi(t)\rangle$ as

$$|\Psi(t)\rangle = \frac{1}{2}\,|A\rangle\,[(1 + \cos(\omega t))\,|I_1\rangle + \sin(\omega t)\,|I_2\rangle]$$
$$+ \frac{1}{2}\,|B\rangle\,[(-1 + \cos(\omega t))\,|I_1\rangle + \sin(\omega t)\,|I_2\rangle]. \quad (41)$$

The probability $p[A, \tau]$ is given by

$$p[A, \tau] = \frac{1}{2} + \frac{1}{2}\cos(\omega \tau) \quad (42)$$

where τ is the time at which the particle P hits the screen.

In conclusion, in a real implementation of this simple idea, it would be interesting to see the appearance and the disappearance of interference (with both fringes and antifringes) as function of the time of flight τ, see Fig. 5. It should be however stressed that the full interaction Hamiltonian does not act on the idler state alone. Indeed, the corresponding Hamiltonian has the form

$$H = \alpha(|R\rangle\,|I_1\rangle\,\langle R|\,\langle I_2| + \text{h.c.}). \quad (43)$$

This is indeed a quite peculiar type of interaction because the idler state rotates only if the particle P is in the state $|R\rangle$ (in the language of § 4, it means: $H_L = 0$, $H_R = \alpha(|I_1\rangle\,\langle I_2| + \text{h.c.}).$). This implies that the spatial trajectory of both states $|I_1\rangle$ and $|I_2\rangle$ must be the same, otherwise the overlap $\langle E_L(t)|E_R(t)\rangle$ would be an extremely small number and the effect that we have described would not take place.

5 Conclusions

We have presented an ideal interference experiment in which we have compared the unitary evolution and the existence of a collapse of the wavefunction. We have

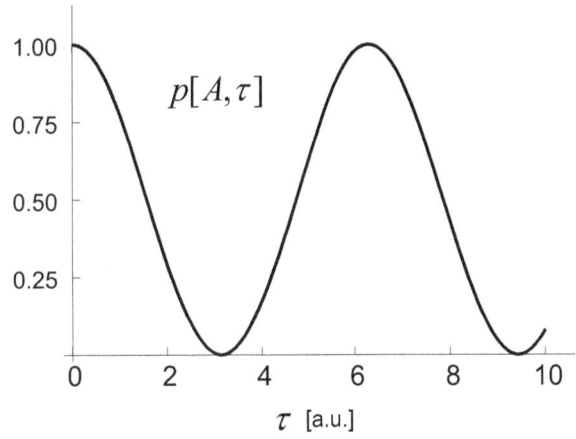

Figure 5: *Quantity $p[A, \tau]$ as function of τ in the case of entanglement with an idler state according to Eq. 41.*

analyzed the case in which a detector measures the which-way information and we have shown that the collapse postulate as well as the no-collapse unitary evolution lead to the same outcome: the disappearance of interference on the screen. In the unitary (no-collapse) evolution, this is true *only* if the states of the detector are orthogonal. This is surely a very good, but not exact, approximation. It was then possible to describe within the very same Gedankenexperiment two astonishing quantum phenomena: the Elitzur–Vaidman bomb and the delayed-choice experiment.

We have then turned to a description of the entanglement with the environment. The phenomenon of decoherence ensures that the interference smoothly disappears. However, as long as the quantum evolution is unitary, it never disappears completely. Conversely, the *real* collapse of the wave function introduces a new kind of dynamics which is not part of the linear Schrödinger equation. While the details differ according to which model is chosen [9], the main features are similar: a quantum state in which one has a delocalized object (superposition of 'here' and 'there') is *not* a stable configuration, but is metastable and decays to a definite position (either 'here' or 'there'). In conclusion, the collapse and the no-collapse views are intrinsically different, as Fig. 4 shows. At a fundamental level, the unitary (no-collapse) evolution leads quite naturally to the many worlds interpretation in which also detectors and observers are included in a superposition. (For a different view see the Bohm interpretation in which an equation describing the trajectories of the particles is added [42, 43]. The positions are the hidden variables of this approach. The Born rule is put in from the very beginning. An extension of the Bohm interpretation to the relativistic framework and to quantum field theories is a difficult task, see Ref. [44] for a critical analysis.)

Even if the distinction between the collapse and the no-collapse alternatives is probably still too difficult to be detected at the moment, the demonstration of decoherence in an experiment with two final states would be an interesting outcome on its own (see the dashed curve in Fig. 4). Also a situation in which an entangled particle is emitted in such a way that an 'oscillating interference' takes place (see Fig. 5) might be an interesting possibility.

A further promising line of research to test the existence of the collapse of the wavefunction is the theoretical and experimental study of unstable quantum systems. The non-exponential behavior of the survival probability for short times renders the so-called Zeno and Anti-Zeno effects possible [45–54]: these are modifications of the survival probability due to the effect of the measurement, which have been experimentally observed [55, 56]. The measurement of an unstable system (for instance, the detection of the decay products) can be modelled as a series of ideal measurements in which the collapse of the wavefunction occurs, but can also be modelled through a unitary evolution in which the wave function of the detector is taken into account and no collapse takes place [57–60]. Then, if differences between these types of measurement appear, one can test how a detector is performing a certain measurement [61]. Quite remarkably, such effects are not restricted to nonrelativistic quantum mechanics, but hold practically unchanged also in the context of relativistic quantum field theory [62–65] and are therefore applicable in the realm of elementary particles.

In conclusion, quantum mechanics still awaits for better understanding in the future. It is surely of primary importance to test the validity of (unitary) standard quantum mechanics for larger and heavier bodies. In this way the new collapse dynamics, if existent, may be discovered.

Acknowledgments

These reflections arise from a series of seminars on *"Interpretation and New Developments of Quantum Mechanics"* and lectures *"Decays in Quantum Mechanics and Quantum Field Theory"*, which took place in Frankfurt over the last 4 years.

The author thanks Francesca Sauli, Stefano Lottini, Giuseppe Pagliara, and Giorgio Torrieri for useful discussions. Stefano Lottini is also acknowledged for a careful reading of the manuscript and for help in the preparation of the figures.

References

[1] Sakurai JJ. Modern Quantum Mechanics. Reading: Addison Wesley, 1994.

[2] Omnès R. The Interpretation of Quantum Mechanics. Princeton Series in Physics, Princeton: Princeton University Press, 1994.

[3] Tegmark M, Wheeler JA. 100 Years of the Quantum. Scientific American 2001; 284: 68–75. arXiv:quant-ph/0101077.

[4] Bell JS. Bertlmann's socks and the nature of reality. In: Speakable and Unspeakable in Quantum Mechanics. Collected papers on quantum philosophy, Cambridge: Cambridge University Press, 1981, pp.139–158. CERN:142461.

[5] Bell JS. On the Einstein-Podolsky-Rosen Paradox. Physics 1964; 1 (3): 195–200. CERN:111654.

[6] Bell JS. On the problem of hidden variables in quantum mechanics. Reviews of Modern Physics 1966; 38 (3): 447–452. doi:10.1103/RevModPhys.38.447.

[7] Bassi A, Ghirardi G. Dynamical reduction models. Physics Reports 2003; 379 (5-6): 257–426. doi:10.1016/S0370-1573(03)00103-0, arXiv:quant-ph/0302164.

[8] Penrose R. The Road to Reality: A Complete Guide to the Laws of the Universe. London: Vintage, 2004.

[9] Bassi A, Lochan K, Satin S, Singh TP, Ulbricht H. Models of wave-function collapse, underlying theories, and experimental tests. Reviews of Modern Physics 2013; 85 (2): 471–527. doi:10.1103/RevModPhys.85.471, arXiv:1204.4325.

[10] Ghirardi GC, Nicrosini O, Rimini A, Weber T. Spontaneous localization of a system of identical particles. Il Nuovo Cimento B 1988; 102 (4): 383–396. doi:10.1007/BF02728509.

[11] Pearle P. Reduction of the state vector by a nonlinear Schrödinger equation. Physical Review D 1976; 13 (4): 857–868. doi:10.1103/PhysRevD.13.857.

[12] Penrose R. On gravity's role in quantum state reduction. General Relativity and Gravitation 1996; 28 (5): 581–600. doi:10.1007/BF02105068.

[13] Diósi L. Quantum stochastic processes as models for state vector reduction. Journal of Physics A: Mathematical and General 1988; 21 (13): 2885–2898. doi:10.1088/0305-4470/21/13/013.

[14] Singh TP. Quantum measurement and quantum gravity: many-worlds or collapse of the wavefunction? Journal of Physics: Conference Series 2009; 174 (1): 012024. doi:10.1088/1742-6596/174/1/012024, arXiv:0711.3773.

[15] Elitzur A, Vaidman L. Quantum mechanical interaction-free measurements. Foundations of Physics 1993; 23 (7): 987–997. doi:10.1007/BF00736012.

[16] Kim Y-H, Yu R, Kulik SP, Shih Y, Scully MO. Delayed "choice" quantum eraser. Physical Review Letters 2000; 84 (1): 1–5. doi:10.1103/PhysRevLett.84.1, arXiv:quant-ph/9903047.

[17] Walborn SP, Cunha MOT, Padua S, Monken CH. Double-slit quantum eraser. Physical Review A 2002; 65 (3): 033818. doi:10.1103/PhysRevA.65.033818, arXiv:quant-ph/0106078.

[18] Everett III H. "Relative state" formulation of quantum mechanics. Reviews of Modern Physics 1957; 29 (3): 454–462. doi:10.1103/RevModPhys.29.454.

[19] Wheeler JA. Assessment of Everett's "relative state" formulation of quantum theory. Reviews of Modern Physics 1957; 29 (3): 463–465. doi:10.1103/RevModPhys.29.463.

[20] DeWitt BS. Quantum mechanics and reality. Physics Today 1970; 23 (9): 30–35. doi:10.1063/1.3022331.

[21] Tegmark M. Many lives in many worlds. Nature 2007; 448 (7149): 23–24. doi:10.1038/448023a, arXiv:0707.2593.

[22] Deutsch D. Quantum theory of probability and decisions. Proceedings of the Royal Society of London. Series A: Mathematical, Physical and Engineering Sciences 1999; 455 (1988): 3129–3137. doi:10.1098/rspa.1999.0443, arXiv:quant-ph/9906015.

[23] Saunders S. Derivation of the Born rule from operational assumptions. Proceedings of the Royal Society of London. Series A: Mathematical, Physical and Engineering Sciences 2004; 460 (2046): 1771–1788. doi:10.1098/rspa.2003.1230, arXiv:quant-ph/0211138.

[24] Wallace D. Everettian rationality: defending Deutsch's approach to probability in the Everett interpretation. Studies in History and Philosophy of Science Part B: Studies in History and Philosophy of Modern Physics 2003; 34 (3): 415–439. doi:10.1016/S1355-2198(03)00036-4, arXiv:quant-ph/0303050, PhilSci:1030.

[25] Zurek WH. Wave-packet collapse and the core quantum postulates: Discreteness of quantum jumps from unitarity, repeatability, and actionable information. Physical Review A 2013; 87 (5): 052111. doi:10.1103/PhysRevA.87.052111, arXiv:1212.3245.

[26] Rae AIM. Everett and the Born rule. Studies in History and Philosophy of Science Part B: Studies in History and Philosophy of Modern Physics 2009; 40 (3): 243–250. doi:10.1016/j.shpsb.2009.06.001, arXiv:0810.2657.

[27] Hsu SDH. On the origin of probability in quantum mechanics. Modern Physics Letters A 2012; 27 (12): 1230014. doi:10.1142/S0217732312300145, arXiv:1110.0549.

[28] Marquardt F, Püttmann A. Introduction to dissipation and decoherence in quantum systems. 2008. arXiv:0809.4403.

[29] Hornberger K. Introduction to decoherence theory. In: Entanglement and Decoherence, vol.768. Lecture Notes in Physics, Buchleitner A, Viviescas C, Tiersch M (editors), Berlin: Springer, 2009, pp.221–276. doi:10.1007/978-3-540-88169-8_5, arXiv:quant-ph/0612118.

[30] Zurek WH. Decoherence, einselection, and the quantum origins of the classical. Reviews of Modern Physics 2003; 75 (3): 715–775. doi:10.1103/RevModPhys.75.715, arXiv:quant-ph/0105127.

[31] Schlosshauer M. Decoherence, the measurement problem, and interpretations of quantum mechanics. Reviews of Modern Physics 2005; 76 (4): 1267–1305. doi:10.1103/RevModPhys.76.1267, arXiv:quant-ph/0312059.

[32] Schlosshauer M. Experimental observation of decoherence. In: Compendium of Quantum Physics. Greenberger D, Hentschel K, Weinert F (editors), Berlin: Springer, 2009, pp.223–229. doi:10.1007/978-3-540-70626-7_70.

[33] Kwiat P, Weinfurter H, Herzog T, Zeilinger A, Kasevich MA. Interaction-free measurement. Physical Review Letters 1995; 74 (24): 4763–4767. doi:10.1103/PhysRevLett.74.4763.

[34] Zehnder L. Ein Neuer Interferenzrefraktor. Zeitschrift für Instrumentenkunde 1891; 11: 275–285. `https://archive.org/details/zeitschriftfrin11gergoog`

[35] Mach L. Ueber einen Interferenzrefraktor. Zeitschrift für Instrumentenkunde 1892; 12: 89–93. `https://archive.org/details/zeitschriftfrin14gergoog`

[36] Klepp J, Sponar S, Hasegawa Y. Fundamental phenomena of quantum mechanics explored with neutron interferometers. Progress of Theoretical and Experimental Physics 2014; 2014 (8): 082A001. `doi:10.1093/ptep/ptu085`, `arXiv:1407.2526`.

[37] Svensson BEY. Pedagogical review of quantum measurement theory with an emphasis on weak measurements. Quanta 2013; 2 (1): 18–49. `doi:10.12743/quanta.v2i1.12`.

[38] Namiki M, Pascazio S. Wave-function collapse by measurement and its simulation. Physical Review A 1991; 44 (1): 39–53. `doi:10.1103/PhysRevA.44.39`.

[39] Adler SL. Why decoherence has not solved the measurement problem: a response to P. W. Anderson. Studies in History and Philosophy of Science Part B: Studies in History and Philosophy of Modern Physics 2003; 34 (1): 135–142. `doi:10.1016/S1355-2198(02)00086-2`, `arXiv:quant-ph/0112095`.

[40] Gerlich S, Eibenberger S, Tomandl M, Nimmrichter S, Hornberger K, Fagan PJ, Tüxen J, Mayor M, Arndt M. Quantum interference of large organic molecules. Nature Communications 2011; 2 (4): 263. `doi:10.1038/ncomms1263`.

[41] Sonnentag P, Hasselbach F. Measurement of decoherence of electron waves and visualization of the quantum-classical transition. Physical Review Letters 2007; 98 (20): 200402. `doi:10.1103/PhysRevLett.98.200402`.

[42] Bohm D. A suggested interpretation of the quantum theory in terms of "hidden" variables. I. Physical Review 1952; 85 (2): 166–179. `doi:10.1103/PhysRev.85.166`.

[43] Bohm D. A suggested interpretation of the quantum theory in terms of "hidden" variables. II. Physical Review 1952; 85 (2): 180–193. `doi:10.1103/PhysRev.85.180`.

[44] Passon O. Why isn't every physicist a Bohmian? 2004. `arXiv:quant-ph/0412119`.

[45] Misra B, Sudarshan ECG. The Zeno's paradox in quantum theory. Journal of Mathematical Physics 1977; 18 (4): 756–763. `doi:10.1063/1.523304`.

[46] Degasperis A, Fonda L, Ghirardi GC. Does the lifetime of an unstable system depend on the measuring apparatus? Il Nuovo Cimento A 1974; 21 (3): 471–484. `doi:10.1007/BF02731351`.

[47] Fonda L, Ghirardi GC, Rimini A. Decay theory of unstable quantum systems. Reports on Progress in Physics 1978; 41 (4): 587–631. `doi:10.1088/0034-4885/41/4/003`.

[48] Khalfin LA. Contribution to the decay theory of a quasi-stationary state. Soviet Physics JETP 1958; 6: 1053–1063. `CERN:424878`.

[49] Koshino K, Shimizu A. Quantum Zeno effect by general measurements. Physics Reports 2005; 412 (4): 191–275. `doi:10.1016/j.physrep.2005.03.001`, `arXiv:quant-ph/0411145`.

[50] Kofman AG, Kurizki G. Acceleration of quantum decay processes by frequent observations. Nature 2000; 405 (6786): 546–550. `doi:10.1038/35014537`.

[51] Facchi P, Nakazato H, Pascazio S. From the quantum Zeno to the inverse quantum Zeno effect. Physical Review Letters 2001; 86 (13): 2699–2703. `doi:10.1103/PhysRevLett.86.2699`.

[52] Giacosa F. Non-exponential decay in quantum field theory and in quantum mechanics: the case of two (or more) decay channels. Foundations of Physics 2012; 42 (10): 1262–1299. `doi:10.1007/s10701-012-9667-3`, `arXiv:1110.5923`.

[53] Giacosa F. Energy uncertainty of the final state of a decay process. Physical Review A 2013; 88 (5): 052131. `doi:10.1103/PhysRevA.88.052131`, `arXiv:1305.4467`.

[54] Giacosa F, Pagliara G. Oscillations in the decay law: a possible quantum mechanical explanation of the anomaly in the experiment at the GSI facility. Quantum Matter 2013; 2 (1): 54–59. `doi:10.1166/qm.2013.1025`, `arXiv:1110.1669`.

[55] Wilkinson SR, Bharucha CF, Fischer MC, Madison KW, Morrow PR, Niu Q, Sundaram B, Raizen MG. Experimental evidence for non-exponential decay in quantum tunnelling. Nature 1997; 387 (6633): 575–577. `doi:10.1038/42418`.

[56] Fischer MC, Gutiérrez-Medina B, Raizen MG. Observation of the quantum Zeno and anti-Zeno effects in an unstable system. Physical Review Letters 2001; 87 (4): 040402. doi:10.1103/PhysRevLett.87.040402, arXiv:quant-ph/0104035.

[57] Schulman LS. Continuous and pulsed observations in the quantum Zeno effect. Physical Review A 1998; 57 (3): 1509–1515. doi:10.1103/PhysRevA.57.1509.

[58] Facchi P, Pascazio S. Quantum Zeno phenomena: pulsed versus continuous measurement. Fortschritte der Physik 2001; 49 (10-11): 941–947. arXiv:quant-ph/0106026.

[59] Koshino K, Shimizu A. Quantum Zeno effect for exponentially decaying systems. Physical Review Letters 2004; 92 (3): 030401. doi:10.1103/PhysRevLett.92.030401, arXiv:quant-ph/0307075.

[60] Koshino K, Shimizu A. Quantum Zeno effect by general measurements. Physics Reports 2005; 412 (4): 191–275. doi:10.1016/j.physrep.2005.03.001, arXiv:quant-ph/0411145.

[61] Giacosa F, Pagliara G. Pulsed and continuous measurements of exponentially decaying systems. Physical Review A 2014; 90 (5): 052107. doi:10.1103/PhysRevA.90.052107, arXiv:1405.6882.

[62] Giacosa F, Pagliara G. Deviation from the exponential decay law in relativistic quantum field theory: the example of strongly decaying particles. Modern Physics Letters A 2011; 26 (30): 2247–2259. doi:10.1142/S021773231103670X, arXiv:1005.4817.

[63] Giacosa F, Pagliara G. Spectral function of a scalar boson coupled to fermions. Physical Review D 2013; 88 (2): 025010. doi:10.1103/PhysRevD.88.025010, arXiv:1210.4192.

[64] Giacosa F, Pagliara G. Spectral functions of scalar mesons. Physical Review C 2007; 76 (6): 065204. doi:10.1103/PhysRevC.76.065204, arXiv:0707.3594.

[65] Giacosa F, Wolkanowski T. Propagator poles and an emergent stable state below threshold: general discussion and the E(38) state. Modern Physics Letters A 2012; 27 (39): 1250229. doi:10.1142/S021773231250229X, arXiv:1209.2332.

7

Accommodating Retrocausality with Free Will

Yakir Aharonov [1,2], *Eliahu Cohen* [1,3] *& Tomer Shushi* [4]

[1] *School of Physics and Astronomy, Tel Aviv University, Tel Aviv, Israel. E-mail: eliahuco@post.tau.ac.il*
[2] *Schmid College of Science, Chapman University, Orange, California, USA. E-mail: yakir@post.tau.ac.il*
[3] *H. H. Wills Physics Laboratory, University of Bristol, Bristol, UK. E-mail: eliahu.cohen@bristol.ac.uk*
[4] *University of Haifa, Haifa, Israel. E-mail: tomer.shushi@gmail.com*

Editors: *Kunihisa Morita, Danko Georgiev & Kelvin McQueen*

Retrocausal models of quantum mechanics add further weight to the conflict between causality and the possible existence of free will. We analyze a simple closed causal loop ensuing from the interaction between two systems with opposing thermodynamic time arrows, such that each system can forecast future events for the other. The loop is avoided by the fact that the choice to abort an event thus forecasted leads to the destruction of the forecaster's past. Physical law therefore enables prophecy of future events only as long as this prophecy is not revealed to a free agent who can otherwise render it false. This resolution is demonstrated on an earlier finding derived from the two-state vector formalism, where a weak measurement's outcome anticipates a future choice, yet this anticipation becomes apparent only after the choice has been actually made. To quantify this assertion, *weak information* is described in terms of Fisher information. We conclude that an *already existing* future does not exclude free will nor invoke causal paradoxes. On the quantum level, particles can be thought of as weakly interacting according to their past and future states, but causality remains intact as long as the future is masked by quantum indeterminism.

Quanta 2016; 5: 53–60.

1 Introduction

Time-symmetric formulations of quantum mechanics are gaining growing interest. Using two boundary conditions rather than the customary one, they offer novel twists to several foundational issues. Such are the Wheeler–Feynman electromagnetic absorber theory [1], Hoyle and Narlikar's theory of gravitation [2], and Cramer's transactional interpretation [3]. Among these, however, the Aharonov–Bergmann–Lebowitz rule [4] and Aharonov's two-state vector formalism [5] are distinct, in that they even predict some novel effects for a combination of forwards and backwards evolving wave functions. When performing a complete post-selection of the quantum state, otherwise counterfactual questions can be intriguingly answered with regard to the state's previous time evolution.

These advances, however, might seem to come with a price that even for adherents is too heavy, namely, dismissing free will. While quantum indeterminism seemed to offer some liberation from the chains imposed on our choices by classical causality, time-symmetric quantum mechanics somewhat undermines quantum indeterminism, as it renders future boundary conditions the missing source of possible causes. This might eventually reveal causality to be just as strict and closed as classical causality. If the future is, in some sense, *already there* to the point of being causally equal to the past, free will (which is defined in Section 2) might appear to be as illusory as it has appeared within the classical framework. We aim to show this is not necessarily the case. The two-state vector formalism is no worse off than classical physics, or other formulations of quantum mechanics as it pertains to the incorporation of free will. In other words, free will is not precluded even when discussing a quantum world having both past and future boundary.

In this special issue of *Quanta*, dedicated to Richard Feynman and discussing time-symmetry in quantum mechanics, we examine what might seem to be a problem in these formulations, namely the notion of free will [6]. Discussion of this kind might at first be regarded as philosophical in character, but we hope to formulate the problem rigorously enough to yield nontrivial physical insights.

2 The Problem

Following Russell and Deery [6], we propose defining free will as follows. Let a physical system be capable of initiating complex interactions with its environment, gaining information about it and predicting its future states, as well as their effects on the system itself. This grants the system purposeful behavior, which nevertheless fully accords with *classical* causality. Now let there be more than one course of action that the system can take in response to a certain event, which in turn lead to different future outcomes that the system can predict. *Free will* then denotes the system's taking one out of various courses of action, independently (at least to some extent) of past restrictions. This definition is very close in spirit to the one employed in [7], i.e. the ability to make choices. It should be emphasized that even in our time-symmetric context, free will means only freedom from the past, not from the future (see also [5]).

In classical physics, conservation laws oblige any event to be strictly determined by earlier causes. In our context, this might apparently leave only one course of action for the system in question, and hence no real choices. When moving to the quantum realm, free will might be recovered [7], but then again, if one adds a final boundary condition to the description of the quantum system, can free will exist? We shall answer both classical and quantum questions on the affirmative, employing statistical and quantum fluctuations, respectively.

In what follows, we analyze a classical causal paradox avoided by the past's instability. We subsequently consider a more acute variant of this paradox and discuss a few possible resolutions. Then we present the

quantum counterpart of these two paradoxes where inherent indeterminism saves causality. We show that within the two-state vector formalism, although both future and past states of the system are known, genuine freedom is not necessarily excluded. We then define and quantify *weak information* that is the kind of information coming from the future that can be encrypted in the past without violating causality.

3 Interaction between Two Systems with Opposing Time-Arrows

To demonstrate the possibility of knowing one's future and its consequences, we discuss a highly simplified classical gedanken experiment. Naturally, there are immediate difficulties with such a setup. For example, can two regions in space have opposite time arrows to begin with? Can observers inside them communicate? These and other questions deserve further probing, but we focus here only on what would happen if several conditions are met, rather than whether and how they can be achieved.

Consider, then, a universe comprised of only two closed, non-interacting laboratories located at some distance from one another. Suppose further that their thermodynamic time arrows are opposite to one another, such that each system's *future* time direction is the other's *past*. Finally let each laboratory host a free agent, henceforth Alice and Bob, capable of free choice.

It is challenging to create a communication channel between two laboratories of this kind. An exchange of signals is possible in the following form. A light beam is sent from the exterior part of one laboratory to the other's boundary, where a static message is posted. The beam is then reflected back to its origin. If the labs are massive enough, the beam imparts only a negligible momentum transfer.

The gedanken experiment proceeds as follows (Fig. 1):

$t_1^{(b)}$: Bob sends a light-beam (red arrow) to Alice's lab.

$t_2^{(b)}$: He receives through his returning beam a message from Alice saying: "Let me know if you see this message" (dotted blue world-line).

$t_3^{(b)}$: Bob posts a confirmation saying: "I saw your message" (red world-line).

Then there are the following events in Alice's lab:

$t_1^{(a)}$: Alice sends a light-beam (blue arrow) to Bob's lab.

$t_2^{(a)}$: Alice receives through her returning beam of particles that scattered off Bob's message, i.e. she gets the information from Bob through this beam reflected from Bob's system to her system.

$t_3^{(a)}$: Alice, realizing that this confirmation comes from her *future*, chooses not to post a message.

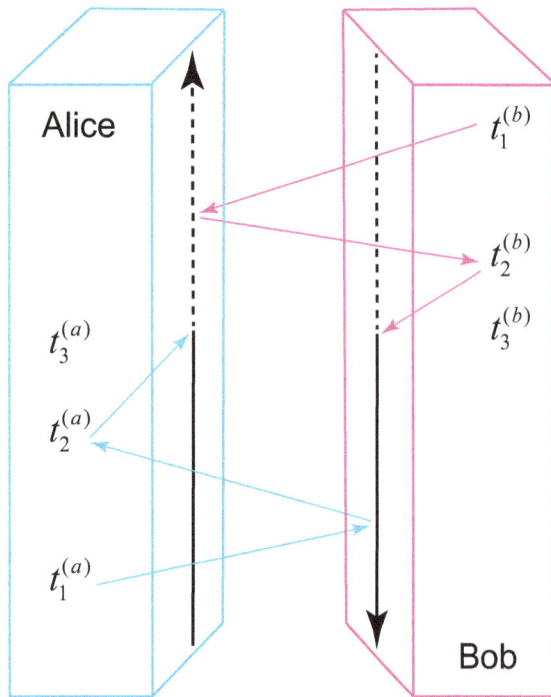

Figure 1: *An illustration of the two labs gedanken experiment with two free agents, Alice and Bob.*

The Causal Paradox is obvious: The dotted blue world-line represents an absent message. How, then, could Bob reply to a message which was removed before he was supposed to see it?

We note that alongside with this formulation of the paradox, one can equivalently describe the complementary scenario: Bob finds through his returning beam that Alice did not post a message. Therefore, he sends no confirmation, but eventually Alice, having free will, decides to post a message in contrast to Bob's observation.

4 The Suggested Resolution

A key element in this causal paradox's resolution is the following well-known fact: Entropy-increasing processes are highly stable, not sensitive to small changes in their initial conditions or their evolution, whereas entropy-decreasing processes are extremely vulnerable to any interference.

Our question therefore is: Which time direction is affected by Alice's decision to change the *future* that *has been forecasted* by Bob? The simplest and most consistent answer is: Bob's past. Upon Alice's decision to remove her message at $t_3^{(a)}$, Bob's *prophecy*, i.e. the message of Bob to Alice regarding her future choice, turns out to be false. This is clearly inconsistent with his earlier observation of Alice's message, which is understood now to be highly unstable. His observation turns out to be a large (hence very rare) statistical fluctuation.

We can now define the arrow of time of any system as the thermodynamic direction which is stable against changes. While a small change at the large system's present will negligibly affect its future, it can have dramatic effects into its past. Alice's future was coupled in our example to Bob's past. By employing her free will, she could completely alter his previous observations, but the apparent paradox is resolved by taking into account the chaotic nature of the entropy decreasing direction. Indeed, the signals are weak enough, which makes them amenable for this reinterpretation as fluctuations.

5 A More Acute Paradox

We shall now discuss an operationally simpler, yet conceptually harder version of the paradox, which emphasizes the role of free will. Let the two labs with opposing time arrows contain two simple machines rather than free agents (see Fig. 2). One machine, A, posts 0 if it receives 0 as an input, and 1 if it receives 1. The second lab's machine, B, posts 1 if it receives 0 and 0 if it receives 1. The paradox is as follows: In case A receives 0 from the other lab, it posts 0. Then B receives the 0 as an input and posts 1, in contrast to A's earlier input. Alternatively, A receives 1 from B, then posts 1. Then, B receives this 1 and posts 0, again in contrast to the A's initial input. It follows there are no valid initial conditions for this combined system at a given time.

The resolution may be:

(1) Communication is impossible between two such systems.

(2) The past of both systems is symmetrically unstable.

(3) There must be some stochastic element allowing consistency.

(4) The operations of the two machines must be coordinated.

As explained above, we assume that communication of simple static messages is possible, hence we shall avoid the first option (nevertheless, this paradox could actually suggest that a special communication protocol is needed between two such systems with opposite time arrows). Options (2) and (3) complement each other and resonate with the above notion of free will, as well as with the quantum paradox to be presented below. Naturally, this combination is favored by us. We believe this paradoxical situation could have been avoided if a minor degree of freedom (e.g. at the form of free will) were allowed. In contrast, alternative (4) implies superdeterminism (see for instance [8, 9]) or the so called *conspiracy* between the two machines, which is philosophically disturbing (at least in our view), negating free will altogether.

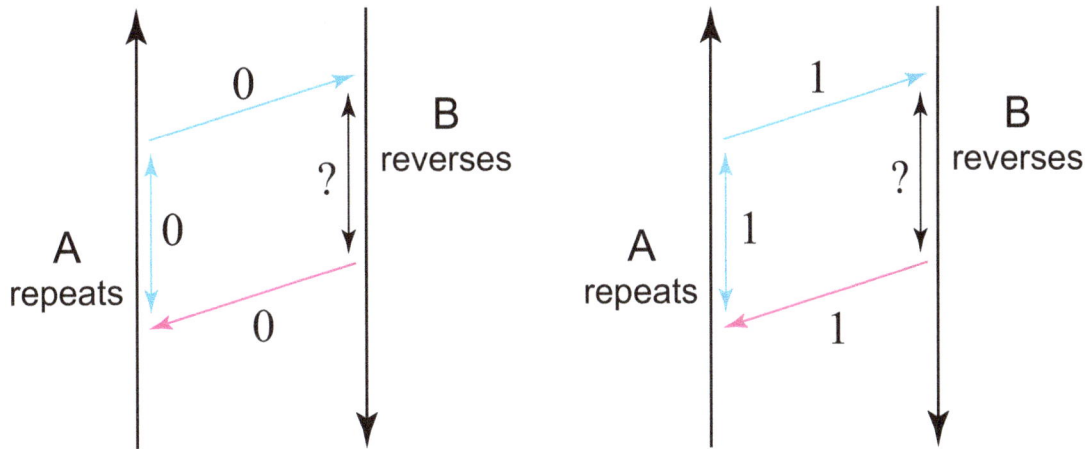

Figure 2: *An illustration of the two machines gedanken experiment. Machine A posts 0 (1) if it receives 0 (1), whereas machine B posts 0 (1) if it receives 1 (0). The paradox is symmetric, but for simplicity it is shown to reside on the B side.*

6 Going Quantum: The Two-State Vector Formalism and Weak Measurements

The possibility for resolving the above problem on classical grounds encourages seeking more interesting avenues at the quantum level. Indeed a similar resolution will be offered, namely the possibility of re-interpreting the past. However, the basic concept on which the resolution relies shifts from thermodynamic to quantum fluctuations which are more suitable for describing small microscopic systems. This is where time-symmetric quantum causality comes in most naturally.

The two-state vector formalism is a time-symmetric formulation of quantum mechanics employing in addition to the forward evolving wave function (pre-selected state) also a backwards evolving wave function (post-selected state). This combination gives rise to the *two-state-vector*, which provides a richer notion of quantum reality between two projective measurements. This world-view has produced several predictions, so far well verified by weak measurements [10–13] which delicately gather information about the quantum state without collapsing it, and thus do not change to post-selection probability.

In an earlier work [13] the following gedanken experiment was proposed. A large ensemble of N spins is prepared in Einstein–Podolsky–Rosen state. Each particle in every pair is weakly measured along the three Bell orientations, before being strongly projected along one of them. As was shown, each weak measurement only slightly disturbs the state and hence the well-known non-local correlations between the strong outcomes are maintained in this experiment. It should be noted that in return each weak measurement provides only a negligible amount of information (to be quantified in Section 7).

However, since all the weak outcomes were classically recorded, upon slicing them according to the projective outcomes, one finds in retrospect, with extreme accuracy, the weak values corresponding to all Bell orientations (not only the ones eventually chosen for the projective measurement). The question is then, how could the values reside in the weak data prior to the final Bell measurements which demonstrated almost perfect non-local correlations? Bell's proof certainly forbids them to be prepared in such a way so the two-state vector formalism answer would be that they came from the future! The important point in this retrocausal interpretation is the weak values could be there, that is, could had causal effect on the pointer's shift, without forcing a specific future outcome.

The resolution is therefore simple: quantum indeterminacy guarantees that, should someone try to abort a future event about which they have received a prophecy, that prophecy would turn out to be a mere error.

Therefore, even in the two-state vector formalism where present is determined by both past and future events, the quantum indeterminism enables free will.

Naturally, more mundane explanations ought to be considered before concluding that results of weak measurement contain information regarding a future event. By normal causality, it should be Alice's measurements which affected Bob's, rather than vice versa. Perhaps, for example, some subtle bias induced by her weak measurements affected his later strong ones.

Such a *past-to-future* effect is considerably strained by the following question: How robust is the alleged bias introduced by the weak measurements? If it is robust enough to oblige the strong measurements, then it is equivalent to full collapse, namely the very local hidden variables already ruled out by Bell's inequality. This is clearly not the case: weakly measured particles remain nearly fully entangled. But then, even the weakest bias, as long as it is expected to show up over a sufficiently large N, is ruled out of the same grounds. The *weak bias* alternative is ruled out also by the robust correlations predicted between all same-spin measurements, whether weak or strong.

Can Alice predict Bob's outcomes on the basis of her own data? To do that, she must feed all her rows of outcomes into a computer that searches for a possible series of spin-orientation choices plus measurement outcomes, such that, when she slices her rows accordingly, she will get the complex pattern of correlations described above.

The number of such possible sequences that she gets from her computation is $\binom{N}{N/2} \propto \frac{2^N}{\sqrt{N}}$. Each such sequence enables her to slice each of her rows into two $N/2$ halves and get the above correlations between her weak measurements and the predicted strong measurements. The distribution of the results is a Gaussian with $\lambda \sqrt{N}/2$ expectation and $\delta \sqrt{N}/2$ standard deviation, so a δ shift in one of the results, or even in $\sqrt{N}$ of them, is very probable. Hence, even if Alice computes all Bob's possible future choices, she still cannot tell which choice he will take, because there are many similar subsets giving roughly the same value. Also, as Aharonov *et al.* pointed out in [13], when Alice finds a subset with a significant deviation from the expected 50%-50% distribution, its origin is much more likely, upon a real measurement by Bob, to turn out to be a measurement error than a genuine physical value. Obviously, then, present data is insufficient to predict the future choice.

7 The Strength of Information Transmission

The information transmission between Alice and Bob can be categorized into two different types with different strength:

Strong information: This type describes the information that, in general, has the potential to interfere with Alice's free choice. This is the classical kind of information transmitted in the first gedanken experiment.

Weak information: In this case the information that Bob sends to Alice will not, in any circumstance, interfere with Alice's free choice because it is buried much below the quantum uncertainty level.

While the strong information transmission was discussed in Sections 3 and 4 and was shown to cast instability into Bob's past, it seems the weak information notion should be further explained and quantified.

We now understand that weak information represents information that does not actively interfere with the Alice's and Bob's systems. Therefore, weak information can be described employing weak measurement outcomes since individually they only provide very partial information that does not interfere with neither Alice's nor Bob's system consistency. Similarly, strong information is related to projective measurement outcomes since they do disturb the systems and provide definite results.

To create a clear distinction between the two kinds of information, we shall discuss a simple thought experiment. Suppose Alice has a spin she wants to measure. To do that, she will use a Stern–Gerlach magnet with a non-homogeneous magnetic field along some direction. Bob, having an opposite time arrow, already knows that Alice will choose the z-axis and will find an *up* outcome. If Bob sends this *strong information* to Alice, she may choose the y-axis instead and find a *down* outcome, reproducing the paradoxical situation discussed above. However, if Bob only tells her that she will find an *up* result along *some* direction, no causal paradox will ensue (see also [14]. This is the kind of *weak information* which does not clash with Alice's free will nor with Bob's history.

8 Fisher Information for Strong and Weak measurements

Fisher information is a tool to quantify the hidden information in a random variable Q regarding a parameter it depends on. Using Fisher information we can now quantitatively define the strong and weak information concepts that were qualitatively introduced in Section 7.

Suppose that there is an unknown parameter θ which we want to estimate (for example, θ can stand for the relative phase between two superposed wave-packets). We define a density function f of Q and another auxiliary parameter Δ which describes the type of information, strong or weak. In probabilistic terms, it is called the *scale parameter* of Q. In this case, Fisher information as a function of Δ, $I_\Delta(\theta)$, is given by

$$I_\Delta(\theta) := E\left(\left[\frac{\partial}{\partial \theta} \ln f\left(\Delta Q; \theta\right)\right]^2 \middle| \theta\right). \tag{1}$$

It can be easily shown that $I_\Delta(\theta)$ is in fact the product of Δ^{-1} and $I(\theta)$:

$$
\begin{aligned}
I_\Delta(\theta) &= \int \left[\frac{\partial}{\partial \theta} \ln f\left(\Delta Q; \theta\right)\right]^2 f\left(\Delta Q; \theta\right) dQ \\
&= \frac{1}{\Delta} \int \left[\frac{\partial}{\partial \theta} \ln f\left(Q; \theta\right)\right]^2 f\left(Q; \theta\right) dQ \\
&= \Delta^{-1} I(\theta). \tag{2}
\end{aligned}
$$

Now, in case of $\Delta \to 0$, we find

$$\lim_{\Delta \to 0} I_\Delta(\theta) = \lim_{\Delta \to 0} \Delta^{-1} I(\theta) = \infty, \tag{3}$$

hence we conclude that $\Delta \to 0$ indicates strong information.

The opposite case of $\Delta \to \infty$ leads to

$$\lim_{\Delta \to \infty} I_\Delta(\theta) = \lim_{\Delta \to \infty} \Delta^{-1} I(\theta) = 0, \tag{4}$$

which implies weak information, so for sufficient large value of Δ, weak information is described by a negligible Fisher information.

Let us now demonstrate this concept. Suppose that θ is the relative phase between two superposed wavepackets, which we want to measure in some interference experiment. Let us assume that the interference pattern is detected via some coupling $1/\Delta$ to a measuring pointer. If our estimation for the relative phase is described by a Gaussian random variable Q, then the density function of Q will be

$$f(Q; \theta) = \frac{1}{\sqrt{2\pi}\theta} \exp\left(-\frac{1}{2\theta^2} Q^2\right). \tag{5}$$

Depending on the coupling strength, the Fisher information will be

$$I_\Delta(\theta) = \Delta^{-1} \theta^{-2}. \tag{6}$$

9 Cryptography Can Protect Causality

Weak or encrypted information can be used for communication between future and past in a causality preserving manner thanks to quantum indeterminism. The main idea behind this type of communication is quantum cryptography [15]. Suppose Bob somehow knows what Alice will choose in the future. He uses a quantum cryptography scheme to encode Alice's future choice and gives her the encrypted prophecy. However, he does not share with her the key to decode this revelation until she actually makes her choice. In this case, similarly to the example in Section 6, both Bob's past and Alice's future are secured. Due to quantum indeterminism, Alice still has free will.

For example, in the BB84 scheme [16,17], even though Alice and Bob communicate through a public channel, their secret key is secured due to another form of quantum indeterminism, namely, that non-orthogonal states are indistinguishable. This means that even if the generated string contains information regarding Eve's future, it will not create a causal paradox.

10 A Few Alternatives

In addition to the proposed resolution for the above paradoxes, there exist some other well-known possibilities. The parallel universe resolution suggests that if one goes back in time and kills his grandfather he actually does it in a parallel universe and therefore he does not interfere with the laws of nature [18, 19]. A different approach to solve this is by postulating another time dimension in which such disagreements can be solved before being recorded in our history [20, 21].

These two resolutions clearly lack simplicity and oblige an excessive ontology to our existing theories. Moreover, detailed work is needed to refute each and every paradox.

Therefore, bearing in mind Occam's razor as a tool for denying complex theories, it seems these alternative solutions are unfavorable.

Another solution simply dictates that one cannot create paradoxes in the universe and therefore cannot, for instance, kill his grandfather. This approach implies a universe guided by global consistency condition such as in [22, 23] and was shown to naturally arise in postselected closed timelike curves [24].

11 Free will and Becoming

Classical physics treats time as a purely geometrical ingredient of the universe, alongside the three spatial dimensions. Against the perfect logical rigor and experimental support that make relativity so powerful, many physicists find the *block universe* picture emerging from it manifestly awkward. In fact, the very notion of space-time implies that, just as all locations have the same degree of reality in space, so do all past, present, and future moments exist along the temporal dimension without any moment being unique as the privileged *now*.

Against this mainstream view, there are alternative accounts [25]. They suspect that, if we experience time so differently from space, this difference may be objective. They provide some models to capture this notion of dynamic time.

Bob's access to Alice's future in the classical gedanken experiment above and the double boundary condition on the wave function proposed by the two-state vector formalism may seem at first sight to resonate with a block universe approach. However, as we have just seen, statistical and quantum fluctuations may provide us with freedom to define the present. As was shown in Sections 4 and 5, this freedom, and also the notion of becoming, is subjective and system-dependent.

Within the two-state vector formalism, while both backward and forward states evolve deterministically, they have limited physical significance on their own — physical reality is created by the product of the causal chains extending in both temporal directions. The past does not determine the future, yet the future is set, and only together do they form the present. However, the existence of a future boundary condition, and its deterministic effect, do not deny our freedom of choice. It is allowed due to the inaccessibility of the data (which is a requirement of causality, as discussed in Section 5). Examining the concept of free will from a physical point of view, we find that it must contain at least partial freedom from past causal constraints, and such freedom is duly manifested in the two-state vector formalism, where a juxtaposition of freedom and determinacy is epitomized.

12 Conclusions

We examined the possibility of free will in a retrocausal theory. Closed causal loops, which arise due to the interaction between two systems with opposing time arrows were discussed. The suggested resolution of the ensuing paradoxes relies on the thermodynamic instability of the past.

Moving to the quantum realm, a similar paradox can be solved via the quantum indeterminism, which is understood to protect free will. This resonates with previous findings of Georgiev [7]. Furthermore, we discussed the strength of information transmission, where the terms *strong* and *weak* are related to strong (projective) and weak values, respectively. When information about a future event is buried under quantum indeterminism it cannot violate free will. Similarly, encrypted information, such as the one available through weak measurements, does not violate causality. The existence of free will in these time symmetric models was conjectured to resonate with a dynamical notion of time.

Acknowledgements

We thank Avshalom C. Elitzur and Nissan Itzhaki for very helpful discussions. Yakir Aharonov and Eliahu Cohen acknowledge support of the Israel Science Foundation Grant No. 1311/14. Yakir Aharonov acknowledges support from ICORE Excellence Center "Circle of Light". Eliahu Cohen was also supported by ERC-AD NLST.

References

[1] Wheeler JA, Feynman RP. Interaction with the absorber as the mechanism of radiation. Reviews of Modern Physics 1945; 17 (2–3): 157–181. `doi:10.1103/RevModPhys.17.157`

[2] Hoyle F, Narlikar JV. A new theory of gravitation. Proceedings of the Royal Society of London A: Mathematical, Physical and Engineering Sciences 1964; 282 (1389): 191–207. `doi:10.1098/rspa.1964.0227`

[3] Cramer JG. The transactional interpretation of quantum mechanics. Reviews of Modern Physics 1986; 58 (3): 647–687. `doi:10.1103/RevModPhys.58.647`

[4] Aharonov Y, Bergmann PG, Lebowitz JL. Time symmetry in the quantum process of measurement. Physical Review B 1964; 134 (6): 1410–1416. `doi:10.1103/PhysRev.134.B1410`

[5] Aharonov Y, Cohen E, Gruss E, Landsberger T. Measurement and collapse within the two-state vector formalism. Quantum Studies: Mathematics and Foundations 2014; 1 (1–2): 133–146. `doi:10.1007/s40509-014-0011-9`

[6] Russell P, Deery O. The Philosophy of Free Will: Essential Readings from the Contemporary Debates. New York: Oxford University Press, 2013.

[7] Georgiev D. Quantum no-go theorems and consciousness. Axiomathes 2013; 23 (4): 683–695. `doi:10.1007/s10516-012-9204-1`

[8] 't Hooft G. The free-will postulate in quantum mechanics. 2007: `arXiv:quant-ph/0701097`

[9] 't Hooft G. The cellular automaton interpretation of quantum mechanics. A view on the quantum nature of our universe, compulsory or impossible? 2014: `arXiv:1405.1548`

[10] Aharonov Y, Albert DZ, Vaidman L. How the result of a measurement of a component of the spin of a spin-$\frac{1}{2}$ particle can turn out to be 100. Physical Review Letters 1988; 60 (14): 1351–1354. `doi:10.1103/PhysRevLett.60.1351`

[11] Tamir B, Cohen E. Introduction to weak measurements and weak values. Quanta 2013; 2 (1): 7–17. `doi:10.12743/quanta.v2i1.14`

[12] Aharonov Y, Cohen E, Elitzur AC. Foundations and applications of weak quantum measurements. Physical Review A 2014; 89 (5): 052105. doi: 10.1103/PhysRevA.89.052105

[13] Aharonov Y, Cohen E, Elitzur AC. Can a future choice affect a past measurement's outcome? Annals of Physics 2015; 355: 258-268. arXiv:1206.6224, doi:10.1016/j.aop.2015.02.020

[14] Elitzur AC, Cohen E, Shushi T. The too-late-choice experiment: Bell's proof within a setting where the nonlocal effect's target is an earlier event. 2015; arXiv:1512.08275

[15] Gisin N, Ribordy G, Tittel W, Zbinden H. Quantum cryptography. Reviews of Modern Physics 2002; 74 (1): 145–195. arXiv:quant-ph/0101098, doi: 10.1103/RevModPhys.74.145

[16] Bennett CH, Brassard G. Quantum cryptography: Public key distribution and coin tossing. Proceedings of the IEEE International Conference on Computers, Systems and Signal Processing, Bangalore, India, December 10–12, 1984, pp. 175–179.

[17] Bennett CH, Brassard G. Quantum cryptography: Public key distribution and coin tossing. Theoretical Computer Science 2014; 560 (1): 7–11. doi:10.1016/j.tcs.2014.05.025

[18] Bousso R, Susskind L. Multiverse interpretation of quantum mechanics. Physical Review D 2012; 85 (4): 045007. arXiv:1105.3796, doi:10.1103/PhysRevD.85.045007

[19] Deutsch D. The structure of the multiverse. Proceedings of the Royal Society of London A: Mathematical, Physical and Engineering Sciences 2002; 458 (2028): 2911–2923. arXiv:quant-ph/0104033, doi:10.1098/rspa.2002.1015

[20] Bars I. Survey of two-time physics. Classical and Quantum Gravity 2001; 18 (16): 3113–3130. arXiv:hep-th/0008164, doi:10.1088/0264-9381/18/16/303

[21] Bars I, Terning J. Two-time physics. In: Extra Dimensions in Space and Time. Multiversal Journeys, Nekoogar F (editor), New York: Springer, 2010, pp. 67–87. doi:10.1007/978-0-387-77638-5_7

[22] Carlini A, Frolov VP, Mensky MB, Novikov ID, Soleng HH. Time machines: the principle of self-consistency as a consequence of the principle of minimal action. International Journal of Modern Physics D 1995; 4 (5): 557–580. arXiv:gr-qc/9506087, doi:10.1142/S0218271895000399

[23] Novikov ID. Time machine and self-consistent evolution in problems with self-interaction. Physical Review D 1992; 45 (6): 1989–1994. doi:10.1103/PhysRevD.45.1989

[24] Lloyd S, Maccone L, Garcia-Patron R, Giovannetti V, Shikano Y. Quantum mechanics of time travel through post-selected teleportation. Physical Review D 2011; 84 (2): 025007. arXiv:1007.2615, doi:10.1103/PhysRevD.84.025007 http://hdl.handle.net/1721.1/66971

[25] Aharonov Y, Popescu S, Tollaksen J. Each instant of time a new universe. In: Quantum Theory: A Two-Time Success Story. Yakir Aharonov Festschrift. Struppa DC, Tollaksen JM (editors), Milan: Springer, 2014, pp. 21–36. doi:10.1007/978-88-470-5217-8_3

8

The Phase Space Formulation of Time-Symmetric Quantum Mechanics

Charlyne de Gosson & Maurice A. de Gosson

Numerical Harmonic Analysis Group, Faculty of Mathematics, University of Vienna, Vienna, Austria.
E-mails: charlyne.degosson@gmail.com, maurice.de.gosson@univie.ac.at

Editors: *Danko Georgiev & Eliahu Cohen*

Time-symmetric quantum mechanics can be described in the Weyl–Wigner–Moyal phase space formalism by using the properties of the cross-terms appearing in the Wigner distribution of a sum of states. These properties show the appearance of a strongly oscillating interference between the pre-selected and post-selected states. It is interesting to note that the knowledge of this interference term is sufficient to reconstruct both states.
Quanta 2015; 4: 27–34.

1 Introduction

Time-symmetric quantum mechanics is an alternative formulation of quantum mechanics exhibiting fascinating and unconventional features whose potentialities have not yet been fully exploited; see [1–5], or the book [6] by Aharonov and Rohrlich. The present paper is a first step towards a formulation of time-symmetric quantum mechanics in terms of phase space concepts such as the Wigner distribution, and the ambiguity transform (the latter is essentially a Fourier transform of the Wigner distribution and is very much used in radar theory). To the best of our knowledge there are very few papers dis-

cussing the phase space approach (which is well-known in conventional quantum mechanics) in the context of time-symmetric quantum mechanics; exceptions to this state of affairs are our previous works [7, 8], and Gray's Conference Proceedings note [9]. The advantage of the phase space approach is that it allows to calculate weak values using the classical observable; a problem that then arises (and which we will study in a forthcoming paper) is that the correspondence between a classical observable a and its *quantization* $\hat{A}$ is by no means obvious: while it is true that most physicists rely on the Weyl scheme, there might be other physically meaningful ways to quantize a classical observable; for instance in [10, 11] we are advocating the use of Born–Jordan quantization, which predates Weyl quantization.

We will also focus on the reconstruction problem, which can roughly be stated as follows: knowing the interference between the pre-selected and post-selected states, can we reconstruct these states? We will see that knowing the cross-Wigner distribution of the pre-selected and post-selected states, suffices to uniquely determine both states. While this result is at first sight surprising, it is well-known in time-frequency analysis [12, 13] that it is possible to reconstruct a signal from the knowledge of its short-time Fourier transform with arbitrary window; the latter is closely related to the cross-Wigner transform.

Parts of this work (in particular the reconstruction formula Eq. 53) have been announced without motivations and proofs in previous work [7, 8]. We also mention that

Lobo and Ribeiro [14] discussed weak values in the quantum phase space using methods that are very different from the Weyl–Wigner–Moyal formalism employed here.

1.1 Notation

We will work with systems having n degrees of freedom. Position or momentum variables are denoted $x = (x_1, ..., x_n)$ and $p = (p_1, ..., p_n)$, respectively. The corresponding phase space variable is (x, p). The scalar product $p_1 x_1 + \cdots + p_n x_n$ is denoted by px. When integrating we will use, where appropriate, the volume elements $d^n x = dx_1 \cdots dx_n$, $d^n p = dp_1 \cdots dp_n$. The unitary $\hbar$-Fourier transform of a square-integrable function $\Psi(x)$ is

$$\tilde{\Psi}(p) = \left(\tfrac{1}{2\pi\hbar}\right)^{\frac{n}{2}} \int e^{-\frac{i}{\hbar}px}\Psi(x)d^n x. \tag{1}$$

We denote by $\hat{x} = (\hat{x}_1, ..., \hat{x}_n)$ and $\hat{p} = (\hat{p}_1, ..., \hat{p}_n)$ the operators defined by $\hat{x}_j\Psi = x_j\Psi$, $\hat{p}_j\Psi = -i\hbar\partial_{x_j}\Psi$.

1.2 The notion of weak value

In time-symmetric quantum mechanics the state of a system is represented by a two-state vector $\langle\Phi| \, |\Psi\rangle$ where the state $\langle\Phi|$ evolves backwards from the future and the state $|\Psi\rangle$ evolves forwards from the past. To make things clear, assume that at a time t_i an observable $\hat{A}$ is measured and a non-degenerate eigenvalue was found: $|\Psi(t_i)\rangle = |\hat{A} = \alpha\rangle$; similarly at a later time t_f a measurement of another observable $\hat{B}$ yields $|\Phi(t_f)\rangle = |\hat{B} = \beta\rangle$. Such a two-time state $\langle\Phi| \, |\Psi\rangle$ can be created as follows [1, 15]: Alice prepares a state $|\Psi(t_i)\rangle$ at initial time t_i. She then sends the system to an observer, Bob, who may perform any measurement he wishes to. The system is returned to Alice, who then performs a strong measurement with the state $|\Phi(t_f)\rangle$ as one of the outcomes. Only if this outcome is obtained, does Bob keep the results of his measurement.

Let now t be some intermediate time: $t_i < t < t_f$. Following the time-symmetric approach to quantum mechanics at this intermediate time the system is described by the *two* wavefunctions

$$\Psi = U_i(t, t_i)\Psi(t_i) \ , \ \Phi = U_f(t, t_f)\Phi(t_f) \tag{2}$$

where $U_i(t, t') = e^{-i\hat{H}_i(t-t')/\hbar}$ and $U_f(t, t') = e^{-i\hat{H}_f(t-t')/\hbar}$ are the unitary operators governing the evolution of the state before and after time t. Consider now the superposition of the two states $|\Psi\rangle$ and $|\Phi\rangle$ (which we suppose normalized); the expectation value

$$\langle\hat{A}\rangle_{\Psi+\Phi} = \frac{\langle\Psi + \Phi|\hat{A}|\Psi + \Phi\rangle}{\langle\Psi + \Phi|\Psi + \Phi\rangle} \tag{3}$$

of the observable $\hat{A}$ in this superposition is obtained using the equality

$$N\langle\hat{A}\rangle_{\Psi+\Phi} = \langle\hat{A}\rangle_\Phi + \langle\hat{A}\rangle_\Psi + 2\,\mathrm{Re}\langle\Phi|\hat{A}|\Psi\rangle \tag{4}$$

where $N = \langle\Psi + \Phi|\Psi + \Phi\rangle$. By definition, if $\langle\Phi|\Psi\rangle \neq 0$, the complex number

$$\langle\hat{A}\rangle_{\Phi,\Psi} = \frac{\langle\Phi|\hat{A}|\Psi\rangle}{\langle\Phi|\Psi\rangle} \tag{5}$$

is the *weak value* of $\hat{A}$.

1.3 What we will do

In the discussion above we have been working directly in terms of the wavefunctions Ψ and Φ; now, a different kind of state description which is very fruitful, particularly in quantum optics, is provided by the Wigner distribution [11, 16–21]

$$W_\Psi(x, p) = \left(\tfrac{1}{2\pi\hbar}\right)^n \int e^{-\frac{i}{\hbar}py}\Psi\left(x + \tfrac{1}{2}y\right)\Psi^*\left(x - \tfrac{1}{2}y\right)d^n y; \tag{6}$$

the latter is directly related to the mean value of the observable $\langle\hat{A}\rangle_\Psi = \langle\Psi|\hat{A}|\Psi\rangle$ by Moyal's formula [11, 17–19, 22]

$$\langle\hat{A}\rangle_\Psi = \iint a(x, p)W_\Psi(x, p)d^n p\,d^n x \tag{7}$$

where $a(x, p)$ is the classical observable whose Weyl quantization is given by the Weyl–Moyal formula

$$\hat{A} = \left(\tfrac{1}{2\pi\hbar}\right)^n \iint \hat{a}(x, p)e^{\frac{i}{\hbar}(x\hat{x}+p\hat{p})}d^n p\,d^n x. \tag{8}$$

Here, we use the terminology *classical observable* in a very broad sense; a can be any complex integrable function, or even a tempered distribution that is an element of $\mathcal{S}'(\mathbb{R}^{2n})$, dual of the Schwartz space $\mathcal{S}(\mathbb{R}^{2n})$ of rapidly decreasing functions. A direct calculation shows that we have

$$W_{\Psi+\Phi} = W_\Phi + W_\Psi + 2\,\mathrm{Re}\,W_{\Psi,\Phi} \tag{9}$$

where the cross-term $W_{\Psi,\Phi}$ is given by

$$W_{\Psi,\Phi}(x, p) = \left(\tfrac{1}{2\pi\hbar}\right)^n \int e^{-\frac{i}{\hbar}py}\Psi\left(x + \tfrac{1}{2}y\right)\Phi^*\left(x - \tfrac{1}{2}y\right)d^n y. \tag{10}$$

The appearance of the term $W_{\Psi,\Phi}$ shows the emergence at time t of a strong interference between the pre-selected and the post-selected states $|\Psi\rangle$ and $|\Phi\rangle$. It is called the cross-Wigner distribution of Ψ, Φ, see [17, 18, 23] and the references therein. We are going to exploit the properties of $W_{\Psi,\Phi}$ to give an alternative working definition of the weak value $\langle\hat{A}\rangle_{\Phi,\Psi}$, namely

$$\langle\hat{A}\rangle_{\Phi,\Psi} = \frac{1}{\langle\Phi|\Psi\rangle} \iint a(x, p)W_{\Psi,\Phi}(x, p)d^n p\,d^n x \tag{11}$$

(see Eq. 20); here $a(x, p)$ is the classical observable whose Weyl quantization is the operator $\hat{A}$. Eq. 11 is justified by an extension of the averaging formula (Eq. 7) to pairs of states: see Eq. 19, well-known in harmonic analysis. This allows us to interpret the function

$$\rho_{\Phi,\Psi}(x,p) = \frac{W_{\Psi,\Phi}(x,p)}{\langle\Phi|\Psi\rangle} \quad (12)$$

as a *complex* probability distribution. We thereafter notice that the cross-Wigner distribution can itself be seen, for fixed (x,p), as a weak value, namely that of Grossmann and Royer's parity operator $\hat{T}_{GR}(x,p)$:

$$W_{\Psi,\Phi}(x,p) = (\pi\hbar)^n \langle\hat{T}_{GR}(x,p)\rangle_{\Psi,\Phi} \langle\Phi|\Psi\rangle \quad (13)$$

(see Eq. 36). Using this approach we prove the following Theorem 2: if $W_{\Psi,\Phi}$ is known, we can reconstruct (up to an unessential phase factor) the wave function Ψ (and hence the state $|\Psi\rangle$) with the use of

$$\Psi(x) = \frac{2^n}{\langle\Phi|\Lambda\rangle} \iint W_{\Psi,\Phi}(y,p)\hat{T}_{GR}(y,p)\Lambda(x)d^n p d^n y \quad (14)$$

where Λ is an arbitrary square-integrable function such that $\langle\Phi|\Lambda\rangle \neq 0$.

2 Weak Values in the Wigner Picture

2.1 The cross-Wigner transform

The cross-Wigner distribution is defined for all square-integrable functions Ψ, Φ; it satisfies the generalized marginal conditions

$$\int W_{\Psi,\Phi}(x,p)d^n p = \Psi(x)\Phi^*(x) \quad (15)$$

$$\int W_{\Psi,\Phi}(x,p)d^n x = \tilde{\Psi}(p)\tilde{\Phi}^*(p) \quad (16)$$

provided that Ψ and Φ are in $L^1(\mathbb{R}^n) \cap L^2(\mathbb{R}^n)$; these formulas reduce to the usual marginal conditions for the Wigner distribution when $\Psi = \Phi$. While W_Ψ is always real (though not non-negative, unless Ψ is a Gaussian), $W_{\Psi,\Phi}$ is a complex function, and we have $W_{\Psi,\Phi}^* = W_{\Phi,\Psi}$. The cross-Wigner distribution is widely used in signal theory and time-frequency analysis [17, 23]; its Fourier transform is the cross-ambiguity function familiar from radar theory [17, 24, 25]. Zurek [26] has studied $W_{\Psi,\Phi}$ when $\Psi + \Phi$ is a Gaussian cat-like state, and has shown that it is accountable for sub-Planck structures in phase space due to interference.

We now make the following elementary, but important remark: multiplying both sides of Eq. 9 by the classical observable $a(x,p)$ and integrating with respect to the x, p variables, we get, using Moyal's formula (Eq. 7),

$$\|\Phi + \Psi\|\langle\hat{A}\rangle_{\Psi+\Phi} = \langle\hat{A}\rangle_\Phi + \langle\hat{A}\rangle_\Psi$$

$$+2 \iint a(x,p)\,\mathrm{Re}\,W_{\Psi,\Phi}(x,p)d^n p d^n x. \quad (17)$$

Comparing with Eq. 4 we see that

$$\mathrm{Re}\langle\Phi|\hat{A}|\Psi\rangle = \iint a(x,p)\,\mathrm{Re}\,W_{\Psi,\Phi}(x,p)d^n p d^n x. \quad (18)$$

It turns out that in the mathematical theory of the Wigner distribution [17, 18] one shows that the equality above actually holds not only for the real parts, but also for the purely imaginary parts, hence we always have

$$\langle\Phi|\hat{A}|\Psi\rangle = \iint a(x,p)W_{\Psi,\Phi}(x,p)d^n p d^n x. \quad (19)$$

An immediate consequence of this equality is that we can express the weak value $\langle\hat{A}\rangle_{\Phi,\Psi}$ in terms of the cross-Wigner distribution and the classical observable $a(x,p)$ corresponding to $\hat{A}$ in the *Weyl quantization scheme*

$$\langle\hat{A}\rangle_{\Phi,\Psi} = \frac{1}{\langle\Phi|\Psi\rangle} \iint a(x,p)W_{\Psi,\Phi}(x,p)d^n p d^n x. \quad (20)$$

We emphasize that one has to be excessively careful when using formulas of the type (Eq. 20) (as we will do several times in this work): the function a crucially depends on the quantization procedure which is used (here Weyl quantization); we will come back to this essential point later, but here is a simple example which shows that things can get wrong if this rule is not observed: let $\hat{H} = \frac{1}{2}(\hat{x}^2 + \hat{p}^2)$ be the quantization of the normalized harmonic oscillator $H(x,p) = \frac{1}{2}(x^2 + p^2)$ (we assume $n = 1$). While it is true that

$$\langle\hat{H}\rangle_{\Phi,\Psi} = \frac{1}{\langle\Phi|\Psi\rangle} \iint H(x,p)W_{\Psi,\Phi}(x,p)dpdx \quad (21)$$

it is in contrast *not true* that

$$\langle\hat{H}^2\rangle_{\Phi,\Psi} = \frac{1}{\langle\Phi|\Psi\rangle} \iint H(x,p)^2 W_{\Psi,\Phi}(x,p)dpdx. \quad (22)$$

Suppose for instance that $\Psi = \Phi$ is the ground state of the harmonic oscillator: $\hat{H}\Psi = \frac{1}{2}\hbar\Psi$. We have

$$\langle\hat{H}^2\rangle - \langle\hat{H}\rangle^2 = 0;$$

however use of Eq. 22 yields the wrong result

$$\langle\hat{H}^2\rangle - \langle\hat{H}\rangle^2 = \frac{1}{4}\hbar^2.$$

The error comes from the inobservance of the prescription above: $\hat{H}^2$ is not the Weyl quantization of $H(x,p)^2$, but that of $H(x,p)^2 - \frac{1}{4}\hbar^2$ as is easily seen using the McCoy [27] rule

$$\widehat{x^r p^s} = \frac{1}{2^s}\sum_{k=0}^{s}\binom{s}{k}\hat{p}^{s-k}\hat{x}^r\hat{p}^k \quad (23)$$

and Born's canonical commutation relation $[\hat{x},\hat{p}] = i\hbar$ (see Shewell [28] for a discussion of related examples).

2.2 A complex phase space distribution

Let us now set

$$\rho_{\Phi,\Psi}(x, p) = \frac{W_{\Psi,\Phi}(x, p)}{\langle\Phi|\Psi\rangle}; \qquad (24)$$

using the marginal conditions given by Eqs. 15,16 we get

$$\int \rho_{\Phi,\Psi}(x, p)d^n p = \frac{\Phi^*(x)\Psi(x)}{\langle\Phi|\Psi\rangle} \qquad (25)$$

$$\int \rho_{\Phi,\Psi}(x, p)d^n x = \frac{\tilde{\Phi}^*(p)\tilde{\Psi}(p)}{\langle\Phi|\Psi\rangle} \qquad (26)$$

hence the function $\rho_{\Phi,\Psi}$ is a complex probability distribution

$$\int \rho_{\Phi,\Psi}(x, p)d^n p d^n x = 1. \qquad (27)$$

The weak value is given in terms of $\rho_{\Phi,\Psi}$ by

$$\langle\hat{A}\rangle_{\Phi,\Psi} = \int a(x, p)\rho_{\Phi,\Psi}(x, p)d^n p d^n x \qquad (28)$$

which reduces to Eq. 7 in the case of an ideal measurement, namely $\Phi = \Psi$. The practical meaning of these relations is the following [5]: the readings of the pointer of the measuring device will cluster around the value

$$\text{Re}\langle\hat{A}\rangle_{\Phi,\Psi} = \int \text{Re}(a(x, p)\rho_{\Phi,\Psi}(x, p))d^n p d^n x \qquad (29)$$

while the quantity

$$\text{Im}\langle\hat{A}\rangle_{\Phi,\Psi} = \int \text{Im}(a(x, p)\rho_{\Phi,\Psi}(x, p))d^n p d^n x \qquad (30)$$

measures the shift in the variable conjugate to the pointer variable. In an interesting paper [29] Feyereisen discusses some aspects of the complex distribution $\rho_{\Phi,\Psi}$.

2.3 The cross-Wigner transform as a weak value

Let $\hat{T}(x_0, p_0) = e^{-\frac{i}{\hbar}(p_0\hat{x} - x_0\hat{p})}$ be the Heisenberg operator; it is a unitary operator whose action on a wavefunction Ψ is given by

$$\hat{T}(x_0, p_0)\Psi(x) = e^{\frac{i}{\hbar}\left(p_0 x - \frac{1}{2}p_0 x_0\right)}\Psi(x - x_0). \qquad (31)$$

It has the following simple dynamical interpretation [18, 21]: $\hat{T}(z_0)$ is the time-one propagator for the Schrödinger equation corresponding to the translation Hamiltonian $\hat{H}_0 = x_0\hat{p} - p_0\hat{x}$. An associated operator is the Grossmann–Royer reflection operator (or displacement parity operator) [18, 30, 31] given by

$$\hat{T}_{\text{GR}}(x_0, p_0) = \hat{T}(x_0, p_0)R^\vee\hat{T}(x_0, p_0)^\dagger \qquad (32)$$

where $R^\vee$ changes the parity of the function to which it is applied: $R^\vee\Psi(x) = \Psi(-x)$; the explicit action of $\hat{T}_{\text{GR}}(z_0)$ on wavefunctions is easily obtained using Eq. 31 and one finds

$$\hat{T}_{\text{GR}}(x_0, p_0)\Psi(x) = e^{\frac{2i}{\hbar}p_0(x - x_0)}\Psi(2x_0 - x). \qquad (33)$$

Now, a straightforward calculation shows that the Wigner distribution W_Ψ is (up to an unessential factor), the expectation value of $\hat{T}_{\text{GR}}(x_0, p_0)$ in the state $|\Psi\rangle$; in fact (dropping the subscripts 0)

$$W_\Psi(x, p) = \left(\frac{1}{\pi\hbar}\right)^n\langle\hat{T}_{\text{GR}}(x, p)\Psi|\Psi\rangle. \qquad (34)$$

More generally, a similar calculation shows that the cross-Wigner transform is given by

$$W_{\Psi,\Phi}(x, p) = \left(\frac{1}{\pi\hbar}\right)^n\langle\hat{T}_{\text{GR}}(x, p)\Phi|\Psi\rangle \qquad (35)$$

and can hence be viewed as a transition amplitude. Taking Eq. 5 into account we thus have

$$W_{\Psi,\Phi}(x, p) = (\pi\hbar)^n\langle\hat{T}_{\text{GR}}(x, p)\rangle_{\Psi,\Phi}\langle\Phi|\Psi\rangle; \qquad (36)$$

this relation immediately implies, using definition (24) of the complex probability distribution $\rho_{\Phi,\Psi}$, the important equality

$$\rho_{\Phi,\Psi}(x, p) = (\pi\hbar)^n\langle\hat{T}_{\text{GR}}(x, p)\rangle_{\Psi,\Phi} \qquad (37)$$

which can in principle be used to determine $\rho_{\Phi,\Psi}$.

As already mentioned, the cross-ambiguity function $A_{\Psi,\Phi}$ is essentially the Fourier transform of $W_{\Psi,\Phi}$; in fact

$$A_{\Psi,\Phi} = \mathcal{F}_\sigma W_{\Psi,\Phi} \ , \ W_{\Psi,\Phi} = \mathcal{F}_\sigma A_{\Psi,\Phi} \qquad (38)$$

where $\mathcal{F}_\sigma$ is the symplectic Fourier transform: if $a = a(x, p)$ then $\mathcal{F}_\sigma a(x, p) = \tilde{a}(p, -x)$ where $\tilde{a}$ is the ordinary $2n$-dimensional $\hbar$-Fourier transform of a; explicitly

$$\mathcal{F}_\sigma a(x, p) = \left(\frac{1}{2\pi\hbar}\right)^n \iint e^{-\frac{i}{\hbar}(xp' - p'x)}a(x', p')d^n p'd^n x'. \qquad (39)$$

Both equalities in Eq. 38 are equivalent because the symplectic Fourier transform is involutive, and hence its own inverse. While the cross-Wigner distribution is a measure of *interference*, the cross-ambiguity function is rather a measure of *correlation*. One shows [11, 17, 18, 23] that $A_{\Psi,\Phi}$ is explicitly given by

$$A_{\Psi,\Phi}(x, p) = \left(\frac{1}{2\pi\hbar}\right)^n \int e^{-\frac{i}{\hbar}py}\Psi\left(y + \frac{1}{2}x\right)\Phi^*\left(y - \frac{1}{2}x\right)d^n y. \qquad (40)$$

The cross-ambiguity function is easily expressed using the Heisenberg operator instead of the Grossmann–Royer operator as

$$A_{\Psi,\Phi}(x, p) = \left(\frac{1}{2\pi\hbar}\right)^n\langle\hat{T}(x, p)\Phi|\Psi\rangle. \qquad (41)$$

The following important result shows that the knowledge of the classical observable a allows us to determine the weak value of the corresponding Weyl operator using the weak value of the Grossmann–Royer (respectively the Heisenberg) operator:

Theorem 1. *Let $\hat{A}$ be the Weyl quantization of the classical observable a. We have*

$$\langle\hat{A}\rangle_{\Phi,\Psi} = \left(\frac{1}{\pi\hbar}\right)^n \iint a(x,p)\langle\hat{T}_{GR}(x,p)\rangle_{\Phi,\Psi}d^npd^nx \quad (42)$$

and

$$\langle\hat{A}\rangle_{\Phi,\Psi} = \left(\frac{1}{2\pi\hbar}\right)^n \iint \mathcal{F}_\sigma a(x,p)\langle\hat{T}(x,p)\rangle_{\Phi,\Psi}d^npd^nx. \quad (43)$$

Proof. In view of Moyal's formula (Eq. 19) we have

$$\langle\Phi|\hat{A}|\Psi\rangle = \iint a(x,p)W_{\Psi,\Phi}(x,p)d^npd^nx \quad (44)$$

that is, taking Eq. 35 into account

$$\langle\Phi|\hat{A}|\Psi\rangle = \left(\frac{1}{\pi\hbar}\right)^n \iint a(x,p)\langle\hat{T}_{GR}(x,p)\Phi|\Psi\rangle d^npd^nx \quad (45)$$

hence Eq. 42; Eq. 43 is obtained in a similar way, first applying the Plancherel formula to the right-hand side of Eq. 44, then applying the first identity given by Eq. 38, and finally using Eq. 41. □

Notice that the formulas above immediately yield the well-known [11, 17, 18, 21] representations of the operator $\hat{A}$ in terms of the Grossmann–Royer and Heisenberg operators:

$$\hat{A} = \left(\frac{1}{\pi\hbar}\right)^n \iint a(x,p)\hat{T}_{GR}(x,p)d^npd^nx \quad (46)$$

and

$$\hat{A} = \left(\frac{1}{2\pi\hbar}\right)^n \iint \mathcal{F}_\sigma a(x,p)\hat{T}(x,p)d^npd^nx. \quad (47)$$

3 The Reconstruction Problem

3.1 Lundeen's experiment

In 2012, Lundeen and his co-workers [32] determined the wavefunction by weakly measuring the position, and thereafter performing a strong measurement of the momentum. They considered the following experiment on a particle: a weak measurement of x is performed which amounts to applying the projection operator $\hat{\Pi}_x = |x\rangle\langle x|$ to the pre-selected state $|\Psi\rangle$; thereafter they perform a strong measurement of momentum, which yields the value p_0,

that is $\Phi(x) = e^{\frac{i}{\hbar}p_0x}$. The result of the weak measurement is thus

$$\langle\hat{\Pi}_x\rangle_{\Psi,\Phi} = \frac{\langle p_0|x\rangle\langle x|\Psi\rangle}{\langle p_0|\Psi\rangle} = \left(\frac{1}{2\pi\hbar}\right)^{\frac{n}{2}} \frac{e^{-\frac{i}{\hbar}p_0x}\Psi(x)}{\tilde{\Psi}(p_0)} \quad (48)$$

where $\tilde{\Psi}$ the Fourier transform of Ψ. Since the value of p_0 is known we get

$$\Psi(x) = \frac{1}{k}e^{\frac{i}{\hbar}p_0x}\langle\hat{\Pi}_x\rangle_{\Psi,\Phi} \quad (49)$$

where $k = (2\pi\hbar)^{\frac{n}{2}}\tilde{\Psi}(p_0)$; Eq. 49 thus allows to determine $\Psi(x)$ by scanning through the values of x. Thus, by reducing the disturbance induced by measuring the position and thereafter performing a sharp measurement of momentum we can reconstruct the wavefunction pointwise. In [33] Lundeen and Bamber generalize this construction to mixed states and arbitrary pairs of observables. Using the complex distribution $\rho_{\Psi,\Phi}(x,p)$ defined above it is easy to recover Eq. 49 of Lundeen *et al.* In fact, choose $a(x,p) = \Pi_{x_0}(x,p) = \delta(x - x_0)$; its Weyl quantization

$$\hat{\Pi}_{x_0}\Psi(x) = \Psi(x_0)\delta(x - x_0)$$

is the projection operator: $\hat{\Pi}_{x_0}|\Psi\rangle = \Psi(x_0)|x_0\rangle$. Using the elementary properties of the Dirac delta function together with the marginal property 25, Eq. 28 becomes

$$\langle\hat{\Pi}_{x_0}\rangle_{\Phi,\Psi} = \int \delta(x - x_0)\rho_{\Phi,\Psi}(x,p)d^npd^nx$$

$$= \int \rho_{\Phi,\Psi}(x_0,p)d^np$$

$$= \frac{\Phi^*(x_0)\Psi(x_0)}{\langle\Phi|\Psi\rangle}$$

which is Eq. 48 since $\Phi(x_0) = e^{\frac{i}{\hbar}p_0x_0}$; Eq. 49 follows.

3.2 Reconstruction: the Weyl–Wigner–Moyal approach

It is well-known [17, 18] that the knowledge of the Wigner distribution W_Ψ uniquely determines the state $|\Psi\rangle$; this is easily seen by noting that W_Ψ is essentially a Fourier transform and applying the Fourier inversion formula, which yields

$$\Psi(x)\Psi^*(x') = \int e^{\frac{i}{\hbar}p(x-x')}W_\Psi\left[\tfrac{1}{2}(x+x'),p\right]d^np; \quad (50)$$

one then chooses x' such that $\Psi(x') \neq 0$, which yields the value of $\Psi(x)$ for arbitrary x. The same procedure applies to the cross-Wigner transform (Eq. 10); one finds that

$$\Psi(x)\Phi^*(x') = \int e^{\frac{i}{\hbar}p(x-x')}W_{\Psi,\Phi}\left[\tfrac{1}{2}(x+x'),p\right]d^np. \quad (51)$$

Notice that if we choose $x' = x$ we recover the generalized marginal condition (Eq. 15) satisfied by the cross-Wigner distribution.

Thus, the knowledge of $W_{\Psi,\Phi}$ and Φ is in principle sufficient to determine the wavefunction Ψ. Here is a stronger statement which shows that the state $|\Psi\rangle$ can be reconstructed from $W_{\Psi,\Phi}$ using an *arbitrary* auxiliary state $|\Lambda\rangle$ non-orthogonal to $|\Phi\rangle$:

Theorem 2. *Let Λ be an arbitrary vector in $L^2(\mathbb{R}^n)$ such that $\langle\Phi|\Lambda\rangle \neq 0$. We have*

$$\Psi(x)\langle\Phi|\Lambda\rangle = 2^n \iint e^{\frac{2i}{\hbar}p(x-y)} W_{\Psi,\Phi}(y,p)\Lambda(2y-x)d^npd^ny \tag{52}$$

that is

$$\Psi(x) = \frac{2^n}{\langle\Phi|\Lambda\rangle} \iint W_{\Psi,\Phi}(y,p)\hat{T}_{GR}(y,p)\Lambda(x)d^npd^ny; \tag{53}$$

equivalently,

$$\Psi(x) = 2^n \frac{\langle\Psi|\Phi\rangle}{\langle\Phi|\Lambda\rangle} \iint \rho_{\Psi,\Phi}(y,p)\hat{T}_{GR}(y,p)\Lambda(x)d^npd^ny. \tag{54}$$

Proof. By a standard continuity and density argument it is sufficient to assume that Ψ, Φ, Λ are in $\mathcal{S}(\mathbb{R}^n)$. Using Eq. 51 we have

$$\Psi(x)\langle\Phi|\Lambda\rangle = \iint e^{\frac{i}{\hbar}p(x-x')} W_{\Psi,\Phi}(\tfrac{1}{2}(x+x'),p)\Lambda(x')d^npd^nx'.$$

Setting $y = \frac{1}{2}(x + x')$ we get Eq. 52 and hence Eq. 53 in view of the explicit formula for the Grossmann–Royer parity operator (Eq. 33). □

Here is an example: viewing $\langle\Phi|\Lambda\rangle$ as the distributional bracket $\langle\Lambda, \Phi^*\rangle$ we may choose $\Lambda(x) = \delta(x - x_0)$. This yields $\langle\Lambda, \Phi^*\rangle = \Phi^*(x_0)$ and the right-hand side of Eq. 52 is just the integral

$$\int e^{\frac{i}{\hbar}p(x-x')} W_{\Psi,\Phi}(\tfrac{1}{2}(x + x'),p)d^np$$

hence we recover Eq. 51 as a particular case.

4 Discussion

We have been able to give a complete characterization of the notion of weak value in terms of the Wigner distribution, which is intimately related to the Weyl quantization scheme through Moyal's formula (Eq. 7). There are however other possible physically meaningful quantization schemes; the most interesting is certainly that of Born–Jordan [34, 35] mentioned in the introduction; the latter

plays an increasingly important role in quantum mechanics and in time-frequency analysis [7, 8, 10, 11, 36–38], and each of these leads to a different phase space formalism, where the Wigner distribution has to be replaced by more general element of the *Cohen class* [39, 40]. Unexpected difficulties however arise, especially when one deals with the reconstruction problem; these difficulties have a purely mathematical origin, and are related to the division of distributions (for a mathematical analysis of the nature of these difficulties, see [38]). The reconstruction problem for general phase space distributions will be addressed in a forthcoming publication. It should also be mentioned that Hiley and Cohen have proposed in [41] an approach to retrodiction from the perspective of the Einstein–Podolsky–Rosen–Bohm experiment; it is possible that this approach could be studied from the point of view of the techniques developed here.

Acknowledgements

Maurice de Gosson has been funded by the grant P-27773 of the FWF Austrian Science Fund.

References

[1] Aharonov Y, Bergmann PG, Lebowitz JL. Time symmetry in the quantum process of measurement. Physical Review B 1964; 134 (6): 1410–1416. doi:10.1103/PhysRev.134.B1410

[2] Aharonov Y, Cohen E, Elitzur AC. Can a future choice affect a past measurement's outcome? Annals of Physics 2015; 355: 258–268. arXiv:1206.6224, doi:10.1016/j.aop.2015.02.020

[3] Aharonov Y, Vaidman L. Properties of a quantum system during the time interval between two measurements. Physical Review A 1990; 41 (1): 11–20. doi:10.1103/PhysRevA.41.11

[4] Aharonov Y, Vaidman L. Complete description of a quantum system at a given time. Journal of Physics A: Mathematical and General 1991; 24 (10): 2315–2328. doi:10.1088/0305-4470/24/10/018

[5] Aharonov Y, Vaidman L. The two-state vector formalism: an updated review. In: Time in Quantum Mechanics. Lecture Notes in Physics, vol.734, Muga JG, Mayato RS, Egusquiza ÍL (editors), Berlin: Springer, 2008, pp.399–447. arXiv:quant-ph/0105101, doi:10.1007/978-3-540-73473-4_13

[6] Aharonov Y, Rohrlich D. Quantum Paradoxes: Quantum Theory for the Perplexed. Weinheim: Wiley-VCH, 2005.

[7] de Gosson MA, de Gosson SM. The reconstruction problem and weak quantum values. Journal of Physics A: Mathematical and Theoretical 2012; 45 (11): 115305. arXiv:1112.5773, doi:10.1088/1751-8113/45/11/115305

[8] de Gosson MA, de Gosson SM. Weak values of a quantum observable and the cross-Wigner distribution. Physics Letters A 2012; 376 (4): 293–296. arXiv:1109.3665, doi:10.1016/j.physleta.2011.11.007

[9] Gray JE. An interpretation of Woodward's ambiguity function and its generalization. Proceedings of the Radar 2010: IEEE International Radar Conference, Arlington, Virginia, May 10-14, 2010, Institute of Electrical and Electronics Engineers, pp.859–864. doi:10.1109/RADAR.2010.5494499

[10] de Gosson MA. Born–Jordan quantization and the equivalence of the Schrödinger and Heisenberg pictures. Foundations of Physics 2014; 44 (10): 1096–1106. doi:10.1007/s10701-014-9831-z

[11] de Gosson MA. Born–Jordan Quantization: Theory and Applications. Fundamental Theories of Physics, vol.182, Berlin: Springer, 2016.

[12] Gröchenig K. Foundations of Time-Frequency Analysis. Applied and Numerical Harmonic Analysis, Boston: Birkhäuser, 2001. doi:10.1007/978-1-4612-0003-1

[13] Hlawatsch F, Auger F. Time-Frequency Analysis: Concepts and Methods. Digital Signal and Image Processing Series, Hoboken, New Jersey: Wiley-ISTE, 2008.

[14] Lobo AC, Ribeiro CA. Weak values and the quantum phase space. Physical Review A 2009; 80 (1): 012112. arXiv:0903.4810, doi:10.1103/PhysRevA.80.012112

[15] Silva R, Guryanova Y, Brunner N, Linden N, Short AJ, Popescu S. Pre- and postselected quantum states: density matrices, tomography, and Kraus operators. Physical Review A 2014; 89 (1): 012121. arXiv:1308.2089, doi:10.1103/PhysRevA.89.012121

[16] Wigner EP. On the quantum correction for thermodynamic equilibrium. Physical Review 1932; 40 (5): 749–759. doi:10.1103/PhysRev.40.749

[17] Folland GB. Harmonic Analysis in Phase Space. Princeton, New Jersey: Princeton University Press, 1989.

[18] de Gosson MA. Symplectic Methods in Harmonic Analysis and in Mathematical Physics. Pseudo-Differential Operators, vol.7, Boston: Birkhäuser, 2011. doi:10.1007/978-3-7643-9992-4

[19] Lee H-W. Theory and application of the quantum phase-space distribution functions. Physics Reports 1995; 259 (3): 147–211. doi:10.1016/0370-1573(95)00007-4

[20] Hillery M, O'Connell RF, Scully MO, Wigner EP. Distribution functions in physics: fundamentals. Physics Reports 1984; 106 (3): 121–167. doi:10.1016/0370-1573(84)90160-1

[21] Littlejohn RG. The semiclassical evolution of wave packets. Physics Reports 1986; 138 (4): 193–291. doi:10.1016/0370-1573(86)90103-1

[22] Moyal JE. Quantum mechanics as a statistical theory. Mathematical Proceedings of the Cambridge Philosophical Society 1949; 45 (01): 99–124. doi:10.1017/S0305004100000487

[23] Hlawatsch F, Flandrin P. The interference structure of the Wigner distribution and related time-frequency signal representations. In: The Wigner Distribution: Theory and Applications in Signal Processing. Mecklenbräuker W, Hlawatsch F (editors), Amsterdam: Elsevier, 1997, pp.59–133.

[24] Woodward PM. Probability and Information Theory, with Applications to Radar. International Series of Monographs on Electronics and Instrumentation, vol.3, Oxford: Pergamon Press, 1953.

[25] Szu HH, Blodgett JA. Wigner distribution and ambiguity function. AIP Conference Proceedings 1980; 65 (1): 355–381. doi:10.1063/1.32325

[26] Zurek WH. Sub-Planck structure in phase space and its relevance for quantum decoherence. Nature 2001; 412 (6848): 712–717.

[27] McCoy NH. On the function in quantum mechanics which corresponds to a given function in classical mechanics. Proceedings of the National Academy of Sciences 1932; 18 (11): 674–676. doi:10.1073/pnas.18.11.674

[28] Shewell JR. On the formation of quantum-mechanical operators. American Journal of Physics 1959; 27 (1): 16–21. doi:10.1119/1.1934740

[29] Feyereisen MR. How the weak variance of momentum can turn out to be negative. Foundations of Physics 2015; 45 (5): 535–556. doi:10.1007/s10701-015-9885-6

[30] Grossmann A. Parity operator and quantization of δ-functions. Communications in Mathematical Physics 1976; 48 (3): 191–194. doi:10.1007/bf01617867 http://projecteuclid.org/euclid.cmp/1103899886

[31] Royer A. Wigner function as the expectation value of a parity operator. Physical Review A 1977; 15 (2): 449–450. doi:10.1103/PhysRevA.15.449

[32] Lundeen JS, Sutherland B, Patel A, Stewart C, Bamber C. Direct measurement of the quantum wavefunction. Nature 2011; 474 (7350): 188–191. arXiv:1112.3575, doi:10.1038/nature10120

[33] Lundeen JS, Bamber C. Procedure for direct measurement of general quantum states using weak measurement. Physical Review Letters 2012; 108 (7): 070402. arXiv:1112.5471, doi:10.1103/PhysRevLett.108.070402

[34] Born M, Jordan P. Zur Quantenmechanik. Zeitschrift für Physik 1925; 34 (1): 858–888. doi:10.1007/bf01328531

[35] Born M, Heisenberg W, Jordan P. Zur Quantenmechanik. II. Zeitschrift für Physik 1926; 35 (8–9): 557–615. doi:10.1007/bf01379806

[36] Boggiatto P, De Donno G, Oliaro A. Time-frequency representations of Wigner type and pseudo-differential operators. Transactions of the American Mathematical Society 2010; 362 (9): 4955–4981. doi:10.1090/S0002-9947-10-05089-0

[37] Boggiatto P, Cuong BK, De Donno G, Oliaro A. Weighted integrals of Wigner representations. Journal of Pseudo-Differential Operators and Applications 2010; 1 (4): 401–415. doi:10.1007/s11868-010-0018-x

[38] Cordero E, de Gosson MA, Nicola F. On the invertibility of Born–Jordan quantization. 2015: arXiv:1507.00144

[39] Cohen L. Generalized phase-space distribution functions. Journal of Mathematical Physics 1966; 7 (5): 781–786. doi:10.1063/1.1931206

[40] Cohen L. Can quantum mechanics be formulated as a classical probability theory? Philosophy of Science 1966; 33 (4): 317–322. doi:10.1086/288104 http://www.jstor.org/stable/186635

[41] Cohen O, Hiley BJ. Retrodiction in quantum mechanics, preferred Lorentz frames, and nonlocal measurements. Foundations of Physics 1995; 25 (12): 1669–1698. doi:10.1007/bf02057882

9

Was Albert Einstein Wrong on Quantum Physics?

Mani L. Bhaumik

Department of Physics and Astronomy, University of California, Los Angeles, California, USA
E-mail: bhaumik@physics.ucla.edu

Editors: Danko Georgiev & Chris C. King

Albert Einstein is considered by many physicists as the father of quantum physics in some sense. Yet there is an unshakable view that he was wrong on quantum physics. Although it may be a subject of considerable debate, the core of his allegedly wrong demurral was the insistence on finding an objective reality underlying the manifestly bizarre behavior of quantum objects. The uncanny wave-particle duality of a quantum particle is a prime example. In view of the latest developments, particularly in quantum field theory, the objections of Einstein are substantially corroborated. Careful investigation suggests that a travelling quantum particle is a holistic wave packet consisting of an assemblage of irregular disturbances in quantum fields. It acts as a particle because only the totality of all the disturbances in the wave packet yields the energy-momentum with the mass of a particle, along with its other conserved quantities such as charge and spin. Thus the wave function representing a particle is not just a fictitious mathematical construct but embodies a reality of nature as asserted by Einstein.
Quanta 2015; 4: 35–42.

1 Introduction

This year we celebrate with much aplomb the centenary of Einstein's unveiling of his ingenious General Theory of Relativity, although its seed was sown in 1905. In the same *Annus Mirabilis*, he also seeded the other seminal breakthrough of the 20th century: quantum mechanics. He is granted undisputed credit for the theory of relativity, but receives only guarded recognition for his essential contribution to the quantum revolution. In fact, there is a general impression that Einstein lost the debate on quantum physics. As we honor him for relativity, it is fitting to ask whether the legendary star of relativity was indeed wrong on quantum physics.

Einstein was the first physicist to support the veracity of Max Planck's radical postulate of quanta of energy [1, 2]. Although the energy quanta were proposed in 1900, after years of frustration in formulating his law of black body radiation, Planck himself did not seem to believe in their actual existence. Even more than a decade later in 1913, in a petition recommending Einstein to be a member of the Prussian Academy of Sciences, Planck (together with Walther Nernst, Heinrich Rubens, and Emil Warburg) made a patronizing remark

> That [Einstein] may sometimes have missed the target in his speculations, as, for example, in his hypothesis of light quanta, cannot really be held too much against him, for it is not possible to introduce fundamentally new ideas, even in the most exact sciences, without occasionally taking a risk. [3, p.44]

Nernst referred to the light quanta as "probably the strangest thing ever thought up." But Einstein daringly peered through the veil.

Essentially, as early as in 1909 in his Salzburg address [4, p.321], Einstein had predicted that physics would have to reconcile itself to a duality in which light could be regarded as both wave and particle. And at the first Solvay Conference in 1911, he had declared that

> these discontinuities, which we find so distasteful in Planck's theory, seem really to exist in nature. [4, p.608]

So, it was in fact Einstein who fostered the innovative notion of the wave-particle duality by asserting the real existence of quanta of radiation or photons, which eventually would open the door for him to his sole Nobel Prize for the photoelectric effect. Following his elicitation, young Louis de Broglie in his PhD thesis extended the concept to matter particles with crucial and enthusiastic support from Einstein.

To de Broglie's thesis advisor Langevin, the idea of a matter wave seemed far-fetched. So, he sent a skeptical note to his friend Einstein requesting that, "although the thesis is a bit strange, could he see if it was still worth something." Einstein replied with a glowing recommendation

> Louis de Broglie's work has greatly impressed me. *He has lifted a corner of the great veil.* In my work I [have recently] obtained results that seem to confirm his. [5, p.242]

Later Einstein admitted to Isidor Isaac Rabi that he indeed thought about the equation for matter waves before de Broglie but did not publish it since there was no experimental evidence for it [5, p.252]. De Broglie expressed his appreciation by writing

> As M. Langevin had great regard for Einstein, he counted this opinion greatly, and this changed a bit his opinion with regard to my thesis. [5, p.250]

Shortly after reading de Broglie's dissertation, Einstein began suggesting to physicists to look in earnest for an evidence of the matter wave. Soon, Clinton Davisson and Lester Germer furnished proof with the accidental discovery of electron waves in observing a diffraction pattern in a nickel crystal [6].

In the meantime Erwin Schrödinger, "inspired by L. de Broglie . . . and by brief, yet infinitely far-seeking remarks of A. Einstein" [7, p.211], formulated the wave mechanics of quantum physics, which turned out to be equivalent to the rather abstract matrix mechanics devised by Werner

Heisenberg at about the same time. Is it then any wonder that eminent physicists like Leonard Susskind [8, p.xi] consider Einstein to be the father of quantum physics in some sense?

Yet, volumes have been written on Einstein's objection to the implications of quantum physics, particularly to the elements of uncertainty, probability, and nonlocality associated with it. There is no question that, as a true scientist, Einstein accepted the extraordinary success and the spectacular results of quantum physics. Can we discern, then, from the very extensive debates and discussions, what was the primary concern of Einstein in his objection to the interpretation of quantum physics? While there can be endless deliberations on this point, why not accept Einstein's own pronouncement on the subject? "At the heart of the problem," Einstein said of quantum mechanics, "is not so much the question of causality but the question of realism" [4, p.460].

Niels Bohr was content with his postulate of complementarity of wave-particle duality, emphasizing there is *no single underlying reality* that is independent of our observation. "It is wrong to think that the task of physics is to find out how nature is," Bohr declared. "Physics concerns what we can say about nature " [4, p.333]. Einstein derided this pronouncement as an almost religious delirium. He firmly believed there was an objective *reality* that existed whether or not we could observe it [4, p.334].

Most contemporary physicists part company with Einstein invoking that it would be futile to look for reality, which becomes totally obscure under the thick smoke of the heavy artillery of Hilbert space necessary to deal with particles in quantum mechanics. It is a daunting task indeed to discern any reality in the thickets of a configuration space! However, if each single particle comprising the ensemble of Hilbert spaces (Fock space) can be shown to have an objective reality individually, would not it be reasonable to infer that the ensemble in Fock space will also have realism even though one may not be able to decipher it?

Here, we present a credible allocution in favor of the existence of a physical reality behind the wave function at the core of quantum physics. This is primarily anchored on the incontrovertible physical evidence that all electrons in the universe are exactly alike. We provide reasonable support to show that the wave function of quantum mechanics is not just a conjured mathematical paradigm, but there is an objective reality underlying it, thus justifying Einstein's primary concern of *the question of realism*.

The answer to the longstanding puzzle of why all electrons are exactly identical in all respects, a feature eventually found to be shared by all the other fundamental particles as well, was finally provided by the quantum field theory of the Standard Model of particle physics constructed by combining Einstein's special theory of relativity with quantum physics, which evolved from his innovative contributions.

Quantum field theory has successfully explained almost all experimental observations in particle physics and correctly predicted a wide range of phenomena with impeccable precision. By way of many experiments over the years, the quantum field theory of the Standard Model has become recognized as a well-established theory of physics. Although one might argue that the Standard Model accurately describes the phenomena within its domain, it is still incomplete since it does not include gravity, dark matter, dark energy, neutrino oscillations and others. However, because of its astonishing success so far, whatever deeper physics may be necessary for its completion would very likely extend its scope without retracting the current fundamental depiction.

2 Nature of Primary Reality Portrayed by Quantum Field Theory

In order to fully grasp the deeper nature of the relativistic quantum fields, one has to go through the rather esoteric mathematical formalism of the quantum field theory. Nevertheless, its essence can be understood in terms of a narrative. Quantum field theory has uncovered a fundamental nature of reality, which is radically different from our daily perception. Our customary ambient world is very palpable and physical. But quantum field theory asserts this is not the primary reality. The fundamental particles involved at the underpinning of our daily physical reality are only secondary. Each fundamental particle, whether it is a boson or a fermion, has its corresponding underlying quantum field from which it originates [9–12]. The particles are excitations of their respective underlying quantum fields possessing propagating states of discrete energies, and it is these which constitute the primary reality. For example, a photon is a quantum of excitation of the photon field (aka electromagnetic field), the electron is a quantum of the electron quantum field, a quark is a quantum of the quark quantum field, and so on for all the fundamental particles of the universe. Inherent quantum fluctuations are also a distinct characteristic of a quantum field. Thus, quantum field theory substantiates the profoundly counter intuitive departure from our normal perception of reality to reveal that the foundation of our tangible physical world is something totally abstract, comprising of continuous quantum fields that create discrete local excitations we call particles.

By far, the most phenomenal step forward made by quantum field theory is the stunning prediction that the primary ingredient of *everything* in this universe is present in *each element of spacetime* (x, y, z, t) of this immensely vast universe [13, p.74]. These ingredients are the underlying quantum fields. We also realize that the quantum fields are alive with quantum activity. These activities have the unique property of being completely spontaneous and utterly unpredictable as to exactly when a particular event will occur [13, p.74]. Furthermore, some of the quantum fluctuations occur at *mind-boggling speeds* with a typical time period of 10^{-21} seconds or less. In spite of these infinitely dynamic, wild fluctuations, the quantum fields have remained immutable, as evinced by their Lorentz invariance, essentially since the beginning and throughout the entire visible universe encompassing regions, which are too far apart to have any communication even with the speed of light. This is persuasively substantiated by the experimental observation that a fundamental particle such as an electron has exactly the same physical properties, be its rest mass, charge or spin, irrespective of when or where the electron has been created, whether in the early universe, through astrophysical processes over the eons or in a laboratory today anywhere in the world. Such a precise match between theory and observation infuses immense confidence on our approach.

3 Quantum Particle in Motion

As elucidated above, an electron represents a propagating discrete quantum of the underlying continuous electron field. In other words, an electron is a quantized wave (or a ripple) of the electron quantum field, which acts as a particle because of its well-defined energy, momentum, and mass, which are conserved fundamentals of the electron. However, even a single electron, in its reference frame, is never alone. It is unavoidably subjected to the perpetual fluctuations of the quantum fields.

When an electron is created instantaneously from the electron quantum field, its position would be indefinite since a regular ripple with a very well-defined energy and momentum is represented by a delocalized periodic function. But the moment the electron comes into existence, it starts to interact with all the other quantum fields facilitated by quantum fluctuations of the fields. For example, the presence of the electron creates a disturbance in the photon (electromagnetic) quantum field. Assisted by a fleeting quantum fluctuation, the disturbance in the photon field can momentarily appear as what is commonly known as a spontaneously emitted virtual photon.

To conserve momentum, the electron would recoil with momentum equal and opposite to that of the photon. A quantum fluctuation of energy ΔE will provide the kinetic energy for the recoil of the electron as well as the

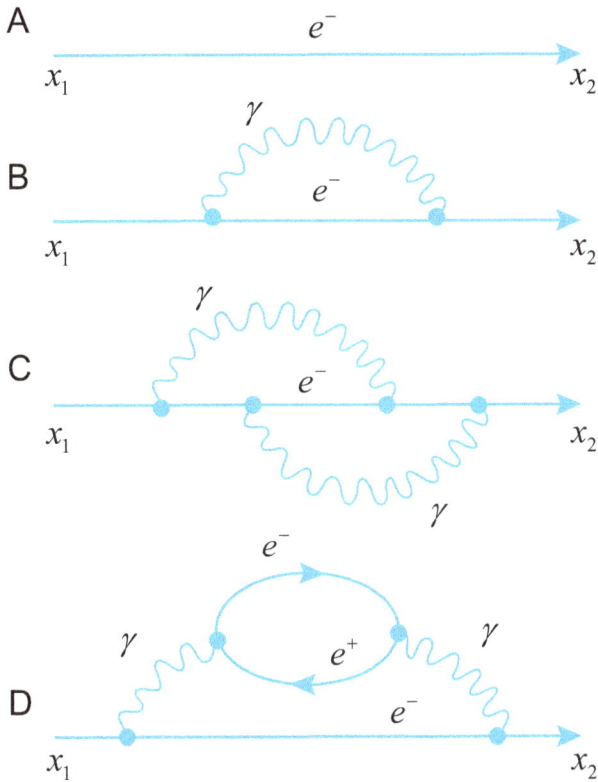

Figure 1: Feynman diagrams showing some of the various interactions between quantum fields during transit of a quantum particle like an electron e^- from x_1 to x_2 (A). (B) Interaction of an electron with the photon field, which is commonly described as the emission of a virtual photon γ by the electron and then reabsorbing it. (C) Emission of two photons and re-absorption by the electron. The photon in turn can create disturbances in the various quantum fields involving a charge. The virtual photon can emit an electron-positron pair $e^- + e^+$ as shown in (D), a muon-anti muon pair, a quark-antiquark pair, etc.

energy of the photon for a time $\Delta T \sim \hbar/\Delta E$. During this transitory moment, the electron by creating a disturbance in the photon field becomes a disturbed ripple itself and therefore ceases to be a normal particle on its own.

All these disturbances are elegantly depicted by Feynman diagrams (see Fig. 1), which also aid in calculating the interaction energies among the various quantum fields. The disturbance in the photon (electromagnetic) field in turn can cause disturbances in all the electrically charged quantum fields, like the electron, muon and the various quark fields. Generally speaking, in this manner, every quantum particle spends some time as a mixture of other virtual particles in all possible ways.

The quantum fluctuations continually and prodigiously create virtual electron-positron pairs in a volume surrounding the electron. "Each pair passes away soon after it comes into being, but new pairs are consistently boiling up to establish an equilibrium distribution" [14, p.404]. Although each pair has a fleeting existence, on average

there is a very significant amount of these pairs to impart a remarkably sizable screening of the bare charge of the electron.

Likewise, though any individual disturbances in the fields or the virtual particles due to quantum fluctuations have an ephemeral existence, there ought to be an equilibrium distribution of such disturbances present at any particular time affecting other aspects of the electron. The effect of these disturbances is very well established in phenomena such as the Lamb shift and the anomalous g-factor of the electron spin.

The electron spin g-factor has been measured to a precision of better than one part in a trillion, compared to the theoretically calculated value that includes Feynman diagrams up to four loops [15]. Therefore it would be reasonable to assume that the equilibrium distribution of disturbances present at any particular time due to all quantum fields involved will be very stable in spite of their fleeting existence.

Let us recall that an electron is a quantized ripple of the electron quantum field, which acts as a particle because it travels holistically with its conserved quantities always sustained as a unit. However, due to interactions of the particle with all the other quantum fields, substantially equivalent to those involved in the Lamb shift and the observed spin g-factor, the ripple in fact becomes very highly distorted immediately after its creation since the quantum fluctuations prompting the interactions of the quantum fields have a typical time period of 10^{-21} seconds. Consequently, the electron ceases to be a ripple of single frequency and becomes a highly deformed *localized* travelling pulse.

It is well known that such a pulse, no matter how deformed, can be expressed by a Fourier integral with weighted linear combinations of simple periodic wave forms like trigonometric functions, briefly mentioned by the author in a previous work [16]. The result would be a wave packet or a wave function that represents a fundamental reality of the universe. Such a wave function would be smooth and continuously differentiable, especially using imaginary numbers in the weighted amplitude coefficients. The wave function $\psi(x)$ will be given by the Fourier integral

$$\psi(x) = \frac{1}{\sqrt{2\pi}} \int_{-\infty}^{+\infty} \tilde{\psi}(k)e^{ikx}dk \tag{1}$$

where $\tilde{\psi}(k)$ is a function that determines the amount of each wave number component $k = 2\pi/\lambda$ that gets added to the combination.

From Fourier analysis, we also know that the spatial wave function $\psi(x)$ and the wave number function $\tilde{\psi}(k)$ are a Fourier transform pair. Therefore, we can find the wave number function through the Fourier transform of $\psi(x)$ as

$$\tilde{\psi}(k) = \frac{1}{\sqrt{2\pi}} \int_{-\infty}^{+\infty} \psi(x)e^{-ikx}dx \qquad (2)$$

Thus the Fourier transform relationship between $\psi(x)$ and $\tilde{\psi}(k)$, where x and k are known as conjugate variables, can help us determine the frequency or the wave number content of any spatial wave function.

4 The Uncertainty Principle

The Fourier transform correlations between conjugate variable pairs have powerful consequences since these variables obey the uncertainty relation

$$\Delta x \Delta k \geq \frac{1}{2} \qquad (3)$$

where Δx and Δk relate to the standard deviations σ_x and σ_k of the wave packet. This is a completely general property of a wave packet with a reality of its own and is in fact inherent in the properties of all wave-like systems. It becomes important in quantum mechanics because of de Broglie's introduction of the wave nature of particles by the relationship $p = \hbar k$, where p is the momentum of the particle. Substituting this in the general uncertainty relationship of a wave packet, the intrinsic uncertainty relation in quantum mechanics becomes

$$\Delta x \Delta p \geq \frac{1}{2}\hbar \qquad (4)$$

This uncertainty relationship has been misunderstood with a rather analogous observer effect, which posits that measurement of certain systems cannot be made without affecting the system. In fact, Heisenberg offered such an observer effect in the quantum domain as a "physical explanation" of quantum uncertainty, a maxim that now popularly goes by the name Heisenberg's uncertainty principle. But the uncertainty principle actually states a fundamental property of quantum systems, and is not a statement about the observational indeterminacy as was emphasized by Heisenberg. In fact, some recent studies [17–19] highlight important fundamental difference between uncertainties in quantum systems and the limitation of measurement in quantum mechanics.

Einstein's fundamental objection to the Copenhagen interpretation was its assertion that any underlying reality of the uncertainties was irrelevant and should be acceptable under the veil of complementarity. We have established that there is indeed an intrinsic uncertainty induced by the wave behavior that is as much a fact of nature as the electron itself, and that it traces its origin back to the wave–particle duality first envisioned by Einstein as a reality.

5 Role of Probability in Measurement

Having been an expert on statistical mechanics, Einstein was no stranger to probability. In fact, he was not opposed to the probabilistic implication of quantum physics. As Wolfgang Pauli reported to Max Born

> Einstein does not consider the concept of 'determinism' to be as fundamental as it is frequently held to be (as he told me emphatically many times) ... In the same way, he *disputes* that he uses as criterion for the admissibility of a theory the question: 'Is it rigorously deterministic?' [20, p.221]

As always, he was essentially searching for realism behind the probabilistic outcome in quantum physics.

In Section 3, we have argued that quantum fields are the primary objective reality. A quantum, or more specifically a quantized ripple of the field, has the characteristics of a particle. But at the very instant of creation, the energy-momentum being fixed, the ripple has only a single frequency and is therefore totally delocalized, which is not a characteristic of a particle. Immediately after the instant of creation, though, the ripple starts to interact with all the other quantum fields. The distorted ripple is equivalent to a localized wave packet that starts to look like a particle, but only the totality of all the disturbances possesses the properties of the particle. The disturbances in the wave packet travel holistically as a unit and thereby acts as a particle. When the wave function is reduced by measurement or otherwise, again the totality of all the field disturbances must be taken as a unit because of the conserved quantities of the particle. The wave function disappears everywhere else, except where it is reduced.

It should now be evident that the random disturbances caused by the intrinsic quantum fluctuations of the underlying field are the reason that a quantum particle such as an electron is always associated with a wave function. Such a wave function is by no means simply a mathematical construct as currently assumed by many physicists. It represents the totality of all the interactions in the various quantum fields caused by the presence of the electron and facilitated by quantum fluctuations. In other words, a quantum particle like an electron in motion is a travelling holistic wave packet consisting of the irregular disturbances of the various quantum fields. It is holistic in the sense that only the *combination of the disturbances* in the electron field together with those in all the other fields *always* maintains a well-defined energy and momentum with an electron mass, since they are conserved quantities for the electron as a particle.

Since a particle like an electron in motion is represented by a wave function, its kinematics cannot be described by the classical equations of motion. Instead, it requires the use of an equation like the Schrödinger equation for a non-relativistic particle

$$i\hbar\frac{\partial}{\partial t}\psi(x,t) = -\frac{\hbar}{2m}\nabla^2\psi(x,t) + V(x)\psi(x,t) \qquad (5)$$

where $V(x)$ is the classical potential and the wave function $\psi(x,t)$ is normalized

$$\int_{-\infty}^{+\infty}\psi^*(x,t)\psi(x,t)dx = 1 \qquad (6)$$

The wave function evolves impeccably in a unitary way. However, when the particle inevitably interacts with a classical device such as a measuring apparatus, the wave function undergoes a sudden discontinuous change known as the *wave function collapse*. Although it is an essential postulate of the Copenhagen interpretation of quantum mechanics, the phenomenon has long been *perplexing* to the physicists [21, p.786]. However, a behavior like this would be a natural consequence of the distinctive nature of a quantum particle described in Section 3. In support of this notion, the holistic nature of the wave function is presented as evidence. In a measurement, this holistic nature becomes obvious since the appearance of the particle in one place prevents its appearance in any other place.

Contrary to the waves of classical physics, the wave function cannot be subdivided during a measurement. This is *specifically* because *the combination of all the disturbances* comprising the wave function possesses a well-defined energy and momentum with the mass of the particle. Consequently, only the totality of the wave function must be taken for detection, causing its disappearance everywhere except where the particle is measured. This inescapable fact could hint at a solution to the well-known measurement paradox.

It has been indeed very difficult to understand why, after a unitary evolution, the wave function suddenly collapses upon measurement or a similar other reductive interaction. The holistic nature of the wave function described above seems to offer a plausible explanation. Parts of the wave function that might spread to a considerably large distance can also terminate instantaneously by the process involved in a plausible quantum mechanical Einstein–Rosen bridge [16, 22] and experimentally demonstrated in quantum entanglement of a single photon [23].

Thus, the very weave of our universe appears to support the objective reality of the wave function, which represents a natural phenomenon and not just a mathematical construct. We also observe that while the wave nature predominates as a very highly disturbed ripple of the quantum field before a measurement, the particle aspect becomes paramount upon measurement.

Because of the wave nature of the particle, the position where the wave packet would land is guided by the *probability density* $|\psi|^2$ given by the *Born rule*. It is only fitting to note that Born followed Einstein in this regard as he stated in his Nobel lecture

> Again an idea of Einstein's gave me the lead. He had tried to make the duality of particles–light quanta or photons–and waves comprehensible by interpreting the square of the optical wave amplitudes as probability density for the occurrence of photons. This concept could at once be carried over to the ψ-function: $|\psi|^2$ ought to represent the probability density for electrons (or other particles). [24, p.262]

The exact mechanism by which the wave function collapses is still hotly debated. The most popular version envisions the wave function becoming entangled with the constituents of the detector and decohering very quickly due to the irreversible thermal motion. One of the principal contributors to the theory of decohernce, Wojciech Zurek contends [25] that the Born rule can actually be derived from the theory of decoherence as opposed to being a mere postulate of quantum theory. There is indeed some support for his contention [26].

The Copenhagen interpretation also requires a conscious observer as an essential part of its formalism, which posits that the reality of a quantum system does not exist until an observer takes part in its detection, thereby causing the wave function to collapse. Einstein objected to this view with his famous query, does "the moon exist only when I look at it?" Although an observer can indeed bring out a particular reality, the fact that the universe, which is quantum at the core, developed to a mature state eons before any manner of conscious observer could appear supports Einstein's skepticism. His contention was that an objective reality should always be present irrespective of measurement.

In contrast, the supporters of the Copenhagen interpretation did not feel it was necessary to delve any further than acceptance of the wave-particle duality and its consequent uncertainty as a principle of complementarity. In view of the nature of reality discussed in this paper, there is no genuine conflict between Einstein's insistence of an underlying reality and the doctrine of complementarity in the Copenhagen interpretation. The intense debate in the pioneering period of quantum physics would appear to be superfluous in view of the nature of the universe revealed to us today. Then the question of who won the debate would have been redundant.

6 Quantum Entanglement

Much has been said about how Einstein got it wrong in the Einstein–Podolsky–Rosen paper [27], in which he attempted to show that quantum mechanics was incomplete and would need further elucidation in the future. For two entangled particles separated by a great distance, Einstein believed there could be no immediate effect to the second particle as a result of anything that was done to the first, since that would violate special relativity. Quantum mechanics predicted otherwise, which he called, "spooky action at a distance." Contrary to Einstein's expectation, all experimental results so far support nonlocality [28–31]. Experimental evidence consistently shows that when two particles undergo entanglement, whatever happens to one of the particles can instantly affect the other, even if the particles are separated by an arbitrarily large distance!

Has Einstein's dream of an objective reality been shattered by these experiments? Not necessarily. It is hard to imagine Einstein would have given up just yet. He would still believe that some deeper reality, perhaps something stranger, lay behind the "spooky action" and certainly that is a reasonable possibility.

Experts such as Maldacena and Susskind [32] postulate that ER=EPR, namely quantum entangled particles (in Einstein–Podolsky–Rosen state [27]) are connected by a wormhole (Einstein–Rosen bridge [22]), implying there is an as yet unknown quantum version of a classical wormhole that permits quantum mechanical nonlocality. There is also a possibility that the quantum fluctuations of the fields are themselves entangled facilitating a quantum wormhole [16]. So there still could be an element of objective reality behind quantum entanglement.

In any case, quantum entanglement violates neither causality nor special relativity, since no classical bit of information can be sent using it [33,34]. Einstein could still have the ultimate chuckle, notwithstanding the fact that some unexpected, specific form of instantaneous action at a distance has been experimentally demonstrated. More so, because in a serendipitous way, the discovery of quantum entanglement has opened up some groundbreaking applications such as quantum cryptography, quantum computing, and quantum teleportation, which have become areas of very active research. As a consequence, the Einstein–Podolsky–Rosen paper [27] has turned out to be a cornerstone in our understanding of quantum physics. If this represents a misstep, it is a fortuitous one that has yielded and will continue to yield a great bounty.

Acknowledgement

The author wishes to thank Zvi Bern and James Ralston for important discussions.

References

[1] Planck M. Ueber das Gesetz der Energieverteilung im Normalspectrum. Annalen der Physik 1901; 309 (3): 553–563. doi:10.1002/andp.19013090310

[2] Einstein A. Über einen die Erzeugung und Verwandlung des Lichtes betreffenden heuristischen Gesichtspunkt. Annalen der Physik 1905; 17 (6): 132–148. doi:10.1002/andp.19053220607

[3] Jammer M. The Conceptual Development of Quantum Mechanics. New York: McGraw-Hill, 1966.

[4] Isaacson W. Einstein: His Life and Universe. New York: Simon & Schuster, 2008.

[5] Stone AD. Einstein and the Quantum: The Quest of the Valiant Swabian. Princeton, New Jersey: Princeton University Press, 2013.

[6] Davisson CJ, Germer LH. Reflection of electrons by a crystal of nickel. Proceedings of the National Academy of Sciences 1928; 14 (4): 317–322. doi:10.1073/pnas.14.4.317

[7] Moore WJ. Schrödinger: Life and Thought. Cambridge: Cambridge University Press, 1989.

[8] Susskind L, Friedman A. Quantum Mechanics: The Theoretical Minimum. What You Need to Know to Start Doing Physics. New York: Basic Books, 2014.

[9] Susskind L. Particle Physics: 1. Basic Concepts. Stanford Continuing Studies 2009: https://www.youtube.com/view_play_list?p=768E1383EA79C603

[10] Kuhlmann M. Quantum field theory. In: Stanford Encyclopedia of Philosophy, Zalta EN, Nodelman U, Allen C (editors), Stanford, California: Stanford University, 2012, http://plato.stanford.edu/entries/quantum-field-theory/

[11] Srednicki M. Quantum Field Theory. Cambridge: Cambridge University Press, 2006. http://web.physics.ucsb.edu/~mark/qft.html

[12] Zee A. Quantum Field Theory in a Nutshell. Princeton, New Jersey: Princeton University Press, 2003.

[13] Wilczek FA. The Lightness of Being: Mass, Ether, and the Unification of Forces. New York: Basic Books, 2008.

[14] Wilczek FA. Fantastic Realities: 49 Mind Journeys And a Trip to Stockholm. Singapore: World Scientific, 2006. doi:10.1142/6019

[15] Brodsky SJ, Franke VA, Hiller JR, McCartor G, Paston SA, Prokhvatilov EV. A nonperturbative calculation of the electron's magnetic moment. Nuclear Physics B 2004; 703 (1–2): 333–362. arXiv:hep-ph/0406325, doi:10.1016/j.nuclphysb.2004.10.027

[16] Bhaumik ML. Reality of the wave function and quantum entanglement. 2014: arXiv:1402.4764

[17] Rozema LA, Darabi A, Mahler DH, Hayat A, Soudagar Y, Steinberg AM. Violation of Heisenberg's measurement-disturbance relationship by weak measurements. Physical Review Letters 2012; 109 (10): 100404. arXiv:1208.0034, doi:10.1103/PhysRevLett.109.100404

[18] Erhart J, Sponar S, Sulyok G, Badurek G, Ozawa M, Hasegawa Y. Experimental demonstration of a universally valid error-disturbance uncertainty relation in spin measurements. Nature Physics 2012; 8 (3): 185–189. arXiv:1201.1833, doi:10.1038/nphys2194

[19] Ozawa M. Heisenberg's original derivation of the uncertainty principle and its universally valid reformulations. 2015: arXiv:1507.02010

[20] Born M, Einstein A. The Born-Einstein Letters: Correspondence Between Albert Einstein and Max and Hedwig Born from 1916–1955, with Commentaries by Max Born. Born I (translator), London: Macmillan, 1971. http://archive.org/details/TheBornEinsteinLetters

[21] Penrose R. The Road to Reality: A Complete Guide to the Laws of the Universe. London: Jonathan Cape, 2004.

[22] Einstein A, Rosen N. The particle problem in the general theory of relativity. Physical Review 1935; 48 (1): 73–77. doi:10.1103/PhysRev.48.73

[23] Fuwa M, Takeda S, Zwierz M, Wiseman HM, Furusawa A. Experimental proof of nonlocal wave-function collapse for a single particle using homodyne measurements. Nature Communications 2015; 6: 6665. arXiv:1412.7790, doi:10.1038/ncomms7665

[24] Born M. The statistical interpretation of quantum mechanics. Nobel Lecture, December 11, 1954. http://www.nobelprize.org/nobel_prizes/physics/laureates/1954/born-lecture.pdf

[25] Zurek WH. Probabilities from entanglement, Born's rule $p_k = |\psi_k|^2$ from envariance. Physical Review A 2005; 71 (5): 052105. arXiv:quant-ph/0405161, doi:10.1103/PhysRevA.71.052105

[26] Schlosshauer M, Fine A. On Zurek's derivation of the Born rule. Foundations of Physics 2005; 35 (2): 197–213. arXiv:quant-ph/0312058, doi:10.1007/s10701-004-1941-6

[27] Einstein A, Podolsky B, Rosen N. Can quantum-mechanical description of physical reality be considered complete? Physical Review 1935; 47 (10): 777–780. doi:10.1103/PhysRev.47.777

[28] Aspect A, Dalibard J, Roger G. Experimental test of Bell's inequalities using time-varying analyzers. Physical Review Letters 1982; 49 (25): 1804–1807. doi:10.1103/PhysRevLett.49.1804

[29] Gröblacher S, Paterek T, Kaltenbaek R, Brukner Č, Żukowski M, Aspelmeyer M, Zeilinger A. An experimental test of non-local realism. Nature 2007; 446 (7138): 871–875. arXiv:0704.2529, doi:10.1038/nature05677

[30] Peruzzo A, Shadbolt PJ, Brunner N, Popescu S, O'Brien JL. A quantum delayed choice experiment. Science 2012; 338 (6107): 634–637. arXiv:1205.4926, doi:10.1126/science.1226719

[31] Shadbolt P, Mathews JCF, Laing A, O'Brien JL. Testing foundations of quantum mechanics with photons. Nature Physics 2014; 10 (4): 278–286. doi:10.1038/nphys2931

[32] Maldacena J, Susskind L. Cool horizons for entangled black holes. Fortschritte der Physik 2013; 61 (9): 781–811. arXiv:1306.0533, doi:10.1002/prop.201300020

[33] Vedral V. Quantum entanglement. Nature Physics 2014; 10 (4): 256–258. doi:10.1038/nphys2904

[34] Brukner Č. Quantum causality. Nature Physics 2014; 10 (4): 259–263. doi:10.1038/nphys2930

10

Holism and Time Symmetry

Peter J. Lewis

Department of Philosophy, University of Miami, Coral Gables, Florida, USA
E-mail: plewis@miami.edu

Editors: *Ken Wharton & Ovidiu Stoica*

Quantum mechanics is often taken to entail holism. I examine the arguments for this claim, and find that although there is no general argument from the structure of quantum mechanics to holism, there are specific arguments for holism available within the three main realist interpretations (Bohm, Ghirardi-Rimini-Weber and many-worlds). However, Evans, Price and Wharton's sideways Einstein-Podolsky-Rosen-Bell example challenges the holistic conclusion. I show how the symmetry between the sideways and standard Einstein-Podolsky-Rosen-Bell set-ups can be used to argue against holism. I evaluate the prospects for extending this insight to more general quantum systems, with a view to producing a genuinely time-symmetric hidden variable theory. I conclude that, although this extension undermines the analogy between the sideways and standard cases, quantum mechanics without holism remains a live possibility.
Quanta 2016; 5: 85–92.

1 Introduction

Quantum mechanics has seemed to many commentators to entail holism, the existence of properties of compound systems that cannot be reduced to properties of their parts. However, the argument is far from straightforward. In what follows, I rehearse the standard argument from en-

tangled states, consider its defects, and then construct a more indirect argument from the three major interpretations of quantum mechanics to a holistic conclusion. I then show how an analogy between entangled states and ordinary preparation/measurement scenarios can be used to undermine even the indirect argument. The analogy suggests that quantum mechanics should be interpreted as a time-symmetric theory, in the sense that causal influences flow both from earlier in time to later and from later to earlier. However, while the analogy is suggestive, it only holds for maximally entangled states, and hence fails when the time-symmetric approach is applied to general quantum states. I argue that the failure of the analogy is no impediment to the time-symmetric interpretation of quantum mechanics, and hence that holism is not necessarily a consequence of quantum mechanics.

2 Arguments for holism

The canonical argument for holism appeals to entangled states of a pair of particles. For example, consider a pair of spin-1/2 particles in the singlet state $|S\rangle = \frac{1}{\sqrt{2}}(|\uparrow_z\rangle_1 |\downarrow_z\rangle_2 - |\downarrow_z\rangle_1 |\uparrow_z\rangle_2)$, where $|\uparrow_z\rangle_1$ and $|\downarrow_z\rangle_1$ are states in which particle 1 is z-spin-up and z-spin-down respectively, and similarly for particle 2. According to Teller [1], the property described by state $|S\rangle$ is a holistic property of the pair of particles: it describes a relation between them that cannot be reduced to their intrinsic properties. Teller's reason is that the state $|S\rangle$ cannot be factored into the product of a state of particle 1 and a state of particle 2. But as it stands this is just the beginning of an argument. Non-factorizability is a mathematical property of a mathematical object. To make the case that this mathematical property tells us something about the physical properties of the particles themselves, we need to know how the mathematical object $|S\rangle$ functions as a description of the physical world. And this is precisely what is up for grabs in quantum mechanics.

Bell's theorem [2] might seem to provide some hope here [3]. Bell's theorem tells us that (subject to some plausible assumptions) *no* ascription of properties to the individual particles can explain the observed behavior of state $|S\rangle$ on measurement. One might be tempted to claim on this basis that we *have* to postulate a holistic property of the pair to explain what we observe. But what does the holistic property add? Suppose we observe that particle 1 is *z*-spin-up and particle 2 is *z*-spin-down. By itself, a holistic property of the pair corresponding to state $|S\rangle$ provides no explanation for what we observe, as $|S\rangle$ is entirely symmetric between spin-up and spin-down for each particle. We could try to break the symmetry by ascribing spin properties to the two particles that are revealed on measurement, but this is precisely what Bell's theorem precludes. That is, the conclusion of Bell's theorem is not that explaining the outcomes we see requires holistic properties, but that, subject to Bell's assumptions, the outcomes we see are *inexplicable*. Most people, Bell included, find this conclusion unacceptable. Hence Bell [4, p.20] saw his theorem as a *reductio* of his assumptions: one of them must be false. But then nothing follows from Bell's theorem (at least directly) about whether the properties of the individual particles are sufficient to explain the outcomes we observe.

So is there any argument that quantum mechanics entails holism? Perhaps the best hope is to look at the various realist interpretations of quantum mechanics on offer: Bohmian hidden variable theories, Ghirardi-Rimini-Weber-type spontaneous collapse theories, and Everettian many-worlds theories. Each of these theories explains the outcomes of measurements on state $|S\rangle$ by violating one of Bell's assumptions. Bohm's theory and the Ghirardi-Rimini-Weber theory violate locality, the assumption that a measurement performed on one particle cannot affect the properties of the other. Many-worlds theories violate uniqueness, the assumption that each measurement has exactly one outcome. So for none of them does Bell's theorem directly imply that the properties of the individual particles alone are insufficient to explain those outcome. But nevertheless, each of them arguably entails holism: even though Bell's theorem does not in itself require the explanation of the measurement outcomes to involve anything over and above the properties of the individual particles, in fact each of these theories does appeal to an irreducible property of the pair of particles to explain what we see. The reason is that in each case the quantum state is typically taken to be genuinely descriptive of the

properties of an entity in the world, of something like a field. This field is most readily understood as an entity residing in a multi-dimensional configuration space, although it can be argued that the field can be understood in three-dimensional terms [5]. In any event, for an entangled state like $|S\rangle$, the properties of this entity include a special connection between particles 1 and 2 that is crucial to the explanation of measurement outcomes.

Let us see briefly why holism is required in the three major realist interpretations [6]. In Bohm's theory, the field described by $|S\rangle$ pushes around a pair of particles, and the positions of the particles correspond to the outcomes of our measurements. There are initial positions of the particles such that if particle 1 is not measured, particle 2 is found to be *z*-spin-up on measurement, but if particle 1 *is* measured, particle 2 is found to be *z*-spin down. That is, the measurement of particle 1 affects the properties of particle 2. This process is non-local, but more to the point, it constitutes a special connection between just these two particles: the measurement of particle 1 does not affect the properties of any other particle. This connection entails counterfactual conditionals such as "If the two particles were to have their spins measured in the same direction, then they would have opposite spins in that direction". These conditional relations are not reducible to the local properties of the individual particles or of the regions of space they occupy. (The conditional relations *are* reducible to the properties of regions of configuration space, but these regions are not *local* in the relevant sense.)

A similar process is at work in the Ghirardi-Rimini-Weber theory. In this case the quantum state $|S\rangle$ is a complete description of the two-particle system, and there is a small chance per unit time of the state undergoing a collapse in which it becomes highly localized in the coordinates of one particle or the other. This collapse probability only becomes significant when many particles become involved, for example on measurement. When the *z*-spin of particle 1 is measured, state $|S\rangle$ collapses to (a state close to) one or other of its terms, and hence both particles acquire *z*-spin properties. Again, this process is non-local, but the key point is the special connection between particle 1 and particle 2 embodied by the entangled state. This state again supports counterfactual conditionals, and is irreducible to the local properties of the spatial regions associated with the individual particles.

The situation in the many-worlds theory is a little more subtle. As in Ghirardi-Rimini-Weber, a quantum state like $|S\rangle$ is a complete description of the system, but unlike Ghirardi-Rimini-Weber, there is no collapse mechanism. Instead, both spin outcomes occur when a spin measurement is performed, and since the terms in the state describing each outcome hardly interact after

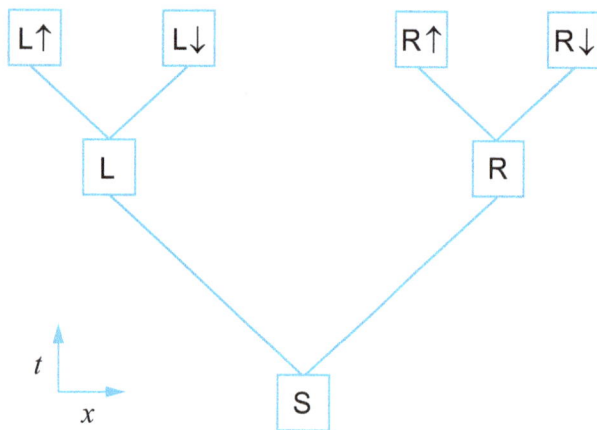

Figure 1: *Spacetime diagram of the standard Einstein-Podolsky-Rosen-Bell experiment.*

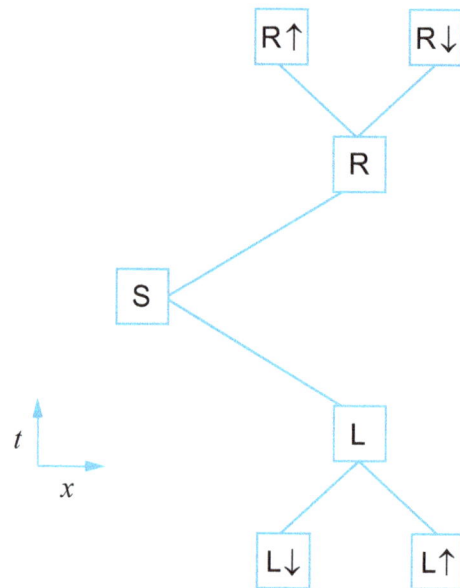

Figure 2: *Spacetime diagram of the sideways Einstein-Podolsky-Rosen-Bell experiment.*

the measurement, they can be treated as separate worlds. When the *z*-spin of particle 1 is measured, two worlds are produced, and relative to a world, both particles acquire *z*-spin properties. There is nothing straightforwardly non-local in the many-worlds case, but the process via which both particles acquire spin properties still embodies a special connection between just these two particles. Again, the state supports counterfactual conditionals, relative to a world: "If the two particles were to have their spins measured in the same direction, then relative to a world, they would have opposite spins in that direction". And again, these conditional relations are not reducible to the local properties of the spatial regions associated with each particle.

So in each case, the explanation of the results we observe appeals to an irreducible property of the pair of particles (or the regions of space associated with each of them) that supports counterfactual relations between them. There are a few caveats, such as attempts by some Bohmians to regard state $|S\rangle$ as describing a law rather than a field [7]. But these aside, the entangled nature of state $|S\rangle$ genuinely corresponds to a holistic property of the ontology associated with the theory in question. So the three main realist interpretations of quantum mechanics all entail holism, and in roughly the same way. One might quite reasonably conclude that holism is an inevitable feature of the quantum world.

But such a conclusion would be premature. Taking my cue from Evans, Price and Wharton [8], I argue that a different perspective on the situation undermines the holistic conclusion.

3 Looking at things sideways

The experiment we have been considering is the standard Einstein-Podolsky-Rosen-Bell experiment. Figure 1 shows a spacetime diagram of the set-up. The particles are emitted at source S and travel outwards to spin detectors at L and R. The spin detectors consist of magnets that deflect particles up or down according to their spin, after which the particles may be run into a fluorescent screen which lights up at the impact point.

Figure 2, on the other hand, shows the Evans-Price-Wharton *sideways* version of the Einstein-Podolsky-Rosen-Bell set-up. Here a single particle is introduced at L and travels to R via S. S is in this case a device that *reflects* the particles, that is, that reverses their direction and their spin. L and R contain the same magnets as before: at R the particle is deflected up or down depending on its spin, and at L the particle is *introduced* either as a spin-up particle through the L-up channel or a spin-down particle through the L-down channel, and is deflected by the magnets onto the trajectory towards S.

These experimental set-ups look rather different: the standard experiment involves two-particles, whereas the sideways version involves a single particle. But as Evans, Price and Wharton point out, the two experiments bear striking symmetry relations to each other. That is, if we reflect the left-hand side of Figure 1 in the $x = 0$ line, and then reflect it again in the $t = 0$ line, we obtain Figure 2. What is more, the probabilistic treatment of the single particle in the sideways version at two times is strongly analogous to the treatment of the two particles in the standard experiment at a single time. That is, if we are completely ignorant about whether the particle in the sideways version is introduced through L-up or L-down, then it is appropriate to ascribe a 50% chance to each, and consequently a 50% chance each to the particle emerging through R-up or R-down. But we know that the spin at L will be the opposite of the spin at R; the spins at the two times are perfectly anticorrelated. This is exactly the probability ascription appropriate for the entangled state $|S\rangle$ in the standard experiment.

But despite the analogies, there is an immediate dis-analogy between the two cases: there is no temptation whatsoever to appeal to holistic properties to explain the particle's behavior in the sideways version. The overall argumentative strategy of Evans, Price and Wharton is to use the analogy between the sideways and standard experiments to dissolve the apparent problems posed by entangled states. Their target is locality: while the standard Einstein-Podolsky-Rosen-Bell experiment is often thought to require non-local action of some kind, nothing of the sort is evident in the sideways version, which casts doubt on the arguments for non-locality in the standard case. My intent is similar: while the standard Einstein-Podolsky-Rosen-Bell experiment is often thought to require holistic properties, nothing of the sort is evident in the sideways version, which casts doubt on the arguments for holism in the standard case. Let us see how that might go.

4 Time symmetry

Consider the sideways Einstein-Podolsky-Rosen-Bell set-up. Suppose a spin-up particle is introduced through the L-up channel. It travels via S to R, where it emerges through the R-down channel. From L to S it is a z-spin-up particle, and from S to R it is a z-spin-down particle: properties ascribed to the particle at a single time suffice to explain what we observe. Similar observations apply if the particle is introduced through L-down and emerges from R-up. Of course, in the standard case, things get more problematic when the magnets at L and R can be rotated from the z-axis: this is the insight behind Bell's proof. But rotating the magnets causes no additional problems for the sideways version. Suppose the magnets at L are rotated so that they make an angle of 120° with the z-axis; call this direction the w-axis. Suppose that a w-spin-up particle is introduced through the L-up channel. It is a w-spin-up particle from L to S and a w-spin-down particle from S to R. But it is measured along the z-axis at R, and we know that empirically speaking there is a 25% chance that it will be measured as z-spin-down and a 75% chance that it will be measured as z-spin-up. We could, if we like, explain this by postulating some process at R by which the particle acquires a z-spin property on measurement, for example a Ghirardi-Rimini-Weber-type collapse.

But in fact this is not required. We can simply stipulate that the w-spin-down particle also has a pre-existing z-spin property, either up or down, distributed statistically in the required 3:1 ratio. One might suspect that Bell's theorem rules out the explanation of the observed outcomes by pre-existing properties, but this is not the case because the sideways set-up explicitly violates one of Bell's assumptions, namely the independence of the particle properties from the device settings at L and R [9]. The properties of the particle clearly depend on the device setting at L: when the magnets are oriented along the w-axis and a particle is introduced through the L-up channel, it has to have the w-spin-up property if it is to be deflected towards S by the magnets. Given that it has this property, it is a straightforward matter to ascribe it spin-properties for any direction in which its spin might be measured at R, without violating the predictions of quantum mechanics. If the angle between the magnets at L and at R is θ, then the required statistical distribution is that the particle is spin-down in the direction of the magnets at R with probability $\cos^2(\theta/2)$, and spin-up with probability $\sin^2(\theta/2)$. So in the sideways version, nothing fancy is required to explain the results we observe: the results just reflect the pre-existing properties of the particles. There is no need for holistic properties, and there is no need for a special mechanism for coordinating the spins of the two particles via a field.

Suppose we carry over this explanation to the standard Einstein-Podolsky-Rosen-Bell case. What is required is that we regard the properties of particle 1 between S and L as dependent on the measurement that is performed at L. The twist, of course, is that in the standard case the particle between S and L is earlier in time than the measurement at L; hence the dependence of its properties on the later measurement event is an example of backwards-in-time causation. The coherence of such accounts has been defended at length by Price [10]. The payoff is that the measurement outcomes in the standard case can be explained entirely by local particle properties that are revealed on measurement. If the magnets at L and R are both oriented along the z-axis, then particle 1 is either z-spin-up or z-spin-down (with a 50% probability of each), and particle 2 has the opposite z-spin property. If the magnets at L are oriented along the w-axis, then particle 1 is either w-spin-up or w-spin-down (with a 50% probability of each), particle 2 has the opposite w-spin, and the z-spin property of particle 2 is statistically distributed accordingly, 3:1 in favor of z-spin-up if particle 2 is w-spin-down, and 3:1 in favor of z-spin-down if particle 2 is w-spin-up. There is no need for holism, no need for the non-local mechanisms of Bohm and Ghirardi-Rimini-Weber, and no need for the world-splitting of the many-worlds theory.

One thing that is strange about the story so far, though, is that it is spatially asymmetric: the measurement performed at L affects the properties of the particles, but the measurement at R simply records them. This asymmetry reflects the obvious temporal asymmetry of the sideways case: L is a preparation event and R is a measurement event. But to inflict a spatial asymmetry on the standard case would be absurd. The obvious alternative, given the time-symmetric causation involved in the current account of correlation, is to deny both the spatial asymmetry in the standard case and the temporal asymmetry in the sideways case. That is, in the standard case particle 1 is affected by the measurement at L, and particle 2 is affected by the measurement at R. Indeed, because of the constraint that their total spin must be zero introduced by the manner of production of the particle pair at S, in fact both measurements affect the properties of both particles. And now that we are in the business of regarding causation as a time-symmetric notion, the same goes for the sideways case: the properties of the single particle are affected both by the past preparation event L and the future measurement event R.

But the details introduce some complications. Consider the standard case first, and suppose that the magnets at L and R are aligned along the z-axis. Then it really does not matter whether we start our causal story at L or at R. That is, we could say that because the magnets at L are along the z-axis, particle 1 is either z-spin-up or z-spin-down (with a 50% probability of each), and because of the way the particles are produced at S, particle 2 has the opposite spin. Alternatively, we could say that because the magnets at R are along the z-axis, particle 2 is either z-spin-up or z-spin-down (with a 50% probability of each), and because of the way the particles are produced at S, particle 1 has the opposite spin. These stories are clearly compatible. But if the magnets at L are aligned along the w-axis the two stories looks different. Starting at L, we say that because of the magnet orientation, the particle is either w-spin-up or w-spin down (with a 50% probability of each), because of the production method at S, particle 2 has the opposite w-spin, and because of particle 2's w-spin, its z-spin properties are distributed 3:1 in favor of one outcome over the other. But starting at R, we say that because of the magnet orientation, the particle is either z-spin-up or z-spin down (with a 50% probability of each), because of the production method at S, particle 2 has the opposite z-spin, and because of particle 2's z-spin, its w-spin properties are distributed 3:1 in favor of one outcome over the other. These two stories look incompatible: they seem to ascribe different probabilities to the various spin properties. For example, one story says that the z-spin of particle 1 is distributed evenly between up and down, and the other story says that it is distributed unevenly.

But this apparent incompatibility dissolves once we untangle what the various probabilities are conditional on. If we start at L, then the probabilities that particle 1 is w-spin-up and w-spin down are *unconditionally* 50% each, and the probabilities that particle 2 is z-spin-up and z-spin down are 75% and 25% respectively, *given* that particle 1 is w-spin-up, and 25% and 75% respectively, *given* that particle 1 is w-spin down. If we start at R, then the probabilities that particle 2 is z-spin-up and z-spin down are *unconditionally* 50% each, and the probabilities that particle 1 is w-spin-up and w-spin down are 75% and 25% respectively, *given* that particle 1 is z-spin-up, and 25% and 75% respectively, *given* that particle 1 is z-spin down. These probability ascriptions are perfectly compatible, and in fact reflect what we observe in an Einstein-Podolsky-Rosen-Bell experiment. That is, if we just look at the results for particle 1 (or just those for particle 2), we see that results are evenly distributed between spin-up and spin-down whichever spin direction is measured. It is only when we compare the results for the two particles that the distinctive entanglement correlations emerge. For example, if we look that the z-spin results for particle 2 only in cases where particle 1 was found to be w-spin-up, we find that they are distributed 3:1 in favor of z-spin-up.

So for a given pair of settings in the standard case, the two stories (the one starting at L and the one starting at R) are just two ways of describing the same distribution of spin properties. And we can do the same for the sideways case. That is, although it is natural to treat L as a preparation event and R as a measurement, we can just as well do the reverse. However, we need to be a little careful, because at the human level preparation and measurement are not the same kind of thing, and there is no obvious sense in which one is the time-reverse of the other. The goal here, though, is just to show that it is possible to regard the quantum process *between* preparation and measurement as time-symmetric. So to that end, let us construe the preparation event so that it is closely analogous to the time-reverse of a measurement event, at least as regards the description of the quantum system. That is, just as a measurement event produces a spin-up or a spin-down outcome at random with a 50/50 statistical distribution, let us suppose that preparation involves introducing a particle at random into the spin-up or spin-down channel with a 50/50 statistical distribution.

Then supposing that the magnets at L are aligned along the w-axis and those at R are aligned along the z-axis, we can describe things like this. Starting at the earlier event L, we say that because of the magnet orientation, the particle at time 1 is either w-spin-up or w-spin down (with a 50% probability of each), because of the reflection method at S, the particle at time 2 has the opposite w-spin, and because of this w-spin, its z-spin properties at R are distributed 3:1 in favor of one outcome over the other. But starting at R, we say that because of the magnet orientation, the particle at time 2 is either z-spin-up or z-spin down (with a 50% probability of each), because of the reflection method at S, the particle at time 1 has the opposite z-spin, and because of this z-spin, its w-spin properties at L are distributed 3:1 in favor of one outcome over the other. These two descriptions are compatible because in each case the 50/50 probabilities are unconditional, and the 75/25 probabilities are conditional on the spin at the other time.

There is something odd about the latter version of the story: it is not the kind we are used to telling. But from the current perspective, the oddity is of a purely pragmatic character that has nothing to do with the structure of the quantum world. That is, given the familiar temporal asymmetries at the macroscopic level, we often prepare a system for later measurement, but we never prepare a system for earlier measurement, so we are interested in the forward-in-time story rather than the backward-in-time story. But these familiar temporal asymmetries (presumably) have an origin in entropic asymmetry that has nothing to do with the behavior of simple quantum systems [10]. So given that backwards-in-time causation is necessary in the standard case, it seems appropriate to propose a temporal symmetry in the sideways case analogous to the spatial symmetry in the standard case.

Thinking back to the arguments in favor of holism in the previous section, what all three realist interpretations of quantum mechanics have in common is that they treat the quantum state as descriptive of the properties of a physical entity. It is here that holism takes root. And now we can see how the time-symmetric account avoids the holistic conclusion: the quantum state is treated merely as a convenient summary of an observer's *knowledge* of the system, as part of a recipe via which the observer can calculate the probability that the system possesses a particular property. But as far as the ontology goes, the time-symmetric account just has particles and their individual properties. The fact that the entangled state $|S\rangle$ cannot be factored into a state of particle 1 and a state of particle 2 just shows that the observer's knowledge is irreducible to knowledge of particle 1 and knowledge of particle 2, since it includes counterfactual conditionals holding between them. But there need be no property of the quantum system itself encoding those conditionals: they are enforced by the causal structures described above.

5 Towards a theory, away from an analogy

Section 4 shows how we can supplement the quantum state with spin properties (hidden variables) to recover the observed measurement results in standard Einstein-Podolsky-Rosen-Bell cases without recourse to holistic properties of the entire system. And since it is such cases that provide the strongest prima facie case for holism, this suggests that holism is not an inevitable consequence of quantum mechanics. But we need to be careful here: this is just a single case, albeit a widely discussed one. There is no guarantee that a general recipe for ascribing properties to systems can be constructed that recovers all the empirical predictions of quantum mechanics. That is, there is no guarantee that the time-symmetric strategy can be turned into a full-fledged hidden variable *theory*.

So let us consider how the property ascription of the previous section might be generalized. The recipe for ascribing probabilities to properties when the w-axis is selected at L and the z-axis at R is summarized in Table 1.

Table 1

		L	
		$\uparrow_w$	$\downarrow_w$
R	$\uparrow_z$	3/8	1/8
	$\downarrow_z$	1/8	3/8

Table 1 shows the unconditional probabilities of the most fine-grained properties: that is, the top left cell shows the probability of the relevant particle (or particles in the standard case) being w-spin-up and z-spin-up. The unconditional probabilities for the two possible properties at L can be obtained by summing the probabilities in a column: for instance, the probability of w-spin-up at L is $3/8 + 1/8 = 1/2$. Similarly, the unconditional probabilities for the two possible properties at R can be obtained by summing the probabilities in a row. The probability of a property at L conditional on a particular property at R can be found by restricting attention to the relevant row and renormalizing. That is, the probability of spin-up on the left given spin-up on the right is $(3/8)/(3/8 + 1/8) = 3/4$. Similarly for the other conditional probabilities.

This probabilistic recipe for ascribing properties to systems can be straightforwardly generalized in a number of ways. First, suppose the w-axis makes an angle θ with the z-axis. Then the probabilities are as in Table 2.

Table 2

		L	
		$\uparrow_w$	$\downarrow_w$
R	$\uparrow_z$	$(1/2)\sin^2(\theta/2)$	$(1/2)\cos^2(\theta/2)$
	$\downarrow_z$	$(1/2)\cos^2(\theta/2)$	$(1/2)\sin^2(\theta/2)$

Table 2 is analogous to a recipe suggested by Price [11] in a very similar context. Second, suppose the state, rather than being the symmetric entangled state $|S\rangle$, is some general state $|\phi\rangle$, that the measurement at L is of a binary observable $\hat{A}$ with two non-degenerate eigenstates $|A_1\rangle$ and $|A_2\rangle$, and that the measurement at at R is of a binary observable $\hat{B}$ with two non-degenerate eigenstates $|B_1\rangle$ and $|B_2\rangle$. Then the recipe for assigning probabilities to the properties A_1, A_2, B_1 and B_2 corresponding to the eigenstates is given in Table 3.

Table 3

		L	
		A_1	A_2
R	B_1	$|\langle A_1 B_1|\phi\rangle|^2$	$|\langle A_2 B_1|\phi\rangle|^2$
	B_2	$|\langle A_1 B_2|\phi\rangle|^2$	$|\langle A_2 B_2|\phi\rangle|^2$

Here $|A_1 B_1\rangle$ is shorthand for $|A_1\rangle_1 |B_1\rangle_2$ (and so on). Finally, we can generalize further to n-ary observables $\hat{A}$ and $\hat{B}$ with non-degenerate eigenstates $|A_i\rangle$ and $|B_i\rangle$ (respectively). Then the recipe for ascribing probabilities to the properties A_i and B_i is given in Table 4.

Table 4

		L			
		A_1	A_2	$\cdots$	A_n
R	B_1	$|\langle A_1 B_1|\phi\rangle|^2$	$|\langle A_2 B_1|\phi\rangle|^2$	$\cdots$	$|\langle A_n B_1|\phi\rangle|^2$
	B_2	$|\langle A_1 B_2|\phi\rangle|^2$	$|\langle A_2 B_2|\phi\rangle|^2$	$\cdots$	$|\langle A_n B_2|\phi\rangle|^2$
	$\vdots$	$\vdots$	$\vdots$	$\ddots$	$\vdots$
	B_n	$|\langle A_1 B_n|\phi\rangle|^2$	$|\langle A_2 B_n|\phi\rangle|^2$	$\cdots$	$|\langle A_n B_n|\phi\rangle|^2$

Table 4 is a fairly general recipe for assigning probabilities to properties based on the settings of the measuring devices. Note that not all potentially measurable properties of the system are assigned probabilities, but only those for which measurements are actually performed. This is what allows the time-symmetric hidden variable approach to avoid the no-go theorems of Bell [2] and Kochen-Specker [12]. Such a recipe can account for the probability distribution of measurement outcomes for a system prepared in a certain state (when L occurs earlier than R, as in the sideways case), and also the measurement outcomes for entangled systems (when L and R are space-like separated, as in the standard case).

This looks like the beginnings of a time-symmetric approach to hidden variables. It is worth noting, though, that the analogy between sideways and standard Einstein-Podolsky-Rosen-Bell cases from which we began does not carry over to this more general context. For maximally entangled states like $|S\rangle$, a "sideways" single-particle analog exists that ascribes exactly the same probabilities to spin properties. But for partially entangled two-particle entangled states, there is no single-particle analog with the same probability ascriptions [13]. This is not a significant problem, though: while the analogy between sideways and standard cases is a useful motivating heuristic, the time-symmetric approach does not depend on any general analogy between two particles at a single time and a single particle at two times. The recipes for ascribing properties to particles in Tables 3 and 4 above still work perfectly well for entangled, partially entangled and unentangled two-particle states in standard Einstein–Podolsky-Rosen-Bell situations, as well as for single-particle states in sideways analogs, where those analogs exist.

There is still more work to be done in developing the time-symmetric approach; there is no dynamical evolution of the state in the simple models considered here. In particular, it is not yet clear how interference should be handled in a time-symmetric hidden variable theory. There are a number of promising approaches: Price [10, pp. 252–257] suggests in general terms how a time-symmetric approach might explain interference phenomena; Spekkens [14] constructs a toy model that exhibits some of the features of interference; Sutherland [15] constructs a time-symmetric version of Bohm's theory that can generate interference effects; and Wharton [16] describes two-slit interference in time-symmetric terms, where the number of slits the particle passes through depends on the later measurement made on it. If one or more of these approaches bears fruit, then the time-symmetric approach will be a notable competitor to the interpretive approaches considered in Section 2.

6 Discussion

Holism comes up repeatedly in the context of quantum mechanics, and while the route from entanglement to holism is not a clear one, a case can be made for holism, in the sense that each of the three main realist interpretations of quantum mechanics appeals to an irreducible property of both subsystems to explain the behavior of an entangled pair. But Evans, Price and Wharton's sideways perspective on entanglement is instructive here. Given the analogy between standard and sideways Einstein-Podolsky-Rosen-Bell experiments, it looks like holism should apply to both or neither. Holism is implausible in the sideways case, and the behavior of the system can be fully explained without appeal to holistic properties. So the idea is that perhaps the same kind of explanation can be applied to the standard case. And indeed it can, as long as we are prepared to accept causal influences from the future to the past. The resulting temporally symmetric account avoids the need for holistic properties, and allows the quantum state to adopt a purely epistemic role, describing the observer's knowledge of the properties of the individual particles.

Evans, Price and Wharton do not give us a full quantum mechanical theory, just an analysis of a couple of examples; other work extends the analogy to cover all maximally entangled states, but not partially entangled states [13, 16]. However, it is straightforward to generalize the recipe for assigning probabilities to properties in the standard case to cover *all* states, maximally entangled or otherwise, provided one is willing to give up the analogy between the standard and sideways cases that was used to motivate the time-symmetric theory. The time-symmetric approach itself is independent of the analogy: once the approach has been recognized, the analogy can be left aside. This gives hope that a time-symmetric hidden variable theory that is provably empirically equivalent to standard quantum mechanics can be constructed. In the meantime, we can at least conclude that the path from quantum mechanics to holism is not a straightforward one.

Acknowledgments

I am grateful for the feedback from participants in the Quantum Foundations Workshop 2015 organized by Shan Gao, especially Mark Stuckey and Ken Wharton. I am also indebted to Ken Wharton for pointing out to me that the analogy between the standard and sideways cases breaks down when the entanglement is not maximal, and for very helpful comments on a draft of this paper.

References

[1] Teller P. Relational holism and quantum mechanics. *British Journal for the Philosophy of Science* 1986; **37**(1): 71–81. JSTOR:686998

[2] Bell JS. On the Einstein–Podolsky–Rosen Paradox. *Physics* 1964; **1**(3): 195–200. CERN:111654

[3] Hawthorne J, Silberstein M. For whom the Bell arguments toll. *Synthese* 1995; **102**(1): 99–138. JSTOR:20117977, doi:10.1007/bf01063901

[4] Bell JS. *Speakable and Unspeakable in Quantum Mechanics, 2nd edition*. Cambridge: Cambridge University Press, 2004.

[5] Wallace D, Timpson CG. Quantum mechanics on spacetime I: Spacetime state realism. *The British Journal for the Philosophy of Science* 2010; **61**(4): 697–727. PhilSci:4621, doi:10.1093/bjps/axq010

[6] Lewis PJ. *Quantum Ontology: A Guide to the Metaphysics of Quantum Mechanics*. Oxford: Oxford University Press, 2016.

[7] Goldstein S, Zanghì N. Reality and the role of the wave function in quantum theory. In: *The Wave Function: Essays on the Metaphysics of Quantum Mechanics*. Ney A, Albert DZ (editors), Oxford: Oxford University Press, 2013, pp. 91-109 arXiv:1101.4575

[8] Evans PW, Price H, Wharton KB. New slant on the EPR-Bell experiment. *British Journal for the Philosophy of Science* 2013; **64**(2): 297–324. arXiv:1001.5057, doi:10.1093/bjps/axr052

[9] Price H. A neglected route to realism about quantum mechanics. *Mind* 1994; **103**(411): 303–336. JSTOR:2253742, arXiv:gr-qc/9406028, doi:10.1093/mind/103.411.303

[10] Price H. *Time's Arrow and Archimedes' Point: New Directions for the Physics of Time*. Oxford: Oxford University Press, 1996.

[11] Price H. Does time-symmetry imply retrocausality? How the quantum world says "Maybe"? *Studies in History and Philosophy of Science Part B: Studies in History and Philosophy of Modern Physics* 2012; **43**(2): 75–83. arXiv:1002.0906, doi:10.1016/j.shpsb.2011.12.003

[12] Kochen SB, Specker EP. The problem of hidden variables in quantum mechanics. *Journal of Mathematics and Mechanics* 1967; **17**(1): 59–87. doi:10.1512/iumj.1968.17.17004

[13] Wharton KB, Miller DJ, Price H. Action duality: a constructive principle for quantum foundations. *Symmetry* 2011; **3**(3): 524–540. doi:10.3390/sym3030524

[14] Spekkens RW. Evidence for the epistemic view of quantum states: A toy theory. *Physical Review A* 2007; **75**(3): 032110. arXiv:0401052, doi:10.1103/PhysRevA.75.032110

[15] Sutherland RI. Causally symmetric Bohm model. *Studies in History and Philosophy of Science Part B: Studies in History and Philosophy of Modern Physics* 2008; **39**(4): 782–805. arXiv:quant-ph/0601095, doi:10.1016/j.shpsb.2008.04.004

[16] Wharton KB. Quantum states as ordinary information. *Information* 2014; **5**(1): 190–208. doi:10.3390/info5010190

11

Quantum Mechanics and Liouville's Equation

Michael Nauenberg

Physics Department, University of California, Santa Cruz, United States
E-mail: mnauenbe@ucsc.edu

Editors: *José N. R. Croca & Danko Georgiev*

In non-relativistic quantum mechanics, the absolute square of Schrödinger's wave function for a particle in a potential determines the probability of finding it either at a position or momentum at a given time. In classical mechanics the corresponding problem is determined by the solution of Liouville's equation for the probability density of finding the joint position and momentum of the particle at a given time. Integrating this classical solution over either one of these two variables can then be compared with the probability in quantum mechanics. For the special case that the force is a constant, it is shown analytically that for an initial Gaussian probability distribution, the solution of Liouville's integrated over momentum is equal to Schrödinger's probability function in coordinate space, provided the coordinate and momentum initial widths of this classical solution satisfy the minimal Heisenberg uncertainty relation. Likewise, integrating Lioville's solution over position is equal to Schrödinger's probability function in momentum space.
Quanta 2017; 6: 53–56.

1 Introduction

In 1926, when Erwin Schrödinger formulated the fundamental non-relativistic equation for quantum mechanics [1], his students rhymed:

> Erwin with his *psi* can do
> Calculations quite a few
> But one thing has not been seen:
> Just what does *psi* really mean [2].

Shortly afterwards, Max Born gave a precise meaning to *psi*, Schrödinger's wave function $\psi(\vec{r}, t)$ for a particle traveling in a potential $V(\vec{r}, t)$, by proposing that $|\psi(\vec{r}, t)|^2$ is the probability density for finding it in the position interval $(\vec{r}, \vec{r} + d\vec{r})$ at time t [3, 4]. Since then, many attempts have been made to derive this interpretation from first principles, but without success, although efforts in this direction have continued up to the present time [5]. To counter early criticisms for his interpretation, Born pointed out that in classical mechanics, the unavoidable uncertainties in initial conditions imply that in practice classical mechanics is also statistical in nature [3, 4]. But recently, in an article entitled "The trouble with quantum mechanics", Steve Weinberg asked "Since Schrödinger's equation is deterministic, how do probabilities get into quantum mechanics?" [6]. The large number of responses by physicists to his article indicates that the answer to this question is still not settled [7].

In the absence of interference effects, the time evolution of the absolute square of Schrödinger's wave function is closely related to the solution of Liouville's equation for the corresponding probability function in classical mechanics. In this note, we demonstrate this correspondence for the motion of a particle moving under the action of a constant force. For the special case that the initial probability distribution is a Gaussian function, the quantum and the classical problem can both be solved analytically. Then, if the initial widths in coordinate and momentum space satisfy the minimal Heisenberg uncertainty relation, it is shown that the evolution of the quantum and classical probability distributions is identically the same.

2 Liouville equation

We consider the Liouville equation in one dimension along the direction x of a constant force. Let $P(x, v, t)$ be the probability of finding a particle at x with velocity v at time t. Then at a later time $t + dt$

$$P(x + dx, v + dv, t + dt) = P(x, v, t), \quad (1)$$

and to first order in the infinitesimals dt, dx and dv,

$$dx\frac{\partial P}{\partial x} + dv\frac{\partial P}{\partial v} + dt\frac{\partial P}{\partial t} = 0. \quad (2)$$

Setting

$$dx = vdt , dv = adt. \quad (3)$$

where $a = dv/dt$ is the acceleration due to an external force that can depend on x, v and t, leads to Liouville's equation

$$\frac{\partial P}{\partial t} = -v\frac{\partial P}{\partial x} - a\frac{\partial P}{\partial v}. \quad (4)$$

For an initial Gaussian dependence of P on x and v,

$$P(x, v, 0) = \frac{1}{2\pi\sigma_x\sigma_v} \exp\left(-\frac{x^2}{2\sigma_x^2} - \frac{v^2}{2\sigma_v^2}\right), \quad (5)$$

where σ_x and σ_v are the widths in coordinate and velocity respectively.

It can be readily shown that for the case that a is a constant, the time dependent solution of Liouville's equation, Eq. 4, is

$$P(x, v, t) = \frac{1}{2\pi\sigma_x\sigma_v} \exp\left[-\frac{\left(x - vt + \frac{at^2}{2}\right)^2}{2\sigma_x^2} - \frac{(v - at)^2}{2\sigma_v^2}\right] \quad (6)$$

Proof.

$$\frac{\partial P}{\partial x} = -\frac{\left(x - vt + \frac{at^2}{2}\right)}{\sigma_x^2}P, \quad (7)$$

$$\frac{\partial P}{\partial v} = \left[\frac{\left(x - vt + \frac{at^2}{2}\right)t}{\sigma_x^2} - \frac{(v - at)}{\sigma_v^2}\right]P. \quad (8)$$

Hence

$$\begin{aligned}\frac{\partial P}{\partial t} &= \left[\frac{\left(x - vt + \frac{at^2}{2}\right)(v - at)}{\sigma_x^2} + \frac{(v - at)a}{\sigma_v^2}\right]P \\ &= -v\frac{\partial P}{\partial x} - a\frac{\partial P}{\partial v} \end{aligned} \quad (9)$$

$\square$

To compare this result with the corresponding solution of the Schrödinger equation for a constant acceleration a, consider the probability $P'(x, t)$ for finding a particle at x at time t independent of its velocity v. Then

$$P'(x, t) = \int_{-\infty}^{+\infty} dvP(x, v, t) = \frac{1}{\sqrt{2\pi}\sigma(t)} \exp\left[-\frac{\left(x - \frac{at^2}{2}\right)^2}{2\sigma^2(t)}\right] \quad (10)$$

where $\sigma(t) = \sqrt{\sigma_x^2 + t^2\sigma_v^2}$.

For the special case that the widths σ_x and σ_v satisfy the minimal Heisenberg uncertainty relation

$$\sigma_x\sigma_v = \frac{\hbar}{2m}, \quad (11)$$

we obtain

$$\sigma(t) = \sqrt{\sigma_x^2 + \left(\frac{\hbar t}{2m\sigma_x}\right)^2}, \quad (12)$$

The corresponding time dependent Schrödinger equation for this problem is

$$\imath\hbar\frac{\partial\psi(x, t)}{\partial t} = -\frac{\hbar^2}{2m}\frac{\partial^2\psi(x, t)}{\partial x^2} + V(x)\psi(x, t), \quad (13)$$

for a potential $V(x) = -amx$. Let the initial state at $t = 0$ be a Gaussian wave function with the same width σ_x as in the classical problem,

$$\psi(x, 0) = \left(2\pi\sigma_x^2\right)^{-\frac{1}{4}} \exp\left(-\frac{x^2}{4\sigma_x^2}\right). \quad (14)$$

Recently, I discussed the solution of Eq. 13 with this initial condition [8], so only the results will be given here. One finds that

$$\psi(x, t) = \phi(x, t)\exp\left[\frac{\imath S(x, t)}{\hbar}\right], \quad (15)$$

where

$$\phi(x, t) = \left[2\pi\left(\sigma_x^2 + \frac{t^2\hbar^2}{4m^2\sigma_x^2}\right)\right]^{-\frac{1}{4}} \exp\left[-\frac{\left(x - \frac{at^2}{2}\right)^2}{4\sigma_x^2 + \frac{2\imath\hbar t}{m}}\right], \quad (16)$$

and

$$S(x, t) = amt\left(x - \frac{at^2}{6}\right). \quad (17)$$

Hence, according to Eq. 10

$$|\psi(x,t)|^2 = P'(x,t), \qquad (18)$$

showing that in coordinate space the probability distribution in quantum mechanics is exactly the same as in classical mechanics. A similar identity is obtained for the probability distribution in momentum space.

3 Conclusion

We have demonstrated analytically that for particle motion under the action of a constant force, the spreading in coordinate space of a quantum mechanical wave packet, and a corresponding classical distribution are exactly the same. In this special case, quantum interference effects do not occur. Originally, Schrödinger interpreted $|\psi(x,t)|^2$ to be the density of a particle like the electron, and concerned with this spreading, he wrote to Lorentz:

> Would you consider it a very weighty objection against the theory if it were to turn out that the electron is incapable of existing in a completely field free space? [9, p. 59]

Even as late as 1946, he wrote to Einstein:

> I am no friend of the probability theory, I have hated it from the first moment when our dear friend Max Born gave it birth. For it could be seen how easy and simple it made everything, in principle, everything ironed out and the true problem concealed. [10, p. 435]

Schrödinger's strong reaction to Born's probability interpretation may partly explain why it continues to be debated up to the present time.

For space dependent forces, the corresponding equations have to be solved numerically, and I have done such a calculation for the important classical and quantum problem of a central inverse square force [11, 12]. In this case the spreading of an initially well localized wave packet occurs around the center of force, and the quantum and classical distribution remain the same until the tip of the distribution catches up with its tail. Afterwards, interference effects occur in quantum mechanics that do not have a classical analog, and recurrences appear that also do not have any classical analog [13]. These recurrences have been verified experimentally for Rydberg atoms [14, 15].

Acknowledgment

I thank David Book for a careful reading of this manuscript and helpful comments.

References

[1] Schrödinger E. An undulatory theory of the mechanics of atoms and molecules. *Physical Review* 1926; **28**(6): 1049–1070. doi:10.1103/PhysRev.28.1049

[2] Bloch F. Heisenberg and the early days of quantum mechanics. *Physics Today* 1976; **29**(12): 23–27. doi:10.1063/1.3024633

[3] Born M. Zur Quantenmechanik der Stoßvorgänge. *Zeitschrift für Physik A* 1926; **37**(12): 863–867. doi:10.1007/BF01397477

[4] Born M. The Statistical Interpretation of Quantum Mechanics. *Nobel Lecture* December 11, 1954. http://www.nobelprize.org/nobel_prizes/physics/laureates/1954/born-lecture.pdf

[5] van Kampen NG. The scandal of quantum mechanics. *American Journal of Physics* 2008; **76**(11): 989–990. doi:10.1119/1.2967702

[6] Weinberg S. The trouble with quantum mechanics. *The New York Review of Books*, January 19, 2017. http://www.nybooks.com/articles/2017/01/19/trouble-with-quantum-mechanics/

[7] Mermin ND, Bernstein J, Nauenberg M, Bricmont J, Goldstein S, Maudlin T. Steven Weinberg and the puzzle of quantum mechanics. *The New York Review of Books*, April 6, 2017. http://www.nybooks.com/articles/2017/04/06/steven-weinberg-puzzle-quantum-mechanics/

[8] Nauenberg M. Einstein's equivalence principle in quantum mechanics revisited. *American Journal of Physics* 2016; **84**(11): 879–882. doi:10.1119/1.4962981

[9] Einstein A, Schrödinger E, Planck M, Lorentz HA. *Letters on Wave Mechanics*. Przibram K (editor), New York: Philosophical Library, 1967.

[10] Moore WJ. *Schrödinger: Life and Thought*. Canto Classics. Cambridge: Cambridge University Press, 2015.

[11] Nauenberg M. Quantum wave packets on Kepler elliptic orbits. *Physical Review A* 1989; **40**(2): 1133–1136. doi:10.1103/PhysRevA.40.1133

[12] Nauenberg M, Keith A. Negative probability and the correspondence between quantum and classical physics. In: *Quantum Chaos—Quantum Measurement*. Cvitanović P, Percival I, Wirzba A (editors), NATO Advanced Science Institutes Series, vol. 358, Dordrecht: Springer, 1992, pp. 265–272. doi:`10.1007/978-94-015-7979-7_22`

[13] Nauenberg M. Autocorrelation function and quantum recurrence of wavepackets. *Journal of Physics B: Atomic, Molecular and Optical Physics* 1990; **23**(15): L385–L390. doi:`10.1088/0953-4075/23/15/001`

[14] Nauenberg M, Stroud C, Yeazell J. The classical limit of an atom. *Scientific American* 1994; **270**(6): 44–49. doi:`10.1038/scientificamerican0694-44`

[15] Nauenberg M. What happened to the Bohr–Sommerfeld elliptic orbits in Schrödinger's wave mechanics? In: *One hundred years of the Bohr atom. Proceedings from a Conference*. Aaserud F, Kragh H (editors), Copenhagen: Royal Danish Academy of Sciences, 2015, pp. 465–477.

12

Schrödinger's Cat: Where Does The Entanglement Come From?

Radu Ionicioiu

Department of Theoretical Physics, Horia Hulubei National Institute of Physics and Nuclear Engineering, Bucharest–Măgurele, Romania. E-mail: r.ionicioiu@theory.nipne.ro

Editors: *Jonas Maziero & Danko Georgiev*

Schrödinger's cat is one of the most striking paradoxes of quantum mechanics that reveals the counterintuitive aspects of the microscopic world. Here, I discuss the paradox in the framework of quantum information. Using a quantum networks formalism, I analyse the information flow between the atom and the cat. This reveals that the atom and the cat are connected only through a classical information channel: the detector clicks → the poison is released → the cat is killed. No amount of local operations and classical communication can entangle the atom and the cat, which are initially in a separable state. This casts a new light on the paradox.
Quanta 2017; 6: 57–60.

1 Introduction

Schrödinger's cat is arguably the most famous and discussed thought experiment in quantum mechanics [1]. Historically, it is also one of the first to reveal the paradoxical nature of quantum theory when applied to macroscopic objects. Consider a box containing a cat, a radioactive atom prepared in the excited state $|0\rangle$ and a detector coupled to the atom. If the atom decays to its ground state $|1\rangle$, the detector triggers a mechanism which releases a poison and kills the cat. We close the box and we let the isolated system evolve. In the textbook description of the experiment, after some time the state of the atom–cat system is [2, pp. 373-374]

$$|\psi_\mathrm{f}\rangle = a|0\rangle|alive\rangle + b|1\rangle|dead\rangle \qquad (1)$$

hence the cat is "simultaneously" alive and dead, thus the paradox; usually one takes $a = b = \frac{1}{\sqrt{2}}$.

The state (1) is entangled, as can be seen from the concurrence $C(|\psi_\mathrm{f}\rangle) = 2|a \cdot b|$; for $|a| = |b| = \frac{1}{\sqrt{2}}$ the state is maximally entangled, $C = 1$. This brings us to the main question of this article:

Where does the entanglement come from?

2 Contradiction

Consider the following statements:

(i) the initial state of the system is separable, $|0\rangle|alive\rangle$. The cat is in a well-defined state $|alive\rangle$ and is not entangled with the atom;

(ii) the atom evolves freely to $a|0\rangle + b|1\rangle$;

(iii) the only coupling between the atom and the cat is via the detector; there is no *entangling quantum interaction* between the two systems;

(iv) the state (1) of the atom–cat system is entangled.

(a)

$$a|0\rangle + b|1\rangle \quad a|0\rangle|alive\rangle + b|1\rangle|dead\rangle$$

$$|alive\rangle$$

(b)

$$a|0\rangle + b|1\rangle \quad m = 0, 1$$

$$|alive\rangle \quad NOT = (1 - m) * alive + m * dead$$

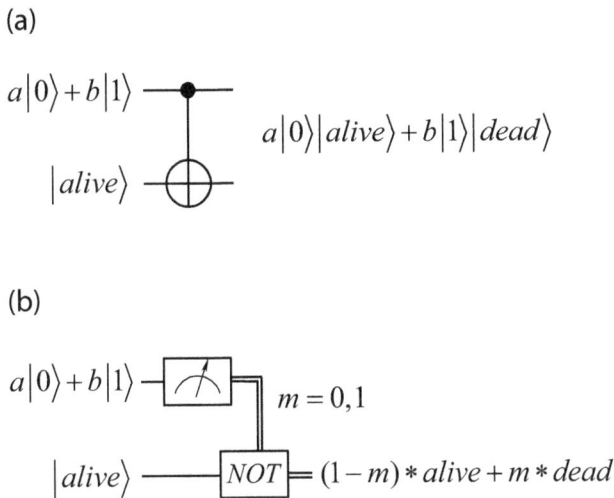

Figure 1: *The quantum network model reveals the information flow in the thought experiment. Single lines represent quantum systems (qubits) and double lines classical systems (bits); blue (red) lines denote the atom (cat). Quantum (classical) information flows along the single (double) lines. (a) Quantum network entangling the atom and the cat. A quantum cat in the initial state |alive⟩ is coupled to the atom via a quantum CNOT gate resulting in the state a|0⟩|alive⟩ + b|1⟩|dead⟩. (b) Quantum network for the standard setup of Schrödinger's cat. The atom evolves freely to a|0⟩ + b|1⟩ and is coupled to the cat via a classical information channel: if the detector clicks (m = 1), a poison is released and kills the cat.*

The purpose of this article is to show that there is a contradiction between the first three statements (i)–(iii) and the last one (iv). Briefly, starting with a separable state and having only a classical communication channel between the atom and the cat one cannot generate the entangled state (1).

Consider first the entangled state (1). In order to entangle two systems, initially in a separable state, we need a quantum interaction, for example an entangling gate or a quantum channel (like in entanglement swapping). One can view this operation in the standard quantum network formalism, Fig.1(a). Therefore we need to apply a quantum CNOT gate between the atom and the cat in order to entangle them. The gate should be fully quantum, meaning it should operate also on quantum superpositions, not only on classical (basis) states.

However, this is not what happens in the usual setup of the thought experiment. The atom evolves freely to $a|0\rangle + b|1\rangle$ and the only interaction between the atom and the cat is through the detector: if the detector clicks, a poison is released and the cat is killed with a *classical device*, Fig.1(b). This is a completely classical information channel between the atom and the cat. There is no coherent quantum CNOT gate acting between the atom (the control) and the cat (the target). No amount of local

operations and classical communication can increase the entanglement between two systems. In particular, two initially separable systems cannot become entangled using only local operations and classical communication. Entanglement is generated only by a quantum interaction, as can be seen in the experimental generation of cat-states [3, 4].

This implies that the final state of the atom–cat system in Fig.1(b) is not described by the entangled state (1), but by a classical statistical mixture

$$\rho = |a|^2|0\rangle\langle 0| \otimes |alive\rangle\langle alive| + |b|^2|1\rangle\langle 1| \otimes |dead\rangle\langle dead| \tag{2}$$

As expected, ρ is separable since the concurrence is $C(\rho) = 0$. This is in contrast to the entangled state (1) which has $C(|\psi_f\rangle) = 2|a \cdot b|$.

To obtain the state (1), which is the crux of the paradox, we need to:

(a) find the appropriate Hilbert space describing the cat. The cat becomes now a quantum system with at least two orthogonal states |alive⟩ and |dead⟩. Here we leave aside the controversial issue of *how to put a cat in a ket*, that is, what is the Hilbert space of a cat? what is its dimension? etc;

(b) apply (experimentally) a quantum CNOT gate between the atom and the quantum cat.

Thus, unless we are able to perform a fully coherent quantum CNOT gate between the atom and the cat, the final state of the system is described by the statistical mixture (2), not by the entangled state (1). Closing the box and refraining from looking inside (in order to prevent the collapse) will not entangle the atom and the cat. The atom still evolves freely and the only coupling between the two subsystems is via a classical channel which does not generate entanglement.

We can better understand the difference between classical and quantum information flow by looking at another thought experiment. In the quantum delayed-choice experiment, we replace the classical control in the Wheeler's delayed-choice [5] with a quantum control [6]. Although the classical and the quantum delayed-choice thought experiments are related, the quantum version gives rise to distinct phenomena which are not present in the classical case. These include a *morphing behaviour* between wave and particle and the ability to measure the control qubit *after* we measure the system [6, 7].

Interestingly, Schrödinger never wrote the entangled state (1), which later became the textbook description of the experiment [2]. The following is the English translation (by J. Trimmer) of Schrödinger's original article [1]:

One can even set up quite ridiculous cases. A cat is penned up in a steel chamber, along with the following diabolical device (which must be secured against direct interference by the cat): in a Geiger counter there is a tiny bit of radioactive substance, *so* small, that *perhaps* in the course of one hour one of the atoms decays, but also, with equal probability, perhaps none; if it happens, the counter tube discharges and through a relay releases a hammer which shatters a small flask of hydrocyanic acid. If one has left this entire system to itself for an hour, one would say that the cat still lives *if* meanwhile no atom has decayed. The first atomic decay would have poisoned it. The ψ-function of the entire system would express this by having in it the living and the dead cat (pardon the expression) mixed or smeared out in equal parts.

It is typical of these cases that an indeterminacy originally restricted to the atomic domain becomes transformed into macroscopic indeterminacy, which can then be *resolved* by direct observation.

3 Discussion

The Schrödinger's cat paradox baffled countless physicists and laymen alike. After more than 80 years, the paradox is still unsolved and generates fierce debates about the measurement problem and the quantum-classical cut. Some of the attempts to solve the paradox [8] invoke Everett's many-worlds interpretation or decoherence. Nevertheless, this still does not explain how one can entangle two separable systems using only a classical channel.

The confusion behind the paradox is well-captured by Jaynes [9, 10]:

> But our present (quantum mechanical) formalism is not purely epistemological; it is a peculiar mixture describing in part realities of Nature, in part incomplete human information about Nature – all scrambled up by Heisenberg and Bohr into an omelette that nobody has seen how to unscramble. Yet we think that the unscrambling is a prerequisite for any further advance in basic physical theory. For, if we cannot separate the subjective and objective aspects of the formalism, we cannot know what we are talking about; it is just that simple.

The (infamous) measurement problem is masterfully discussed in John Bell's (less famous) article entitled *Against 'measurement'* [11]. Here we did not attempt to solve the measurement problem, which would need a (presently missing) quantum ontology [12]. Our goal was more modest: we showed that, in the usual description of the Schrödinger's cat paradox [1, 2], we cannot entangle the cat and the atom via a classical channel, that is, a detector. In order to entangle the two systems we need a *coherent quantum interaction* between the two systems.

Indeed, the experiments preparing cat-states $\frac{1}{\sqrt{2}}(|000\dots0\rangle + |111\dots1\rangle)$ use entangling quantum interactions between the subsystems [3, 4]. However, such an interaction is absent in the usual description of the paradox in which a detector click starts a classical chain of events killing the cat.

In conclusion, Schrödinger's cat is *either* alive *or* dead, not both simultaneously. This kills the paradox (but not the cat).

Acknowledgments

I am grateful to Daniel Terno for comments and suggestions. This work has been funded by the Romanian Ministry of Research and Innovation through grant PN 16420101/2016.

References

[1] Schrödinger E. Die gegenwärtige Situation in der Quantenmechanik. *Die Naturwissenschaften* 1935; 23:807–812; 823–828; 844–849. English translation: Trimmer JD, *Proceedings of the American Philosophical Society* 1980; 124: 323–338. Reprinted in: *Quantum Theory and Measurement*, Wheeler JA, Zurek WH (editors), pp. 152–167. Princeton: Princeton University Press, 1983.

[2] Peres A. *Quantum Theory: Concepts and Methods*. New York: Kluwer, 2002.

[3] Gao WB, Lu CY, Yao XC, Xu P, Gühne O, Goebel A, Chen YA, Peng CZ, Chen ZB, Pan JW. Experimental demonstration of a hyper-entangled ten-qubit Schrödinger cat state. *Nature Physics* 2010; 6:331–335. `arXiv:0809.4277`, `doi:10.1038/nphys1603`.

[4] Sychev DV, Ulanov AE, Pushkina AA, Richards MW, Fedorov IA, Lvovsky AI. Enlargement of optical Schrödinger's cat states. *Nature Photonics* 2017; 11:379–382 (2017). `arXiv:1609.08425`, `doi:10.1038/nphoton.2017.57`.

[5] Wheeler JA. Law without law. In: *Quantum Theory and Measurement*, Wheeler JA, Zurek WH (editors), pp. 182–213. Princeton: Princeton University Press, 1983.

[6] Ionicioiu R, Terno DR. Proposal for a quantum delayed-choice experiment. *Physical Review Letters* 2011; 107:230406. arXiv:1103.0117, doi:10.1103/PhysRevLett.107.230406.

[7] Céleri LC, Gomes RM, Ionicioiu R, Jennewein T, Mann RB, Terno DR. Quantum control in foundational experiments. *Foundations of Physics* 2014; 44:576–587. arXiv:1301.6969, doi:10.1007/s10701-014-9792-2.

[8] Vedral V. Observing the observer. In: Living in a Quantum World. *Scientific American* 2011; 304(6):38–43. arXiv:1603.04583. doi:10.1038/scientificamerican0611-38.

[9] Jaynes ET. Probability in quantum theory. In: *Complexity, Entropy, and the Physics of Information*, Zurek WH (editor), pp. 381–404. Redwood City: Addison-Wesley, 1990.

[10] Pusey MF, Barrett J, Rudolph T. On the reality of the quantum state. *Nature Physics* 2012; 8:475–478. arXiv:1111.3328, doi:10.1038/nphys2309.

[11] Bell JS. Against 'measurement'. *Physics World* 1990; 3(8):33–40. doi:10.1088/2058-7058/3/8/26.

[12] Ionicioiu R. Quantum mechanics: knocking at the gates of mathematical foundations. In: *Romanian Studies in Philosophy of Science*, Pârvu I, Sandu G, Toader ID (editors), Boston Studies in the Philosophy and History of Science, vol. 313, pp. 167–179. Heidelberg: Springer, 2015. arXiv:1506.04511, doi:10.1007/978-3-319-16655-1_11.

13

Is Schrödinger's Cat Alive?

Mani L. Bhaumik

Department of Physics and Astronomy, University of California, Los Angeles, USA. E-mail: bhaumik@physics.ucla.edu

Editors: *Zvi Bern* & *Danko Georgiev*

Erwin Schrödinger is famous for presenting his wave equation of motion that jump-started quantum mechanics. His disenchantment with the Copenhagen interpretation of quantum mechanics led him to unveil the Schrödinger's cat paradox, which did not get much attention for nearly half a century. In the meantime, disappointment with quantum mechanics turned his interest to biology facilitating, albeit in a peripheral way, the revelation of the structure of DNA. Interest in Schrödinger's cat has recently come roaring back making its appearance conspicuously in numerous scientific articles. From the arguments presented here, it would appear that the legendary Schrödinger's cat is here to stay, symbolizing a profound truth that quantum reality exists at all scales; but we do not observe it in our daily macroscopic world as it is masked for all practical purposes, most likely by environmental decoherence with irreversible thermal effects.
Quanta 2017; 6: 70–80.

1 Introduction

Like most cats, this one slipped into the room practically unnoticed. Schrödinger's cat [1–5], which is supposedly both alive and dead at the same time, has by now attained a celebrated status in both scientific and public spheres. In addition to its surprisingly frequent mention in quantum physics articles, the quantum computation community

has bestowed it a permanent berth by coining a cat state in its repute. The appearance of a cat's image adorning the jacket of a popular undergraduate textbook [6] as well as two popular science books on the subject [7,8] speaks volumes about its privileged place in scientists' fancy. It has come to be the common metaphor for superposition so ubiquitous in the quantum domain. Perhaps because of its somewhat melodramatic character, it has also secured a place in popular culture with appearances in literature, television, film, music, cartoons, jokes, and video games. It would be rather fascinating to follow the trail of this captivating story since in a real sense Schrödinger's cat is a conspicuous symbolic entity representing a profound reality, namely: the manifest coexistence of the quantum and classical reality, which appear so particularly different in their attributes.

At the very dawn of the twentieth century, Max Planck initiated the quantum revolution by introducing the idea of a quantum [9]. After years of frustrating failure to work out an accurate formula for thermal radiation of a black body, Planck postulated that radiation energy is emitted and absorbed only in discrete packages he called quanta. But he was, however, a reluctant revolutionary and did not believe in the physical existence of the quantum, which so uniquely contributed in formulating the correct equation for the black body radiation.

It was Albert Einstein who recognized the reality of a quantum and suggested that radiation in space also consists of discrete quanta [10]. With near unanimity, the entire physics community—including Planck himself—remained very skeptical about the postulate since by then Maxwell's theory of electromagnetic radiation in terms of continuous wave motion had, through its numerous confirmations, become ingrained in people's minds. It also evoked the counter intuitive idea of a wave-particle duality. Nevertheless, Einstein unwaveringly persisted almost for two decades as the sole champion keeping the nascent quantum revolution alive. His tenacity was finally vindicated spectacularly by the discovery of the Compton effect in 1923, which demonstrated that X-rays could be deflected like billiard balls by an electron [11].

With an overwhelming support from Einstein, the young French graduate student, Louis de Broglie galvanized the quantum revolution in 1924 by boldly postulating in his doctoral thesis that even matter particles could have wave like properties. His hypothesis gained support from the fact that it could explain the radius of the stationary orbits of the electron in Niels Bohr's audacious model of the hydrogen atom, offered in 1913 [12–14]. The radius of a Bohr's conjectured orbit was an integral multiple of de Broglie's proposed wavelength of the electron. Years later, Einstein confided that he also came up with the idea of matter waves but did not publish it due to a lack of any evidence. After de Broglie submitted his thesis, Einstein ardently appealed to the physicists to look for an experimental proof of the postulated waves. Soon Clinton Davisson and Lester Germer provided the evidence in 1927 by observing a diffraction pattern in the beam of electrons scattered by nickel crystals [15].

In the meantime, through his correspondence with Einstein [16, p. 412], Erwin Schrödinger became fascinated by the radical concept of de Broglie's matter wave. At the request of Peter Debye, Schrödinger gave a seminar at ETH in Zurich toward the end of November 1925, enthusiastically explaining how the matter waves gave support to Niels Bohr's atomic model with stable electronic orbits. In the seminar, Debye expressed some skepticism about how waves, which normally spreads out, could confine itself in stable atomic orbits. He suggested that a relevant wave equation of motion should be formulated to deal with a wave in a proper way [16, pp. 419–421].

Schrödinger took up the challenge, since apparently he was already thinking about the subject himself, and promptly formulated his groundbreaking equation of motion for matter waves in early 1926 [17, 18]. With some assistance from his associate, mathematical physicist Hermann Weyl, Schrödinger found the solutions for the standing waves of the discrete orbits of the hydrogen atom [19, p. 1165]. Schrödinger stated in his first paper

> The essential thing seems to me to be, that the postulation of "whole numbers" no longer enters into the quantum rules mysteriously, but that we have traced the matter a step further back, and found the "integralness" to have its origin in the finiteness and single-valuedness of a certain space function. [18, p. 9]

The astounding success of his equation in giving precise explanation of the quantized orbital motion of an electron in a hydrogen atom initiated a storm of activities and the quantum revolution went into full swing.

2 Trail to the Schrödinger Equation

As we have just described, after the demonstration of the Compton effect, wave-particle duality of massless particles like photons was established beyond a reasonable doubt. Louis de Broglie's profound insight to extend the idea to massive particles was a feat of genius. He correctly surmised that a massive particle like an electron would have a characteristic internal frequency v_0 in compliance with the Planck–Einstein relation

$$E = m_0 c^2 = h v_0 \tag{1}$$

that would obey the rules of special relativity to provide the length of the matter wave $\lambda = h/p$ or equivalently $p = \hbar k$ analogous to that of the photon. However, he struggled with a derivation of the matter wave using the special theory of relativity to such an extent that he portrayed the matter wave to be merely a fictitious guiding wave to be used for the kinematics of the matter particle. Consequently, the matter wave proposition received the moniker de Broglie's hypothesis instead of a theory. We now know that the relationship $p = \hbar k$ can be elegantly comprehended using the four-vector procedure of the special theory of relativity.

In an earlier presentation [20], it was shown that the wave packet of a quantum particle could be deduced from an objective reality to be given by

$$\psi(x,t) = \frac{1}{\sqrt{2\pi}} \int_{-\infty}^{\infty} \tilde{\psi}(k) e^{i(kx - \omega_k t)} dk \tag{2}$$

The group velocity of the wave-packet for a massless particle like a photon is

$$v_g = \frac{d\omega}{dk} = \frac{d(kc)}{dk} = c \tag{3}$$

For a massive particle like an electron, the group velocity of the wave packet in Eq. (2) is no longer equal to the constant c, but a variable velocity v depending upon its energy-momentum. Accordingly, the spacetime dependence $e^{i(kx - \omega_k t)}$ must change frequently, which can be deduced from the special theory of relativity.

The transformation of the momentum four-vector $(\frac{E}{c}, \vec{p})$ and the wave four-vector $(\frac{\omega}{c}, \vec{k})$ that keeps their magnitude invariant is the Lorentz transformation. Considering a laboratory frame S and a rest frame S' with a boost velocity v in the x direction, the Lorentz transformation relations for the momentum four-vector and the wave four-vector are

$$\frac{E'}{c} = \gamma(\frac{E}{c} - \beta p_x) \tag{4}$$

$$p'_x = \gamma(p_x - \beta\frac{E}{c}) \tag{5}$$

$$p'_y = p_y \tag{6}$$

$$p'_z = p_z \tag{7}$$

and

$$\frac{\omega'}{c} = \gamma(\frac{\omega}{c} - \beta k_x) \tag{8}$$

$$k'_x = \gamma(k_x - \beta\frac{\omega}{c}) \tag{9}$$

$$k'_y = k_y \tag{10}$$

$$k'_z = k_z \tag{11}$$

where $\beta = \frac{v}{c}$ and $\gamma = 1/\sqrt{1 - \beta^2}$.

To determine the proportionality constant between the two four-vectors, we correlate their timelike components after multiplying the component of the wave four-vector by $\hbar$:

$$\frac{E'}{c} = \gamma(\frac{E}{c} - \beta p_x) \tag{12}$$

$$\frac{\hbar\omega'}{c} = \gamma(\frac{\hbar\omega}{c} - \beta\hbar k_x) \tag{13}$$

Subtracting Eq. (13) from Eq. (12), we find

$$\frac{E'}{c} - \frac{\hbar\omega'}{c} = \gamma(\frac{E}{c} - \frac{\hbar\omega}{c} - \beta p_x + \beta\hbar k_x) \tag{14}$$

Since the laws of physics are the same in all inertial frames of reference, the Planck's law $E = \hbar\omega$ in frame S should hold true in frame S', giving us $E' = \hbar\omega'$. Thus Eq. (14) reduces to

$$\gamma\beta(p_x - \hbar k_x) = 0 \tag{15}$$

According to the zero product property of algebra, either $\gamma\beta = 0$ or $p_x - \hbar k_x = 0$. Because $\gamma\beta$ is non-zero, we obtain

$$p_x = \hbar k_x \tag{16}$$

Or more generally, $p = \hbar k$ irrespective of the mass of the particle, zero or otherwise. This relationship for a massive particle is the celebrated de Broglie hypothesis. However, we now realize that the relationship can indeed be derived from fundamental considerations and does not need to be a mere hypothesis.

As presented earlier [21], using the relationship in Eq. (9) the group velocity of the wave packet is $v_g = v$, which is the velocity of the "particle" represented by the wave packet. Therefore, the wave packet moves with the velocity of either a massive or a massless particle. Steven Weinberg also confirms this result using a slightly different consideration [22, pp. 11–12].

3 Kinematics of the wave packet

Following Schrödinger, the equation of motion for the wave packet given in Eq. (2) can now be formulated using the relationship in Eq. (16).

Schrödinger's astute intuition was to express the phase of a plane wave utilizing Eq. (16) to achieve

$$\Psi(x, t) = Ae^{i(kx-\omega t)} = Ae^{\frac{i}{\hbar}(px-Et)} \tag{17}$$

and to realize that the first order partial derivatives were: with respect to space

$$\frac{\partial\Psi}{\partial x} = \frac{i}{\hbar}pAe^{\frac{i}{\hbar}(px-Et)} = \frac{i}{\hbar}p\Psi \tag{18}$$

or

$$p\Psi = -i\hbar\frac{\partial\Psi}{\partial x} \tag{19}$$

thus the momentum operator is

$$\hat{p} = -i\hbar\frac{\partial}{\partial x} \tag{20}$$

and with respect to time

$$\frac{\partial\Psi}{\partial t} = -\frac{i}{\hbar}EAe^{\frac{i}{\hbar}(px-Et)} = -\frac{i}{\hbar}E\Psi \tag{21}$$

or

$$E\Psi = i\hbar\frac{\partial\Psi}{\partial t} \tag{22}$$

thus the energy operator is

$$\hat{E} = i\hbar\frac{\partial}{\partial t} \tag{23}$$

for any state of energy E.

Max Born generalized this relationship for any system described by a Hamiltonian $\hat{H}$. The time dependence of any wave function, whether or not for a state of definite energy, is then given by

$$i\hbar\frac{\partial\Psi}{\partial t} = \hat{H}\Psi \tag{24}$$

The equation (24) is known as the general Schrödinger equation. It is indeed quite general and used throughout quantum mechanics, for everything from the Dirac equation to quantum field theory, by utilizing various complicated expressions for the Hamiltonian $\hat{H}$.

If the Hamiltonian itself is not explicitly dependent on time, Eq. (22) shows that

$$\hat{H}\Psi = E\Psi \tag{25}$$

For a single non-relativistic particle of mass m, its energy in a potential $V(x)$ is $\frac{p^2}{2m} + V(x)$ giving us

$$-\frac{\hbar^2}{2m}\nabla^2\Psi(x) + V(x)\Psi(x) = E\Psi(x) \tag{26}$$

This is the time-independent Schrödinger equation for a single non-relativistic particle of energy E. The equation (26) solved for the hydrogen atom, with the three boundary conditions that: $\Psi(x)$ is single-valued, $\Psi(x)$ returns to the same value if x goes around a closed curve, and $\Psi(x)$ vanishes as the magnitude of x goes to infinity, accurately reproduced Bohr's formula for the discrete energy levels but without any assumption.

Einstein wholeheartedly endorsed the paper since he considered the wave nature of particle gave a clearer intuitive picture as compared to the rather abstract matrix mechanics introduced by Werner Heisenberg, Max Born and Pasqual Jordan in 1925. In rapid succession within a few months after his first paper, Schrödinger was able to demonstrate that the analytical solution of his time independent wave equation indeed predicts the discrete energy levels of several other non-relativistic quantum systems, among them, the quantum harmonic oscillator, diatomic molecule, and Stark effect [18,23]. With Einstein's commendation and the immense success of the wave equation, Schrödinger equation became the benchmark for wave mechanics at the time bestowing its inventor an instant renown. The very following year, he was honored with the venerable chair formerly held by Max Planck at the Friedrich Wilhelm University in Berlin.

After the spectacular success of the time independent wave equation, Schrödinger focused his attention on developing a time dependent equation for treating problems in which the quantum system changes with time as in scattering problems. Using the total energy $H = \frac{p^2}{2m} + V(x)$ for a single non-relativistic particle in a potential $V(x)$, and taking the square of the momentum operator, Schrödinger soon unveiled his time dependent equation of motion for a wave packet

$$i\hbar\frac{\partial\Psi(x,t)}{\partial t} = \hat{H}\Psi(x,t) = -\frac{\hbar^2}{2m}\nabla^2\Psi(x,t) + V(x)\Psi(x,t) \tag{27}$$

Clearly, the plane wave in Eq. (17) is a solution of the Schrödinger equation (27). However, the Schrödinger equation is a linear differential equation. So a linear combination of plane waves is also a valid solution. The wave packet in Eq. (2) is just such a linear combination. The momentum wave function $\tilde{\psi}(k)$ appearing in the integrand is an integral of the position wave functions since the position and momentum space wave functions are Fourier transforms of each other. Therefore the kinematics of the wave packet described in Eq. (2) can be processed using the Schrödinger equation (27).

This distinctive aspect of superposition of solutions of the linear Schrödinger equation is its unique signature. It should not be surprising since in our daily lives, we do see water wavelets superpose when they come in contact.

In the quantum domain as well, we see chargeless bosons like photons and gravitons exhibit superposition by producing macroscopic classical waves. But the fermion wave functions, being antisymmetric, in addition to some of them having different charges, are not alike and do not produce classical waves. However, the bizarre implication of superposition, especially as construed by the Copenhagen interpretation for measurement, made both Einstein and Schrödinger, the pioneering leaders of the quantum revolution, enormously unhappy.

4 Probability Amplitude

In his initial studies involving the time independent wave equation to predict energies of discrete quantum states, Schrödinger did not have to be too concerned about what exactly the wave function $\Psi(x,t)$ did represent. It would be tempting to think, as in fact Schrödinger originally did, that the wave function represented a smeared out charge distribution of the electron. In an experiment involving the scattering of electrons, Max Born realized that Schrödinger's contention that the wave function represented a charge distribution could not be sustained [24]. Instead he suggested as it appears in his Nobel lecture

> Again an idea of Einstein's gave me the lead. He had tried to make the duality of particles— light quanta or photons—and waves comprehensible by interpreting the square of the optical wave amplitudes as probability density for the occurrence of photons. This concept could at once be carried over to the ψ-function: $|\psi|^2$ ought to represent the probability density for electrons (or other particles). [25]

This is known as the Born's rule.

But recalling how the wave packet had come to be [20], it would have been obvious that Born was correct. The wave packet consists of irregular disturbances, the sum total of which represents the mass, energy momentum, charge of a particle like electron. Therefore, the wave function is in fact a function of probability amplitude for finding the particle.

It should be noted, however, that although the attributes of the various irregular disturbances are mostly characteristics of their respective quantum fields with different charge, spin, etc., they have one aspect in common. The element of disturbance in energy is identical for all fields. Energy density of a wave is given by the square of its amplitude. Therefore, to get the probability density, we have to take the square of the amplitude of the wave function, which usually involves a complex quantity. Consequently, the square of the amplitude $\Psi^*(x,t)\Psi(x,t)$, which is the probability density function $P(x, t)$, should represent the probability density for finding a particle in position space at time t. As discussed earlier, this is acknowledged as the renowned Born's rule, which is a necessary hypothesis of quantum mechanics. But as we have just presented, it is a natural consequence of the reality of the wave function revealed in a previous article [20] and does not need to be a hypothetical rule.

However, it is of critical importance that the wave function is normalized

$$\int_{-\infty}^{\infty} \Psi^*(x, t)\Psi(x, t)dx = 1 \qquad (28)$$

since the equation describes one particle with the sum of probability to be 1. Born's probabilistic description of the wave function was promptly taken over by Niels Bohr in Copenhagen who in essence became the father of the long-reigning Copenhagen interpretation of quantum mechanics.

According to the Copenhagen interpretation, physical systems in superposed states do not possess real properties prior to being measured, and quantum mechanics can only predict the probabilities that measurements will produce certain results. The act of measurement using a macroscopic device by an observer collapses the wave function, causing the set of probabilities to reduce to only one of the possible values.

It is well known that both pioneers of quantum mechanics, Einstein and Schrödinger, were extremely troubled by the Copenhagen interpretation. Much has been written on the famous Bohr–Einstein debates. The concerns of Einstein have been discussed in greater detail in an earlier communication [20]. His most strident objection was the almost cult-like denial by the proponents of the Copenhagen interpretation of any reality in a quantum system prior to measurement. Among others, in one noteworthy sarcastic comment he states

> The present quantum theory is unable to provide the description of a real state of physical facts, but only of an (incomplete) knowledge of such. Moreover, the very concept of a real factual state is debarred by the 'orthodox' quantum theoreticians. The situation arrived at corresponds almost exactly to that of the good old Bishop Berkeley. [26, pp. 73–74]

He expected the proponents to be at least open, as he himself was, to the possibility that quantum mechanics as currently formulated might be incomplete.

Schrödinger's main objection appears to be the rather fictional depiction of the superposed states and their probability interpretation as well as the collapse of the wave function. He envisioned that after a unitary evolution of

his wave equation, it could be somehow extrapolated up to the macroscopic scale of the measuring device. Sometime later, he sarcastically commented that

> I don't like it, and I'm sorry I ever had anything to do with it. [7, p. v]

5 Quantum Entanglement

Einstein expressed his dismay regarding the implications of quantum mechanics very publicly through the famous Bohr–Einstein debates beginning at the Solvay conference in 1927, of which Bohr was the generally supposed winner. Bohr, meanwhile, was troubled by none of the elements that appalled Einstein. He made his own peace with the contradictions by asserting a principle of complementarity that emphasized the role of the observer over the observed and placed limits on what the observer could know about a quantum system. Given his disenchantment with the Copenhagen interpretation, Einstein's attention shifted from quantum mechanics primarily toward his interest in developing a unified theory of gravity and electro-magnetism. After the well-known turbulent times in Germany, Einstein left Europe and in 1933, permanently settled down at the Institute for Advanced Study in Princeton, where he finally had the time and serenity to resume his thoughts on quantum mechanics.

In 1935, Albert Einstein together with Boris Podolsky and Nathan Rosen published the seminal EPR paper in an attempt to show that the description of reality as given by a wave function was not complete [27]. Einstein tried to illustrate that superposition of the quantum states of two particles cannot be real. Because when the particles are separated by an arbitrarily large distance, measuring a particular property of one of the particles would instantly affect the other. In a letter to Born, Einstein wrote

> I admit, of course, that there is a considerable amount of validity in the statistical approach which you were the first to recognize clearly as necessary given the framework of the existing formalism. I cannot seriously believe in it because the theory cannot be reconciled with the idea that physics should represent a reality in time and space, free from spooky actions at a distance. [28, p. 158]

Because of Einstein's repute, the EPR article generated some impressive newspaper headlines. But the physics community essentially ignored it for quite a while. Later, in the 1950s, David Bohm tried to show that Einstein's desired element of reality could be supported by existence of some local hidden variables. Bohm's ideas met with stiff resistance from the scientists.

It was not until 1964, when John Bell started his seminal analysis known as Bell's inequality that helped to discard the existence of local hidden variables, proposed by Bohm, but indicated the existence of nonlocality revealed in Bohm's investigation. Bell ingenuously proposed an approach to experimentally verify nonlocality between a pair of quantum particles in superposed states, which are now called entangled states as coined by Schrödinger. Numerous experimental verifications of the existence of quantum entanglement beyond any reasonable doubt have now been accomplished establishing that quantum superposition is undoubtedly real. Using a satellite, Chinese scientists have demonstrated entanglement over a distance as large as 1200 kilometers patently attesting that Einstein's "spooky action at a distance" is here to stay [29]. He still might have derived some satisfaction from knowing that in conformity with his special theory of relativity, no meaningful signal can be sent using entanglement and the spookiness of the action is circumscribed.

As explained by Narnhofer and Thirring, in quantum field theory almost everything is entangled in the quantum and the mesoscopic domains [30]. Penrose finds this extremely puzzling, stating

> Since, according to quantum mechanics, entanglement is such a ubiquitous phenomenon—and we recall that the stupendous majority of quantum states are actually entangled ones—why is it something that we barely notice in our direct experience in the world? [31, p. 591]

It is rather ironic that the very pivotal EPR thought experiment in which Einstein attempted in effect to show the impossibility of quantum entanglement would lead to one of the most astounding discoveries of the twentieth century. It has opened a floodgate of activities in quantum physics. The possible uses of quantum entanglement in a variety of novel applications such as quantum cryptography, quantum computation, and quantum teleportation have become areas of very active research.

It has also provided an immense stimulus to basic scientific investigations. Experts such as Maldacena and Susskind postulate that ER=EPR implying there is an as yet unknown quantum mechanical version of a classical wormhole that permits quantum entanglement [32]. There is a possibility that the quantum fluctuations of the fields are themselves entangled facilitating a quantum mechanical Einstein–Rosen bridge [33]. Most exciting perhaps is the intriguing prospect that spacetime itself could be stitched together by entanglement [34, 35].

6 Arrival of the Schrödinger's Cat

Like Einstein, Schrödinger fled the rapidly deteriorating political situation and left Berlin in 1933 for England to join the University of Oxford. Soon after he arrived, he received the Nobel Prize, which should have augured a very comfortable existence for him. Instead, the glare that focused on his personal life and unconventional marital arrangement outshone the Nobel Prize and caused him considerable difficulty in securing a tenured position until 1939, when he became the Director of the Institute for Advanced Studies in Dublin. In the midst of the turmoil of his tenure issues in 1935, during extensive correspondence with Albert Einstein, he proposed what is now called the Schrödinger's cat thought experiment.

As one of the doctrines of the Copenhagen interpretation, a system stops being in a superposition of states and becomes one or the other, only when an observation takes place, which collapses the entire wave function. Both Einstein and Schrödinger thought this to be downright preposterous.

In the same period of time during which Einstein presented his EPR paper, he also wrote to Schrödinger comparing the absurdity of the Copenhagen interpretation to the notion of a keg of gun powder that is simultaneously both exploded and unexploded. In an attempt to take Einstein's thoughts a step further, the famed creator of wave mechanics conjured up the enigmatic Schrödinger's cat, which by virtue of being unobservable inside a box and subject to a random quantum trigger that may or may not kill the kitty, is put into a quantum superposition of being alive and dead at the same time.

Einstein was duly impressed and wrote in a letter to Schrödinger a bit later:

> You are the only contemporary physicist, besides Laue, who sees that one cannot get around the assumption of reality, if only one is honest. Most of them simply do not see what sort of risky game they are playing with reality— reality as something independent of what is experimentally established. They somehow believe that the quantum theory provides a description of reality, and even a *complete* description; this interpretation is, however, refuted most elegantly by your system of radioactive atom + Geiger counter + amplifier + charge of gun powder + cat in a box, in which the ψ-function of the system contains both the cat alive and blown to bits. Nobody really doubts that the presence or absence of the cat is something independent of the act of observation. [36, p. 39]

Although Schrödinger loved animals and apparently had a cat named Milton at the time he devised his paradox, no one knows why he chose a cat and not a dog or something else. Schrödinger's thought experiment, however, had little impact during his lifetime. From the time he proposed it and till his death, it was scarcely mentioned in the literature. Even Schrödinger rarely brought it up. After its rather clandestine appearance, Schrödinger's cat went into a long slumber of nearly half a century.

The absurd scenario portrayed in the thought experiment remained mostly an academic curiosity until the 1980s, when it was realized that, under suitable conditions, a macroscopic object with many microscopic degrees of freedom could behave quantum mechanically, provided that it was sufficiently decoupled from its environment. Such decoupling is required because quantum superposition rapidly decoheres as a result of complicated interactions with the environment and its inherently irreversible thermal processes.

In the 1980s, there were also some other very exciting developments. The French physicist Alain Aspect, in 1982, gave a quite definitive experimental demonstration of quantum entanglement that led to a burst of activities [37, 38]. In 1984, Charles Bennet and Gilles Brassard proposed a theoretical system for quantum cryptography using photons in a superposition state to create a secure key distribution [39]. Achievement of a mesoscopic Schrödinger's cat state allowing potential studies of quantum decoherence generated considerable enthusiasm as well.

The explosion of publication of research articles on these subjects is too voluminous to describe here. An extensive list of publications on decoherence can be found in papers by its prominent investigator, W. H. Zurek [40–43], as well as in the book by M. Schlosshauer and his other papers [44–46]. Only a selection of especially pertinent papers on Schrödinger's cat will now be discussed.

The physics Nobel Laureates Serge Haroche and David Wineland have allowed Schrödinger's cat to roam the most prestigious halls by mentioning it in their Nobel lectures [47, 48]. Wineland shared the Nobel Prize with Haroche in 2012 "for ground breaking experimental methods that enable measuring and manipulation of individual quantum systems." They independently developed methods for measuring and manipulating individual particles while preserving their quantum mechanical nature in ways that were previously thought unachievable. The two researchers have taken different routes to the study of some of the same phenomena.

Wineland and his group used a trapped beryllium ion, laser cooled to its zero-point energy [49]. A Schrödinger's cat state was ingenuously produced by applying a sequence of laser pulses to create superposition of the ion's internal hyperfine electronic states with its external positional states resulting from its zero-point motion in the potential well of the trapping electro-magnetic field. A mesoscopic distance of more than 80 nm separating the individual wave packets of about 7 nm was accomplished creating a mesoscopic Schrödinger's cat state that could allow controlled studies of quantum decoherence and study of the quantum-classical boundary. Quantum decoherence has received great interest to find a realistic solution of the long standing measurement problem and more recently for application to quantum computing [50, 51] and quantum cryptography [52].

Of necessity, to measure the outcome of any quantum system we have to use a macroscopic device composed of many quantum particles interacting in an incredibly complicated way, and also partaking in irreversible thermal processes. Is it then surprising that the transition from the quantum to classical domain is the most difficult as well as a controversial subject in quantum physics?

Fortunately, with the help of many sophisticated contemporary experiments accompanied by highly developed theoretical investigations, significant progress is being achieved although a very substantial difference exists between theoretically anticipated and experimentally observed values of decoherence time. A consensus seems to be developing in favor of the environmental decoherence model pioneered by Zurek [40–43]. According to his theoretical analysis, the decoherence time of a macroscopic object with a mass of 1 g at a temperature of 300 K extended over a distance of 1 cm could be as short as 10^{-23} s. This, however, appears to be far too short compared to some recent experimental investigations.

Observation has confirmed that quantum coherence is sustained for longer than a few picoseconds inside photosynthetic light harvesting pigment-protein complexes at room temperature [53–60]. Even more spectacular are analyses of recent experimental studies of the avian compass [61]. Quantum superposition and entanglement are sustained in this living system for at least tens of microseconds. Similar time frames for sustaining quantum superposition have been conjectured in the tubulin molecules of microtubules in the cytoskeletons of living organisms [62].

These observations are strikingly at variance with the view that life is too warm and wet for such quantum phenomena to endure over such a relatively long time. It would appear that living entities may develop at least in particular locations some sorts of shielding mechanism to protect quantum superposition from environmental decoherence. But this leaves us in a quandary about how long quantum superposition could last in a living cat weighing at least 5 kg and containing some 10^{27} atoms. In the absence of any meaningful data, the best educated guess would be that decoherence will perhaps occur in extremely short time scale under ambient conditions.

The coherence of superposition in large quantum objects is usually demonstrated by a double slit type diffraction experiment. Such studies have been performed with molecules as large as 1 nm, namely carbon 60 Bucky balls [63]. Recently, a diffraction pattern has been observed in experiment of superposed molecules containing 10 000 atomic mass units [64].

Most remarkably, using a rather sophisticated procedure, coherent superposition has been demonstrated in a macroscopic object containing an estimated 10 trillion atoms [65]. For this purpose, the investigators used a 40 micrometer long mechanical resonator, just large enough to be visible with the naked eye. The resonator with a resonant frequency of 6.175 GHz to its first excited phonon state was cooled to a temperature of merely 25 mK over absolute zero and put in a very high vacuum to minimize environmental decoherence. Under these circumstances, the resonator was confirmed to be in its ground state. Then a signal from a coupled qubit possessing the resonance frequency of 6.175 GHz was injected into the resonator thereby transferring the superposition feature of the qubit to the macroscopic object. Superposition of the ground and the first excited phonon state of the macroscopic resonator lasted for the resonator relaxation time of 6.1 ns.

The above demonstration provides strong evidence that quantum mechanics and its attendant aspect of superposition applies to macroscopic objects and can be revealed under appropriate circumstances. Can it apply to Schrödinger's cat? The answer in principle should be yes—at least at an extremely short time scale. But to prove it, the cat will surely perish for other reasons! Because in order to conduct the experiment, it would be necessary to remove all sources of environmental decoherence by exposing the cat to exceptionally low temperatures and high vacuum and perhaps stopping irreversible metabolic processes as well.

7 Conclusion

By now the Schrödinger's cat has been in existence for over eight decades surviving longer than the mythical cat's nine lives. Considering the fact that quantum superposition has been convincingly established in quantum, mesoscopic and a macroscopic domain, Schrödinger's cat is likely to live on, symbolizing a profound truth that quantum reality exists in all scales. We do not observe it in our daily macroscopic world because it is masked for all practical purposes, predominantly by environmental decoherence possessing irreversible thermal effects.

It should be noted however, that the concerns, which brought the Schrödinger's cat into existence are on their way out. We have cogently pointed out that the probabilistic feature of a quantum particle is an inherent aspect arising out of the genesis of the particle as a wave packet comprising a collection of irregular disturbances of the underlying quantum fields [20]. This is in addition to the innate uncertainty relation $\Delta x \Delta p \geq \frac{1}{2}\hbar$ of a real wave packet. Comprehension of the reality of the wave packet acting as a particle was possible only after the development of the standard model in quantum field theory and was thus unavailable to both Einstein and Schrödinger and to the framers of the Copenhagen interpretation.

Bohr's contention that there is no evocative reality before measurement collapses the wave function was not entirely without merit. But his emphatic denial of any reality whatsoever appears unjustifiable. His emphasis on the critical role of a conscious observer for measurement is also untenable, since the universe developed to a fairly mature state obeying quantum rules long before the possibility of emergence of conscious beings. Einstein's insistence on the existence of reality in all scales from quantum to classical would be correct, if only we allow him the understandable failure to see that the quantum reality is not necessarily the same as the classical reality. Had the Bohr–Einstein debate take place today, it probably would have been declared a draw!

Acknowledgement

The author wishes to thank Joseph Rudnick and Zvi Bern for helpful discussions.

References

[1] Schrödinger E. Die gegenwärtige Situation in der Quantenmechanik. I. *Naturwissenschaften* 1935; **23**(48): 807–812. doi:10.1007/bf01491891

[2] Schrödinger E. Die gegenwärtige Situation in der Quantenmechanik. II. *Naturwissenschaften* 1935; **23**(49): 823–828. doi:10.1007/bf01491914

[3] Schrödinger E. Die gegenwärtige Situation in der Quantenmechanik. III. *Naturwissenschaften* 1935; **23**(50): 844–849. doi:10.1007/bf01491987

[4] Trimmer JD. The present situation in quantum mechanics: a translation of Schrödinger's "cat paradox" paper. *Proceedings of the American Philosophical Society* 1980; **124**(5): 323–338. JSTOR:986572

[5]   Schrödinger E. The present situation in quantum mechanics: a translation of Schrödinger's "cat paradox" paper. In: *Quantum Theory and Measurement.* Wheeler JA, Zurek WH (editors), Trimmer JD (translator), New Jersey: Princeton University Press, 1983, pp. 152–167.

[6]   Griffiths DJ. *Introduction to Quantum Mechanics,* 2nd edition. Upper Saddle River, New Jersey: Prentice Hall, 2004.

[7]   Gribbin J. *In Search of Schrödinger's Cat: Quantum Physics and Reality.* New York: Bantam Books, 1984.

[8]   Rowlands P. *How Schrödinger's Cat Escaped the Box.* Singapore: World Scientific, 2015. doi:10.1142/9391

[9]   Planck M. Ueber das Gesetz der Energieverteilung im Normalspectrum. *Annalen der Physik* 1901; **309**(3): 553–563. doi:10.1002/andp.19013090310

[10] Einstein A. Über einen die Erzeugung und Verwandlung des Lichtes betreffenden heuristischen Gesichtspunkt. *Annalen der Physik* 1905; **17**(6): 132–148. doi:10.1002/andp.19053220607

[11] Compton AH. A quantum theory of the scattering of X-rays by light elements. *Physical Review* 1923; **21**(5): 483–502. doi:10.1103/PhysRev.21.483

[12] Bohr N. On the constitution of atoms and molecules. Part I. *Philosophical Magazine* 1913; **26**(151): 1–25. doi:10.1080/14786441308634955

[13] Bohr N. On the constitution of atoms and molecules. Part II. Systems containing only a single nucleus. *Philosophical Magazine* 1913; **26**(153): 476–502. doi:10.1080/14786441308634993

[14] Bohr N. On the constitution of atoms and molecules. Part III. Systems containing several nuclei. *Philosophical Magazine* 1913; **26**(155): 857–875. doi:10.1080/14786441308635031

[15] Davisson C, Germer LH. Diffraction of electrons by a crystal of nickel. *Physical Review* 1927; **30**(6): 705–740. doi:10.1103/PhysRev.30.705

[16] Mehra J, Rechenberg H. *The Historical Development of Quantum Theory. Vol. 5. Erwin Schrödinger and the Rise of Wave Mechanics. Part 2. The Creation of Wave Mechanics: Early Response and Applications 1925–1926.* New York: Springer, 1987.

[17] Schrödinger E. Quantisierung als Eigenwertproblem: Erste Mitteilung. *Annalen der Physik*

1926; **4**(79): 361–376. doi:10.1002/andp.19263840404

[18] Schrödinger E. *Collected Papers on Wave Mechanics.* London: Blackie & Son, 1928. https://archive.org/details/in.ernet.dli.2015.220691

[19] Mehra J. Erwin Schrödinger and the rise of wave mechanics. II. The creation of wave mechanics. *Foundations of Physics* 1987; **17**(12): 1141–1188. doi:10.1007/bf01889592

[20] Bhaumik ML. Was Albert Einstein wrong on quantum physics? *Quanta* 2015; **4**(1): 35–42. arXiv:1511.05098, doi:10.12743/quanta.v4i1.47

[21] Bhaumik ML. Deciphering the enigma of wave-particle duality. *Quanta* 2016; **5**(1): 93–100. arXiv:1611.00226, doi:10.12743/quanta.v5i1.54

[22] Weinberg S. *Lectures on Quantum Mechanics.* Cambridge: Cambridge University Press, 2013.

[23] Schrödinger E. An undulatory theory of the mechanics of atoms and molecules. *Physical Review* 1926; **28**(6): 1049–1070. doi:10.1103/PhysRev.28.1049

[24] Born M. Zur Quantenmechanik der Stoßvorgänge. *Zeitschrift für Physik A* 1926; **37**(12): 863–867. doi:10.1007/BF01397477

[25] Born M. The statistical interpretation of quantum mechanics. *Nobel Lecture,* December 11, 1954: http://www.nobelprize.org/nobel_prizes/physics/laureates/1954/born-lecture.pdf

[26] Jammer M. Einstein and quantum physics. In: *Albert Einstein, Historical and Cultural Perspectives: The Centennial Symposium in Jerusalem.* Holton G, Elkana Y (editors), Princeton: Princeton University Press, 1982, pp. 59–76. doi:10.2307/j.ctt7zvrpt.7

[27] Einstein A, Podolsky B, Rosen N. Can quantum-mechanical description of physical reality be considered complete? *Physical Review* 1935; **47**(10): 777–780. doi:10.1103/PhysRev.47.777

[28] Born M, Einstein A. *The Born-Einstein Letters: Correspondence Between Albert Einstein and Max and Hedwig Born from 1916–1955, with Commentaries by Max Born.* Born I (translator), London: Macmillan, 1971. http://archive.org/details/TheBornEinsteinLetters

[29] Yin J, Cao Y, Li Y-H, Liao S-K, Zhang L, Ren J-G, Cai W-Q, Liu W-Y, Li B, Dai H, Li G-B, Lu Q-M, Gong Y-H, Xu Y, Li S-L, Li F-Z, Yin Y-Y, Jiang Z-Q, Li M, Jia J-J, Ren G, He D, Zhou Y-L, Zhang X-X, Wang N, Chang X, Zhu Z-C, Liu N-L, Chen Y-A, Lu C-Y, Shu R, Peng C-Z, Wang J-Y, Pan J-W. Satellite-based entanglement distribution over 1200 kilometers. *Science* 2017; **356**(6343): 1140–1144. `doi:10.1126/science.aan3211`

[30] Narnhofer H, Thirring W. Entanglement, Bell inequality and all that. *Journal of Mathematical Physics* 2012; **53**(9): 095210. `doi:10.1063/1.4738376`

[31] Penrose R. *The Road to Reality: A Complete Guide to the Laws of the Universe*. London: Jonathan Cape, 2004.

[32] Maldacena J, Susskind L. Cool horizons for entangled black holes. *Fortschritte der Physik* 2013; **61**(9): 781–811. `arXiv:1306.0533, doi:10.1002/prop.201300020`

[33] Bhaumik ML. Reality of the wave function and quantum entanglement. 2014. `arXiv:1402.4764`

[34] van Raamsdonk M. Building up spacetime with quantum entanglement. *General Relativity and Gravitation* 2010; **42**(10): 2323–2329. `arXiv:1005.3035, doi:10.1007/s10714-010-1034-0`

[35] Swingle B. Constructing holographic spacetimes using entanglement renormalization. 2012. `arXiv:1209.3304`

[36] Einstein A, Schrödinger E, Planck M, Lorentz HA. *Letters on Wave Mechanics*. Przibram K (editor), New York: Philosophical Library, 1967.

[37] Aspect A, Grangier P, Roger G. Experimental realization of Einstein–Podolsky–Rosen–Bohm Gedankenexperiment: A new violation of Bell's inequalities. *Physical Review Letters* 1982; **49**(2): 91–94. `doi:10.1103/PhysRevLett.49.91`

[38] Aspect A, Dalibard J, Roger G. Experimental test of Bell's inequalities using time-varying analyzers. *Physical Review Letters* 1982; **49**(25): 1804–1807. `doi:10.1103/PhysRevLett.49.1804`

[39] Bennett CH, Brassard G. Quantum cryptography: public key distribution and coin tossing. Proceedings of the *IEEE International Conference on Computers, Systems and Signal Processing*, Bangalore, India, December 10–12, 1984, pp. 175–179. `doi:10.1016/j.tcs.2014.05.025`

[40] Zurek WH. Decoherence and the transition from quantum to classical. *Physics Today* 1991; **44**(10): 36–44. `doi:10.1063/1.881293`

[41] Zurek WH. Decoherence and the transition from quantum to classical—revisited. *Los Alamos Science* 2002; **27**: 86–109. `arXiv:quant-ph/0306072, http://library.lanl.gov/cgi-bin/getfile?27-09.pdf`

[42] Zurek WH. Decoherence, einselection, and the quantum origins of the classical. *Reviews of Modern Physics* 2003; **75**(3): 715–775. `arXiv:quant-ph/0105127, doi:10.1103/RevModPhys.75.715`

[43] Zurek WH. Probabilities from entanglement, Born's rule $p_k = |\psi_k|^2$ from envariance. *Physical Review A* 2005; **71**(5): 052105. `arXiv:quant-ph/0405161, doi:10.1103/PhysRevA.71.052105`

[44] Schlosshauer M. *Decoherence and the Quantum-To-Classical Transition*. The Frontiers Collection, Elitzur AC, Silverman MP, Tuszynski JA, Vaas R, Zeh H-D (editors), Berlin: Springer, 2007. `doi:10.1007/978-3-540-35775-9`

[45] Schlosshauer M. Decoherence, the measurement problem, and interpretations of quantum mechanics. *Reviews of Modern Physics* 2005; **76**(4): 1267–1305. `arXiv:quant-ph/0312059, doi:10.1103/RevModPhys.76.1267`

[46] Schlosshauer M, Fine A. On Zurek's derivation of the Born rule. *Foundations of Physics* 2005; **35**(2): 197–213. `arXiv:quant-ph/0312058, doi:10.1007/s10701-004-1941-6`

[47] Haroche S. Controlling photons in a box and exploring the quantum to classical boundary. *Nobel Lecture*, December 8, 2012. `https://www.nobelprize.org/nobel_prizes/physics/laureates/2012/haroche-lecture.pdf`

[48] Wineland DJ. Superposition, entanglement, and raising Schrödinger's cat. *Nobel Lecture*, December 8, 2012. `https://www.nobelprize.org/nobel_prizes/physics/laureates/2012/wineland-lecture.pdf`

[49] Monroe C, Meekhof DM, King BE, Wineland DJ. A "Schrödinger cat" superposition state of an atom. *Science* 1996; **272**(5265): 1131–1136. `doi:10.1126/science.272.5265.1131`

[50] Zhang J, Pagano G, Hess PW, Kyprianidis A, Becker P, Kaplan H, Gorshkov AV, Gong ZX, Monroe C. Observation of a many-body dynamical phase transition with a 53-qubit quantum simulator. *Nature* 2017; **551**(7682): 601–604. `doi:10.1038/nature24654`

[51] Song C, Xu K, Liu W, Yang C-P, Zheng S-B, Deng H, Xie Q, Huang K, Guo Q, Zhang L, Zhang P, Xu D, Zheng D, Zhu X, Wang H, Chen YA, Lu CY, Han S, Pan J-W. 10-qubit entanglement and parallel logic operations with a superconducting circuit. *Physical Review Letters* 2017; **119**(18): 180511. `arXiv:1703.10302, doi:10.1103/PhysRevLett.119.180511`

[52] Shenoy-Hejamadi A, Pathak A, Radhakrishna S. Quantum cryptography: key distribution and beyond. *Quanta* 2017; **6**(1): 1–47. `doi:10.12743/quanta.v6i1.57`

[53] Engel GS, Calhoun TR, Read EL, Ahn T-K, Mancal T, Cheng Y-C, Blankenship RE, Fleming GR. Evidence for wavelike energy transfer through quantum coherence in photosynthetic systems. *Nature* 2007; **446**(7137): 782–786. `doi:10.1038/nature05678`

[54] Lee H, Cheng Y-C, Fleming GR. Coherence dynamics in photosynthesis: protein protection of excitonic coherence. *Science* 2007; **316**(5830): 1462–1465. `doi:10.1126/science.1142188`

[55] Calhoun TR, Ginsberg NS, Schlau-Cohen GS, Cheng Y-C, Ballottari M, Bassi R, Fleming GR. Quantum coherence enabled determination of the energy landscape in light-harvesting complex II. *Journal of Physical Chemistry B* 2009; **113**(51): 16291–16295. `doi:10.1021/jp908300c`

[56] Collini E, Wong CY, Wilk KE, Curmi PMG, Brumer P, Scholes GD. Coherently wired light-harvesting in photosynthetic marine algae at ambient temperature. *Nature* 2010; **463**(7281): 644–647. `doi:10.1038/nature08811`

[57] Panitchayangkoon G, Hayes D, Fransted KA, Caram JR, Harel E, Wen J, Blankenship RE, Engel GS. Long-lived quantum coherence in photosynthetic complexes at physiological temperature. *Proceedings of the National Academy of Sciences* 2010; **107**(29): 12766–12770. `doi:10.1073/pnas.1005484107`

[58] Sarovar M, Ishizaki A, Fleming GR, Whaley KB. Quantum entanglement in photosynthetic light-harvesting complexes. *Nature Physics* 2010; **6**(6): 462–467. `arXiv:0905.3787, doi:10.1038/nphys1652`

[59] Ishizaki A, Fleming GR. Quantum superpositions in photosynthetic light harvesting: delocalization and entanglement. *New Journal of Physics* 2010; **12**(5): 055004. `doi:10.1088/1367-2630/12/5/055004`

[60] Whaley KB, Sarovar M, Ishizaki A. Quantum entanglement phenomena in photosynthetic light harvesting complexes. *Procedia Chemistry* 2011; **3**(1): 152–164. `doi:10.1016/j.proche.2011.08.021`

[61] Gauger EM, Rieper E, Morton JJL, Benjamin SC, Vedral V. Sustained quantum coherence and entanglement in the avian compass. *Physical Review Letters* 2011; **106**(4): 040503. `arXiv:0906.3725, doi:10.1103/PhysRevLett.106.040503`

[62] Hameroff SR, Penrose R. Consciousness in the universe: a review of the 'Orch OR' theory. *Physics of Life Reviews* 2014; **11**(1): 39–78. `doi:10.1016/j.plrev.2013.08.002`

[63] Arndt M, Nairz O, Vos-Andreae J, Keller C, van der Zouw G, Zeilinger A. Wave–particle duality of C_{60} molecules. *Nature* 1999; **401**(6754): 680–682. `doi:10.1038/44348`

[64] Eibenberger S, Gerlich S, Arndt M, Mayor M, Tüxen J. Matter-wave interference of particles selected from a molecular library with masses exceeding 10 000 amu. *Physical Chemistry Chemical Physics* 2013; **15**(35): 14696–14700. `doi:10.1039/c3cp51500a`

[65] O'Connell AD, Hofheinz M, Ansmann M, Bialczak RC, Lenander M, Lucero E, Neeley M, Sank D, Wang H, Weides M, Wenner J, Martinis JM, Cleland AN. Quantum ground state and single-phonon control of a mechanical resonator. *Nature* 2010; **464**(7289): 697–703. `doi:10.1038/nature08967`

14

Erwin Schrödinger and Quantum Wave Mechanics

John J. O'Connor & Edmund F. Robertson

School of Mathematics and Statistics, University of St Andrews, North Haugh, St Andrews, Fife, Scotland
E-mails: joc@st-andrews.ac.uk, efr@st-andrews.ac.uk

Editors: *Stefan K. Kolev & Danko Georgiev*

The fathers of matrix quantum mechanics believed that the quantum particles are *unanschaulich* (unvisualizable) and that quantum particles pop into existence only when we measure them. Challenging the orthodoxy, in 1926 Erwin Schrödinger developed his wave equation that describes the quantum particles as a packet of quantum probability amplitudes evolving in space and time. Thus, Schrödinger visualized the unvisualizable and lifted the veil that has been obscuring the wonders of the quantum world.
Quanta 2017; 6: 48–52.

Erwin Schrödinger's father, Rudolf Schrödinger, ran a small linoleum factory which he had inherited from his own father. Erwin's mother, Emily Bauer, was half English, this side of the family coming from Leamington Spa, and half Austrian with her father coming from Vienna.

Schrödinger learnt English and German almost at the same time due to the fact that both were spoken in the household. He was not sent to elementary school, but received lessons at home from a private tutor up to the age of ten. He then entered the Akademisches Gymnasium in the autumn of 1898, rather later than was usual since he spent a long holiday in England around the time he might have entered the school. He wrote later about his time at the Gymnasium:

> I was a good student in all subjects, loved mathematics and physics, but also the strict logic of the ancient grammars, hated only the memorization of incidental dates and facts. Of the German poets, I loved especially the dramatists, but hated the pedantic dissection of their works. [1, §1]

A student in Schrödinger's class at school also wrote:

> Especially in physics and mathematics, Schrödinger had a gift for understanding that allowed him, without any homework, immediately and directly to comprehend all the material during the class hours and to apply it. After the lecture [...] it was possible for [our professor] to call Schrödinger immediately to the blackboard and to set him problems, which he solved with playful facility. For us average students, mathematics and physics were frightful subjects, but they were his preferred fields of knowledge. [1, §1]

Schrödinger graduated from the Akademisches Gymnasium in 1906 and, in that year, entered the University of Vienna. In theoretical physics, he studied analytical mechanics, applications of partial differential equations to dynamics, eigenvalue problems, Maxwell's equations and electromagnetic theory, optics, thermodynamics, and statistical mechanics. It was Fritz Hasenöhrl's lectures on theoretical physics which had the greatest influence on Schrödinger. In mathematics, he was taught calculus and algebra by Franz Mertens, function theory, differential equations and mathematical statistics by Wilhelm Wirtinger (whom he found uninspiring as a lecturer). He also studied projective geometry, algebraic curves and continuous groups in lectures given by Gustav Kohn.

On May 20, 1910, Schrödinger was awarded his doctorate for the dissertation *On the Conduction of Electricity on the Surface of Insulators in Moist Air* [2]. After this he undertook voluntary military service in the fortress artillery. Then he was appointed to an assistantship at Vienna but, rather surprisingly, in experimental physics rather than theoretical physics. He later said that his experiences conducting experiments proved an invaluable asset to his theoretical work since it gave him a practical philosophical framework in which to set his theoretical ideas.

Having completed the work for his habilitation, he was awarded the degree on September 1, 1914. That it was not an outstanding piece of work is shown by the fact that the committee was not unanimous in recommending him for the degree. As Walter J. Moore writes:

> Schrödinger's early scientific development was inhibited by the absence of a group of first-class theoreticians in Vienna, against whom he could sharpen his skills by daily argument and mutual criticism. [1, §2]

In 1914, Schrödinger's first important paper was published developing ideas of Boltzmann. However, with the outbreak of World War I, Schrödinger received orders to take up duty on the Italian border. His time of active service was not wasted as far as research was concerned, however, for he continued his theoretical work, submitting another paper from his position on the Italian front. In 1915, he was transferred to duty in Hungary and from there he submitted further papers for publication. After being sent back to the Italian front, Schrödinger received a citation for outstanding service commanding a battery during a battle.

In the spring of 1917, Schrödinger was sent back to Vienna and assigned to teach a course in meteorology. He was able to continue research and it was at this time that he published his first results on quantum theory. After the end of the war he continued working at Vienna. From 1918 to 1920, he made substantial contributions to the theory of colour vision [3–7].

Schrödinger had worked at Vienna on radioactivity, proving the statistical nature of radioactive decay. He had also made important contributions to the kinetic theory of solids, studying the dynamics of crystal lattices. On the strength of his work he was offered an associate professorship at Vienna in January 1920 but by this time he wished to marry Anny Bertel. They had become engaged in 1919

and Anny had come to work as a secretary in Vienna on a monthly salary which was more than Schrödinger's annual income. Then he was offered an associate professorship, still not at a salary large enough to support a non-working wife, so he declined.

Schrödinger accepted instead an assistantship in Jena and this allowed him to marry Anny on March 24, 1920. After only a short time there, he moved to a chair in Stuttgart where he became friendly with Hans Reichenbach. He then moved to a chair at Breslau, his third move in eighteen months. Soon however he was to move yet again, accepting the chair of theoretical physics in Zürich in late 1921. During these years of changing from one place to another, Schrödinger studied physiological optics, in particular he continued his work on the theory of colour vision [8–12].

Hermann Weyl was Schrödinger's closest colleague in his first years in Zürich and he was to provide the deep mathematical knowledge which would prove so helpful to Schrödinger in his work. The intellectual atmosphere in Zürich suited Schrödinger and Zürich was to be the place where he made his most important contributions. From 1921 he studied atomic structure, then in 1924 he began to study quantum statistics. Soon after this he read de Broglie's thesis which became a turning point in the direction of his research and had a major influence on his thinking. On November 3, 1925 Schrödinger wrote to Albert Einstein:

> A few days ago I read with great interest the ingenious thesis of Louis de Broglie, which I finally got hold of [...] [1, §6]

On November 16, in another letter to Alfred Landé, professor of physics at Tübingen, Schrödinger wrote:

> I have been intensely concerned these days with Louis de Broglie's ingenious theory. It is extraordinarily exciting, but still has some very grave difficulties. [1, §6]

One week later Schrödinger gave a seminar on de Broglie's work and a member of the audience, a student of Sommerfeld's, suggested that there should be a wave equation. Within a few weeks Schrödinger had found his wave equation.

Schrödinger published his revolutionary work relating to wave mechanics and the general theory of relativity in a series of six papers in 1926 [13–18]. In its most general form, the time-dependent Schrödinger equation describes the time evolution of closed quantum physical systems as

$$i\hbar\frac{\partial}{\partial t}\Psi(\mathbf{r}, t) = \hat{H}\Psi(\mathbf{r}, t)$$

where ι is the imaginary unit, $\hbar$ is the reduced Planck constant, the symbol $\frac{\partial}{\partial t}$ indicates a partial derivative with respect to time t, Ψ is the wave function of the quantum system, $\mathbf{r}$ and t are the position vector and time respectively, and $\hat{H}$ is the Hamiltonian operator characterizing the total energy of the quantum system. Wave mechanics, as proposed by Schrödinger [19], was the second formulation of quantum theory, the first being matrix mechanics due to Heisenberg. The relation between the two formulations of wave mechanics and matrix mechanics was understood by Schrödinger immediately as this quotation from one of his 1926 papers shows

To each function of the position- and momentum- coordinates [in wave mechanics] there may be related a matrix in such a manner, that these matrices, in every case satisfy the formal calculation rules of Born and Heisenberg. [...] The solution of the natural boundary-value problem of this differential equation [in wave mechanics] is completely equivalent to the solution of Heisenberg's algebraic problem. [19, p. 46]

The work was indeed received with great acclaim. Max Planck described it as

epoch-making work [1, §6]

Einstein wrote:

the idea of your work springs from true genius [1, §6]

Then, ten days later Einstein wrote again:

I am convinced that you have made a decisive advance with your formulation of the quantum condition [...] [1, §6]

Paul Ehrenfest wrote:

I am simply fascinated by your [wave equation] theory and the wonderful new viewpoints that it brings. Every day for the past two weeks our little group has been standing for hours at a time in front of the blackboard in order to train itself in all the splendid ramifications. [1, §6]

The author of Schrödinger's obituary in *The Times* wrote

The introduction of wave mechanics stands [...] as Schrödinger's monument and a worthy one. [20]

Schrödinger accepted an invitation to lecture at the University of Wisconsin, Madison, leaving in December 1926 to give his lectures in January and February 1927. Before he left he was told he was the leading candidate for Planck's chair in Berlin. After giving a brilliant series of lectures in Madison he was offered a permanent professorship there but:

He was not at all tempted by an American position, and he declined on the basis of a possible commitment to Berlin. [1, §7]

The list of candidates to succeed Planck in the chair of theoretical physics at Berlin was impressive. Arnold Sommerfeld was ranked in first place, followed by Schrödinger, with Max Born as the third choice. When Sommerfeld decided not to leave Munich, the offer was made to Schrödinger. He went to Berlin, taking up the post on October 1, 1927 and there he became a colleague of Einstein's.

Although he was a Catholic, Schrödinger decided in 1933 that he could not live in a country in which persecution of Jews had become a national policy. Alexander Lindemann, the head of physics at Oxford University, visited Germany in the spring of 1933 to try to arrange positions in England for some young Jewish scientists from Germany. He spoke to Schrödinger about posts for one of his assistants and was surprised to discover that Schrödinger himself was interested in leaving Germany. Schrödinger asked for a colleague, Arthur March, to be offered a post as his assistant.

To understand Schrödinger's request for March we must digress a little and comment on Schrödinger's liking for women. His relations with his wife had never been good and he had had many lovers with his wife's knowledge. Anny had her own lover for many years, Schrödinger's friend Weyl. Schrödinger's request for March to be his assistant was because, at that time, he was in love with Arthur March's wife Hilde.

Many of the scientists who had left Germany spent the summer of 1933 in the South Tyrol. Here Hilde became pregnant with Schrödinger's child. On November 4, 1933 Schrödinger, his wife and Hilde March arrived in Oxford. Schrödinger had been elected a fellow of Magdalen College. Soon after they arrived in Oxford, Schrödinger heard that, for his work on wave mechanics, he had been awarded the Nobel prize.

In the spring of 1934, Schrödinger was invited to lecture at Princeton and while there he was made an offer of a permanent position. On his return to Oxford he negotiated about salary and pension conditions at Princeton but in the end he did not accept. It is thought that the fact that he wished to live at Princeton with Anny and Hilde both sharing the upbringing of his child was not found acceptable. The fact that Schrödinger openly had two wives, even if one of them was married to another man, did not go down too well in Oxford either, but his daughter Ruth Georgie Erica was born there on May 30, 1934.

In 1935, Schrödinger published a three-part essay on *The Present Situation in Quantum Mechanics [21–23]* in which his famous Schrödinger's cat paradox appears. This was a thought experiment where a cat in a closed box either lived or died according to whether a quantum event occurred. The paradox was that both universes, one with a dead cat and one with a live one, seemed to exist in parallel until an observer opened the box.

In 1936, Schrödinger was offered the chair of physics at the University of Edinburgh in Scotland. He may have accepted that post but for a long delay in obtaining a work permit from the Home Office. While he was waiting he received an offer from the University of Graz and he went to Austria and spent the years 1936–1938 in Graz. Born was then offered the Edinburgh post which he quickly accepted.

However the advancing Nazi threat caught up with Schrödinger again in Austria. After the Anschluss the Germans occupied Graz and renamed the university Adolf Hitler University. Schrödinger wrote a letter to the University Senate, on the advice on the new Nazi rector, saying:

> I had misjudged up to the last the true will and the true destiny of my country. I make this confession willingly and joyfully. [1, §9]

It was a letter he was to regret for the rest of his life. He explained the reason to Einstein in a letter written about a year later:

> I wanted to remain free – and could not do so without great duplicity. [1, §9]

The Nazis could not forget the insult he had caused them when he fled from Berlin in 1933 and on August 26, 1938 he was dismissed from his post for "political unreliability". He went to consult an official in Vienna who told him that he must get a job in industry and that he would not be allowed to go to a foreign country. He fled quickly with Anny, this time to Rome from where he wrote to Éamon de Valera as President of the League of Nations. De Valera offered to arrange a job for him in Dublin in the new Institute for Advanced Studies he was trying to set up. From Rome, Schrödinger went back to Oxford, and there he received an offer of a one year visiting professorship at the University of Gent.

After his time in Gent, Schrödinger went to Dublin in the autumn of 1939. There he studied electromagnetic theory and relativity and began to publish on a unified field theory. His first paper on this topic was written in 1943. In 1946, he renewed his correspondence with Einstein on this topic. In January 1947, he believed he had made a major breakthrough:

> Schrödinger was so entranced by his new theory that he threw caution to the winds, abandoned any pretence of critical analysis, and even though his new theory was scarcely hatched, he presented it to the Academy and to the Irish press as an epoch-making advance. [1, §11]

The *Irish Times* carried an interview with Schrödinger the next day in which he said:

> This is the generalisation. Now the Einstein Theory becomes simply a special case [...] I believe I am right, I shall look an awful fool if I am wrong. [1, §11]

Einstein, however, realised immediately that there was nothing of merit in Schrödinger's "new theory".

> [Schrödinger] was even thinking of the possibility of receiving a second Nobel prize. In any case, the entire episode reveals a lapse in judgment, and when he actually read Einstein's comment, he was devastated. [1, §11]

Einstein wrote immediately saying that he was breaking off their correspondence on unified field theory. Unified field theory was, however, not the only topic to interest Schrödinger during his time at the Institute for Advanced Study in Dublin. His study of Greek science and philosophy is summarised in *Nature and the Greeks* (1954), which he wrote while in Dublin. Another important book written during this period was *What is Life?* (1944) [24] which led to progress in biology.

On the personal side Schrödinger had two further daughters while in Dublin, to two different Irish women. He remained in Dublin until he retired in 1956 when he returned to Vienna and wrote his last book *My View of the World* (1961) expressing his own metaphysical outlook.

During his last few years Schrödinger remained interested in mathematical physics and continued to work on general relativity, unified field theory and meson physics.

References

[1] Moore WJ. *Schrödinger: Life and Thought*. Canto Classics, Cambridge: Cambridge University Press, 2015.

[2] Schrödinger E. Über die Leitung der Elektrizität auf der Oberfläche von Isolatoren an feuchter Luft. *Sitzungsberichte der kaiserlichen Akademie*

der Wissenschaften in Wien. Mathematisch-naturwissenschaftliche Klasse, Abteilung 2a 1910; **119**: 1215–1222.

[3] Schrödinger E. Theorie der Pigmente von größter Leuchtkraft. *Annalen der Physik* 1920; **367**(15): 603–622. doi:10.1002/andp.19203671504

[4] Schrödinger E. Grundlinien einer Theorie der Farbenmetrik im Tagessehen: I. Mitteilung. *Annalen der Physik* 1920; **4**(63): 397–426. doi:10.1002/andp.19203682102

[5] Schrödinger E. Grundlinien einer Theorie der Farbenmetrik im Tagessehen: II. Mitteilung. *Annalen der Physik* 1920; **4**(63): 427–456. doi:10.1002/andp.19203682103

[6] Schrödinger E. Grundlinien einer Theorie der Farbenmetrik im Tagessehen: III. Mitteilung. *Annalen der Physik* 1920; **4**(63): 481–520. doi:10.1002/andp.19203682202

[7] Schrödinger E. Farbenmetrik. *Zeitschrift für Physik* 1920; **1**(5): 459–466. doi:10.1007/bf01332674

[8] Schrödinger E. Über das Verhältnis der Vierfarben- zur Dreifarbentheorie. *Sitzungsberichte der kaiserlichen Akademie der Wissenschaften in Wien. Mathematisch-naturwissenschaftliche Klasse, Abteilung 2a* 1925; **134**: 471–490.

[9] Schrödinger E. On the relationship of four-color theory to three-color theory, with Commentary by Qasim Zaidi. *Color Research & Application* 1994; **19**(1): 37–47. doi:10.1111/j.1520-6378.1994.tb00059.x

[10] Schrödinger E. Über Farbenmessung. *Physikalische Zeitschrift* 1925; **26**: 349–352.

[11] Schrödinger E. Über die subjektiven Sternfarben und die Qualität der Dämmerungsempfindung. *Naturwissenschaften* 1925; **13**(18): 373–376. doi:10.1007/bf01559096

[12] Schrödinger E. Die geometrische Lösung von Farbenmischungsaufgaben. *Naturwissenschaften* 1926; **14**(8): 146–147. doi:10.1007/bf01507301

[13] Schrödinger E. Quantisierung als Eigenwertproblem: Erste Mitteilung. *Annalen der Physik* 1926; **4**(79): 361–376. doi:10.1002/andp.19263840404

[14] Schrödinger E. Quantisierung als Eigenwertproblem: Zweite Mitteilung. *Annalen der Physik* 1926; **4**(79): 489–527. doi:10.1002/andp.19263840602

[15] Schrödinger E. Über das Verhältnis der Heisenberg–Born–Jordanschen Quantenmechanik zu der meinem. *Annalen der Physik* 1926; **4**(79): 734–756. doi:10.1002/andp.19263840804

[16] Schrödinger E. Quantisierung als Eigenwertproblem: Dritte Mitteilung. *Annalen der Physik* 1926; **4**(80): 437–490. doi:10.1002/andp.19263851302

[17] Schrödinger E. Quantisierung als Eigenwertproblem: Vierte Mitteilung. *Annalen der Physik* 1926; **4**(81): 109–139. doi:10.1002/andp.19263861802

[18] Schrödinger E. Der stetige Übergang von der Mikro- zur Makromechanik. *Naturwissenschaften* 1926; **14**(28): 664–666. doi:10.1007/bf01507634

[19] Schrödinger E. *Collected Papers on Wave Mechanics*. London: Blackie & Son, 1928. https://archive.org/details/in.ernet.dli.2015.211600

[20] Professor Erwin Schrödinger: Physicist and Nobel Prize Winner. *The Times*, 1961. http://www-history.mcs.st-and.ac.uk/Obits/Schrodinger.html

[21] Schrödinger E. Die gegenwärtige Situation in der Quantenmechanik. I. *Naturwissenschaften* 1935; **23**(48): 807–812. doi:10.1007/bf01491891

[22] Schrödinger E. Die gegenwärtige Situation in der Quantenmechanik. II. *Naturwissenschaften* 1935; **23**(49): 823–828. doi:10.1007/bf01491914

[23] Schrödinger E. Die gegenwärtige Situation in der Quantenmechanik. III. *Naturwissenschaften* 1935; **23**(50): 844–849. doi:10.1007/bf01491987

[24] Schrödinger E. *What is Life? The Physical Aspect of the Living Cell & Mind and Matter*. Cambridge: Cambridge University Press, 1977. https://archive.org/details/WhatIsLife_201708

15

Constructive Empiricism, Partial Structures and the Modal Interpretation of Quantum Mechanics

Otávio Bueno

Department of Philosophy, University of Miami, Coral Gables, FL 33124, USA. E-mail: otaviobueno@me.com

Editors: *Eliahu Cohen & Danko Georgiev*

Van Fraassen's modal interpretation of non-relativistic quantum mechanics is articulated to support an anti-realist account of quantum theory. However, given the particular form of van Fraassen's anti-realism (constructive empiricism), two problems arise when we try to make it compatible with the modal interpretation: one difficulty concerns the tension between the need for modal operators in the modal interpretation and van Fraassen's skepticism regarding real modality in nature; another addresses the need for the truth predicate in the modal interpretation and van Fraassen's rejection of truth as the aim of science. After examining these two problems, I suggest a formal framework in which they can be accommodated–using da Costa and French's partial structures approach–and indicate a variant of van Fraassen's modal interpretation that does not face these difficulties.
Quanta 2014; 3: 1–15.

1 Introduction

The quantum-logical approach to the foundations of non-relativistic quantum mechanics has provided rich insights into our understanding of quantum theory. (Here, I will only consider non-relativistic quantum mechanics. Reference to quantum mechanics is restricted to this class of theories and their interpretations.) The quantum-logical approach has indicated mathematical structures that are often used in the formulation of quantum mechanics (such as orthomodular lattices), and clarified the relationship between these structures and logic [1–5]. Just as in classical mechanics the algebra of states of a physical system is a Boolean algebra, in quantum mechanics, the corresponding algebra is a non-distributive, orthomodular lattice. The corresponding logic–the consequence relation defined for the theory's language–is then not *classical*, but *quantum*. (It is a delicate issue to determine the language of quantum mechanics. According to certain authors, there is no such thing. The language of the theory is only a fragment of functional analysis [6, pp.83-85]. This is part and parcel of the peculiarities of quantum mechanics. For example, given that quantum particles lack identity conditions (at least in one interpretation of the theory), the *language* of quantum mechanics cannot be classical, but it has to accommodate *objects* for which identity does not meaningfully apply [7–11].)

The motivation for the shift in logic is the same underlying the construction of other interpretations of quantum mechanics, namely, to solve conceptual difficulties facing the theory; in particular, (1) to provide an account of measurement, (2) to accommodate the well-known paradoxes (such as EPR and Schrödinger's), and (3) to examine the issue of identity and non-individuality of quantum particles. It has to be admitted that, from the quantum-logical point of view, these issues have not been as thoroughly examined as they could (most work concentrates on the issue of measurement), and that exaggerated claims have been made on the basis of quantum logic (for a critical appraisal, see [12, 13]). However, overall, the approach still provides important insights, since it suggests a unified picture to examine the foundations of quantum mechanics.

Independently of the merits of the quantum-logical approach, it is important to distinguish quantum logic as a particular *interpretation* of quantum mechanics from quantum logic as a *methodological tool* to investigate the theory. As an interpretation, Putnam [3] argued that quantum logic provides an argument to support realism in quantum mechanics. Roughly speaking, this is because in this approach all observables can be assumed to have sharp values. (Putnam's move towards realism is correctly criticised in [14]; see also [12].) As a methodological tool, we can use the structures and techniques provided by quantum logic as conceptual resources to explore the quantum domain, without assuming the interpretative moves typically made by quantum logicians. Admittedly, the distinction is somewhat vague, since the interpretative claims from quantum logic are based on the use of (quantum) logical tools. However, as often happens, the (logical) formalism does not single out one unique (quantum-logical) interpretation, but underdetermines several possible interpretations. Thus, using the methodological tools of quantum logic, we are able to provide alternative interpretations of quantum mechanics–resisting, for instance, Putnam's move towards realism.

In this paper, I want to explore one aspect of quantum logic: its role in the formulation of the *modal* interpretation of quantum mechanics. Instead of assuming the interpretative claims of the quantum-logical approach, I shall use it as a method to construct one kind of interpretation. Although the modal interpretation is often associated with realism in quantum mechanics [15], it was first formulated by van Fraassen as an anti-realist proposal [13, 16–18]. And it is in the anti-realist camp that it belongs more comfortably. However, as we shall see, given the particular form of van Fraassen's constructive empiricism [19, 20], two problems arise: one difficulty concerns the tension between the need for modal operators in the modal interpretation and van Fraassen's skepticism with regard to real modality in nature [20]; another addresses the need for the truth predicate in the

modal interpretation and van Fraassen's rejection of truth as the aim of science [19]. After examining these two problems, I will suggest a formal framework in which they can be accommodated–using da Costa and French's partial structures and quasi-truth [21–23]–and which provides a variant of van Fraassen's modal interpretation. Given the role played by structures in this new account, I shall call it a *modal-structural* interpretation.

2 Quantum Logic, Empiricism and the Modal Interpretation of Quantum Mechanics

In order to develop a new formulation of empiricism, van Fraassen articulated a proposal to overcome the major shortcoming of previous empiricist accounts: to accommodate the theoretical aspects of science within the bounds of empiricism. The difficulty is that theoretical talk cannot be eliminated from the scientific description of the world without radically depriving science from its actual content, nor can it be reduced to purely empirical factors. The early attempts, in the hands of logical positivists, to reduce the theoretical content of science to observation reports failed for well-known reasons (ultimately, science cannot be regimented in the restricted framework assumed by positivism). The alternative suggested by van Fraassen consists in broadening the limits of empiricism, and rejecting the reductionism of former approaches.

There are two central features of van Fraassen's approach that I wish to emphasize. First, as is well known, instead of trying to reduce the theoretical aspects of science to empirical phenomena, van Fraassen's strategy is to change the aim of science, but still stressing the role of observation to achieve this aim. According to his proposal (constructive empiricism), the aim of science is not truth, but something weaker: *empirical adequacy* [19, p.12]. Using the resources of the semantic approach (according to which to present a theory is to specify a family of structures, its models [19, p.64]), van Fraassen characterizes this concept in the following way. A theory is empirically adequate if it has a model such that its empirical substructures are isomorphic to the structures that represent empirical phenomena ([19, p.64]; for a discussion, see [24]). In this way, the empirical adequacy of a theory reflects one role that empirical information plays in science: to provide constraints for theory acceptance. Now, empirical adequacy is strictly weaker than truth, and it is reducible to truth only if we are dealing with observable phenomena [19, p.72]. Thus, the search for empirically adequate theories does not entail the commitment to unobservable entities. For suppose that there are no unobservable entities in the world; a false theory, which postulates the existence of these entities, can still be empirically adequate. In this sense, by taking empirical adequacy as the aim of science, van Fraassen is able to provide a proposal in which empirical factors have a crucial role to play in science without collapsing into a reduction of theory to observation.

The avoidance of reductionism allows van Fraassen to put forward a crucial argument against realism: the underdetermination argument. Because there is a gap between certain theories (such as classical mechanics or non-relativistic quantum mechanics and their interpretations) and empirical phenomena, empirically equivalent but conceptually distinct theories may arise. And the empiricist sees no epistemic reason to select one of them; after all, as far as the empirical phenomena are concerned, they are equally adequate [19]. (Note that van Fraassen only uses the argument from underdetermination in very specific contexts, invoking particular theories to make his case. The argument is never used as a feature of every theory, applied to science as a whole [25, pp.346-347].) A great deal can be said, and has been said, about this argument, but for our present purposes it is enough to indicate the role it plays in constructive empiricism. The gap between theoretical models and models of the phenomena allows the empiricist to remain agnostic about the existence of unobservable entities, which are postulated in some descriptions of the phenomena, but not in other, also empirically adequate, descriptions.

This leads to the second empiricist feature of van Fraassen's approach that I wish to examine. Although the constructive empiricist is only agnostic about the existence of unobservable entities (as van Fraassen points out: "I wish merely to be agnostic about the existence of the unobservable aspects of the world described by science" [19, p.72]), he is entirely skeptical about the existence of real modality in nature, such as laws of nature and objective chance. In van Fraassen's own words:

> To be an empiricist is to withhold belief in anything that goes beyond the actual, observable phenomena, and *to recognize no objective modality in nature.* ([19, pp.202-203]; the italics are mine; see also [20]. A formulation of empiricism in terms of a stance rather than a doctrine to be believed is developed in [26].)

To recognize real modality in nature is to recognize the existence of non-actual, merely possible phenomena, and also to recognize necessary connections in reality. These are, of course, anathema to empiricism ever since Hume. Thus van Fraassen rehearses here a characteristic point of the empiricist tradition, although with a new twist. Among the real modality the empiricist is critical of,

laws of nature are prominent, and with the emergence of quantum mechanics, a new alternative could be provided. After all, given the crucial role played by symmetry in the construction of quantum mechanics, van Fraassen has grounds to propose, roughly speaking, that the concept of a law of nature could be replaced, in some contexts, by the concept of symmetry [20]. Objective modality, such as laws, are not only philosophically problematic [20, pp.16-128], but can also be replaced, in some contexts, by a metaphysically less problematic alternative, based on symmetry considerations [20, pp.216-289].

Now, because quantum mechanics provides a crucial case for van Fraassen's approach, and because of its overall importance in science, it is understandable that the empiricist be expected to provide an interpretation of it. In this context, van Fraassen formulates and develops the modal interpretation of quantum mechanics. The interpretation is presented as an implementation of central features of van Fraassen's general empiricist approach, and I now discuss some of its distinctive traits. (In the preface to [20], van Fraassen notes: "This book was originally twice as long. When a general approach is announced and advocated, it remains hand-waving except to the extent that it is implemented. Accordingly, the now missing part was devoted to a detailed study of the structure and interpretation of quantum mechanics. It will appear separately, as *Quantum Mechanics: An Empiricist View* [20, pp.vii-viii].)

The modal interpretation is explicitly articulated as an alternative to von Neumann's interpretation rule [1]. This rule provides a close link between observables and states of a physical system, indicating how to read assignments of values to observables (essentially, value assignments classify the states). The link is as follows:

> Observable B has value b iff a measurement of B is certain to have outcome b. (For a discussion, see [13, pp.274-278].)

The crucial feature, as van Fraassen notes, is the logical form of this interpretation rule: a biconditional linking assignments of values to observables and outcomes of measurements. The (apparent) classical flavor of this link is noticeable. However, what happens if a measurement is not certain to have a given outcome? To this question, von Neumann provides a peculiar answer: in this case, the observable has no value. The classical flavor of the interpretation is therefore deceptive: not all observables have a value after all—unmeasured observables have no value, excluding the case of certainty [13, p.274].

The main feature of the modal interpretation is then to distinguish between two concepts of state: *value* and *dynamic states.* The distinction is motivated by the fact that, in the case of quantum mechanics, (1) we cannot assume that observables have values, 'there to be seen if we look', nor can we (2) assume that the evolution of the physical system is determined completely by what those values are. These assumptions may have been taken for granted in the case of classical mechanics, but the picture changes once we move to the quantum domain. To each of these assumptions, there corresponds a particular concept of state. The *value state* is fully specified by stating which observables have values, and which they are; the *dynamic state* is completely determined by stating how the system will develop if acted upon in a particular way, and how it will develop if isolated [13, p.275].

Corresponding to these two concepts of state, we have two types of proposition: a *value-attributing proposition*, denoted by $\langle m, E \rangle$, which states that observable m actually has a value in E (where E is, typically, a Borel set); and a *state-attributing proposition*, denoted by $[m, E]$, according to which the state is such that a measurement of m must have a value in E [13, p.275]. These propositions have a well-determined body of truth-makers: value-attributing propositions are true (or false) depending on value states, whereas the truth-values of state-attributions depend on dynamic states.

If von Neumann's interpretation rule is rejected, the equivalence between value-attributing and state-attributing propositions is denied. Only one side of the biconditional holds. But what side? Since measurement outcomes are relevant to what state the system is in, it is natural that $[m, E]$ implies $\langle m, E \rangle$; that is, if the state is such that a measurement of m must have a value in E, then m does have a value in E [13, p.276]. In other words, in von Neumann's interpretation rule only the right-to-left conditional holds. This allows van Fraassen to introduce *unsharp* values to observables. After all, if $[m, E]$ is not true (that is, if the state is not such that m must have a value in E), it is still possible that m does have a value in E, although this value may be *unsharp* [13, pp.276;282-283]. Furthermore, crucial information about the physical system is provided by the dynamic state, which still remains the basic state to consider, since it gives information about the system's evolution. In this sense, the importance of the actual values of observables derives from the fact that they provide indications about the dynamic state. And once we know that state, we are in a position to know how the system will evolve (either in isolation or in interaction with another system).

Far more can be said, of course, about the modal interpretation; in particular, how it handles the measurement problem [13, pp.283-299;317-337], the EPR paradox [13, pp.338-374], and the issues of identity and non-individuality of quantum particles [13, pp.375-482]. But instead of going into these issues here, I will consider another pressing point: what are the empiricist features

in the modal interpretation? Two of them deserve notice. The first derives from van Fraassen's examination of the empirical basis of quantum theory [13, §4-5]. As he argues, there are possible phenomena in the domain of quantum mechanics that cannot be accommodated by any causal model (in a minimal sense of causality). (These are common cause models, which spell out at least necessary conditions for causality. A correlation between events A and B has a *common cause* in factor C if: (1) C precedes A and B; (2) $P(A|C) > P(A|\neg C)$ and $P(B|C) > P(B|\neg C)$; and (3) $P(A \wedge B|C) = P(A|C)P(B|C)$, where $P(X|Y)$ is the conditional probability of X given Y [13, pp.53-57;81-85]. If it is argued that common causes do not provide even a necessary condition for causality, it can be replied that the notion of causality under consideration is then decidedly too weak [13, p.487, note 4].) Due to empirical considerations, theories articulated in terms of common cause models are shown to be inadequate, in the sense that the structures describing the data of quantum phenomena cannot be embedded into the theoretical models.

The second empiricist feature comes in with the introduction of modality. As an interpretation of quantum theory, the modal interpretation spells out how the world could be if quantum mechanics were true [13, p.242]. It explores possibilities in the way the theory pictures the world (for instance, in what respect quantum theory is indeterministic, in what respect it is not), extending the account provided by the theory to unobservable factors, in such a way that a fuller picture of the world is presented. In this sense, the modal interpretation increases our understanding of quantum mechanics. Roughly speaking, the empiricist component enters with the claim that, with regard to unobservable entities (such as elementary particles), it is enough to say how they can be. (The epigraph of [13] is precisely this statement of Descartes's: "Que, touchant les choses que nos sens n'aperçoivent point, il suffit d'expliquer comment elles peuvent être" (*Principles*, iv, 204; "That, concerning those things that our senses do not perceive, it is enough to explain how they can be"). This is, of course, a thoroughly empiricist point.) In this way, a commitment to the existence of these entities is not required, since the theory can be interpreted so that what it states about the phenomena are only certain possibilities (probabilities of measurement outcomes, relative possibilities of trajectories of a given system etc.).

However, despite these empiricist features in van Fraassen's interpretation, I think the modal interpretation faces two problems if we try to make it compatible with constructive empiricism. To these issues I should turn now.

3 Two Problems: Truth and Modality

The first problem is concerned with the notion of truth, and its role in the modal interpretation. The second concerns modality: its status and the way it is introduced into van Fraassen's proposal. I argue that there is a tension between these two moves, and suggest a way out using the partial structures framework. It goes without saying that van Fraassen has provided a strong case for understanding quantum mechanics in terms of certain modal operators, and his search for an empiricist view of the theory in a decidedly classical setting (with regard to logic and probability) is significant. (As van Fraassen argues, quantum theory itself does not require us to abandon either classical logic or classical probability theory [13, pp.134-135].) The point is not uncontentious, but as he points out, quantum theory is compatible with the adoption of a non-classical logic (such as quantum logic) as a fragment of a broader logic that is classical.) None of this is in question here. My point is to suggest an alternative empiricist formulation of the modal interpretation, which preserves as much as possible of van Fraassen's positive program, without being subject to the philosophical difficulties found in it.

3.1 Truth

Before spelling out these difficulties, note an important feature of the distinction between state attributions and value attributions of a physical system. First, state attributions are *theoretical constructs*, and part of the worries involved in theory construction arises from the proper representation of these states. The point here is that, in order to accommodate the claim that a given physical system is in a certain state (state attribution), and to distinguish this statement from the system having a certain value (value attribution), *the notion of truth is introduced.* As we saw, the connection between a value state and a value-attributing proposition is that value states are *truth-makers* of value-attributing propositions [13, pp.275-276]. Similarly, dynamic states and state-attributing propositions are connected by the fact that the former are what make the latter *true.*

In other words, the value-state distinction is cashed out in terms of *truth*. As van Fraassen points out [13, pp.280-281], if a physical system X has dynamic state (represented by a density operator) W at a time t, the state-attributions $[M, E]$ which are *true* are those such that $\text{Tr}(WI_E^M) = 1$. (A few comments about the notation: (a) Tr is a linear map, called *trace*, of operators into numbers, which give us the probability that a measurement of m has a value in E. (b) I is a projection operator, which is a Her-

mitian operator I such that $II = I$. (c) I_E^M is then defined as the Hermitian operator I such that Ix is x if $Mx = ax$ for some value a in E, and is the zero vector if $Mx = bx$ for some value b outside E, where E is a Borel set. (d) That the trace function Tr provides a probability is due to the following equations: $P_x^m(E) = (x \cdot I_E^M x) = \text{Tr}(I_x^M I_E^M)$, where $P_x^m(E)$ is the probability that a measurement of m has a value in E; $(x \cdot I_E^M x)$ is the inner product of x and $I_E^M x$; and I_x^M is the projection on the subspace $[x]$ spanned by x. For details, see [13, pp.147-152;157-165;280-281].) As opposed to state-attributions, value-attributions *cannot* be deduced from the dynamic state. But, according to van Fraassen, they are constrained in three ways, which again are spelled out in terms of *truth*: (1) If $[M, E]$ is *true*, so is the value-attribution $\langle M, E \rangle$, that is, observable M has value in E; (2) all *true* value-attributions could have probability 1 together; and (3) the set of *true* value-attributions is maximal with respect to feature (2) [13, p.281]. So, the assignment of truth-conditions to state-attributing and value-attributing propositions is crucial to spell out the difference between them (the former, but *not* the latter, can be deduced from the dynamic state).

But how can we make sense, in empiricist terms, of these truth-conditions? As we saw, for the constructive empiricist, science does not aim at the truth, but at empirical adequacy. In van Fraassen's formulation, a theory T is empirically adequate if there is a model of T such that all the phenomena (properly structured) are isomorphic to the empirical substructures of this model [19, p.64]. The idea is that empirical adequacy, as defined, is reduced to truth if we consider only observable phenomena [13, p.4]. Note, however, that in van Fraassen's formulation, empirical adequacy is a model-theoretic notion, and therefore we are dealing here with the notion of *truth in a structure*. Thus, empirical adequacy can be reduced at best to this notion of truth, and *not* to truth *simpliciter*. It may be argued that we can define the latter in terms of the former by stating that a sentence α is true *simpliciter* if it is true in the structure @ of the actual world (supposing that there is one). The problem is that an empiricist cannot accept this formulation without a severe qualification, since presumably the structure @ incorporates information about unobservables which the empiricist is agnostic about. So the distinction between observable and unobservable entities would have to be drawn with regard to @, and we would be back to the properties of the structure in question, and hence to the notion of truth in a structure. I conclude that if empirical adequacy is reducible to truth in van Fraassen's account, the latter cannot be truth *simpliciter*.

Now I raise this point because, as van Fraassen acknowledges, state attributions are theoretical constructs, and arguably involve reference to unobservables (in fact, a state is represented by a statistical operator, which is typically a particular kind of Hermitian operator defined on a Hilbert space). However, it is not clear how we should understand, in empiricist terms, the claim that state attributions are *true*. In order not to be committed to the existence of unobservables involved in such attributions, the empiricist would have to distinguish, in the structure which describes a state attribution, an *empirical substructure* from a *theoretical superstructure*. But it is clear that this cannot be done, since state attributions, as such, are entirely theoretical constructions, and they do not seem to incorporate an *empirical* part. The point here is that, in order not to be committed to the theoretical content involved in state attributions, the empiricist should avoid talking about their truth, and should run the semantic analysis in terms of a weaker notion. My suggestion, to be spelled out below, is to change here the norm of theoretical talk to quasi-truth, rather than truth. (Someone may argue that van Fraassen does not need a substantial notion of truth to assign truth-values to state-attributing and value-attributing propositions. He can simply adopt a *disquotational* approach. The suggestion is, of course, well motivated, given that, in some writings of van Fraassen, the disquotational account plays some role [27, 28]. But the suggestion faces a serious problem. As we have just seen, in order for the notion of empirical adequacy to be reducible to truth (when we only consider observable phenomena), the empiricist cannot adopt the notion of truth *simpliciter*. After all, as cashed out by van Fraassen, empirical adequacy is a *model-theoretic* notion, and so truth has to be understood in a model-theoretic way. Given that a disquotational account is not model-theoretic, it will not help the empiricist here. For what is needed for an interpretation of state and value attributions is the notion of truth in a structure. However, as argued above, this notion is too strong for the empiricist to use when talking about unobservable entities.)

3.2 Modality

Before discussing the concept of quasi-truth, let me consider the second feature of the modal interpretation: the introduction of modal operators. Since van Fraassen is skeptical about the existence of objective modality, all modal talk involved in his interpretation comes from language, as it were. As he points out, modal talk is talk about the structure of our own ways of representing the phenomena [20, pp.68;92;213-214;223]. It is not taken to correspond to something real, such as a possible world in Lewis's sense [29], nor is it taken to be primitive (else van Fraassen's position would lead to modalism). Indeed, countenancing possible worlds would be incompatible with a thoroughly empiricist outlook, given the increase

in the ontological commitments that they bring. And admitting a primitive notion of modality would be equally problematic, since if this notion has any metaphysical import, it will clash with van Fraassen's skepticism about modality (in particular, with the rejection of the notion of objective chance).

It may be argued that constructive empiricism is compatible with Lewis's modal realism [30]. After all, strictly speaking, constructive empiricism is only a claim about the aim of science–involving the search for empirically adequate theories and restricting belief to the empirical adequacy of the relevant scientific theories [19, p.12]. Modal realism is compatible with that aim for science, just as a number of other metaphysical views are. Although this is literally correct, the point of constructive empiricism is to develop an empiricist account of science, and a significant feature of the empiricist stance is to be suspicious precisely of the kind of postulational metaphysics advanced by the modal realist [26]. As a result, it seems unmotivated and questionable for the constructive empiricist to embrace modal realism. A less metaphysically loaded approach seems in order.

It is not surprising that modal operators are introduced in van Fraassen's interpretation of quantum mechanics as *algebraic* devices; that is, constructions which mathematically have the form of modal operators in modal logic, but which lack any other (metaphysical) warrant. Van Fraassen introduces two kinds of modal operators, one of them based on a transitive relative possibility relation R (between possible situations), and another based on an equivalence relation (namely, equality of dynamic states). Let me briefly spell this out.

The modal interpretation, formulated by van Fraassen in quantum-logical terms, has the notion of possible situation as basic [13, p.302]. A proposition is then identified with a set of situations; intuitively, those situations in which the proposition is true. We say that a proposition q is true in a situation w exactly if w is in q [13, p.302]. Let $\mathbf{V}$ be the set of value-attributions, i.e. $\mathbf{V} = \{\langle m, E \rangle : m$ is an observable and E is a Borel set$\}$, and $\mathbf{P}$ the set of state-attributions: $\mathbf{P} = \{[m, E] : m$ is an observable and E is a Borel set$\}$. Now, according to the modal interpretation, a situation w is *possible relative to a situation w'* (in symbols, wRw') if for all q in $\mathbf{V}$, if w is in q, then w' is in q [13, p.311]. In terms of this notion, modal operators of necessity $\square$ and possibility $\lozenge$ are then defined for all propositions [13, p.314]. Indeed, if q is a proposition and w a possible situation, we say that

$$\square q = \{w : \text{for all } w', \text{ if } wRw', \text{ then } w' \in q\} \quad (1)$$

$$\lozenge q = \{w : \text{for some } w', \text{ if } wRw', \text{ then } w' \in q\} \quad (2)$$

This establishes at once a connection between state-attributing and value-attributing propositions, namely: $[m, E] = \square \langle m, E \rangle$ (a proof can be found in [13, p.316]). The second kind of modal operator introduced is based on an equivalence relation; namely, equality of a dynamic state x with respect to situations w and w'. Indeed, if q is a proposition, we say that $\square q = \{w :$ for all w', if $x(w) = x(w')$, then $w' \in q\}$. As a consequence, a similar result also holds for this necessity operator: $[m, E] = \square \langle m, E \rangle$ [13, pp.316-317].

From this brief exposition, it should be clear that all the modal operators introduced in the modal interpretation are only *algebraic* constructs, called *modal* because of the formal similarities that they bear to the *necessity* and *possibility* operators in modal logic. This may well be all right given van Fraassen's skepticism about modality. After all, as noted, van Fraassen is not willing to be committed to the existence of a *modal reality* to which these operators refer. But is this *algebraic* similarity enough to characterize these operators as *modal*?

There is certainly a whole story to be told here, but perhaps it suffices to note a few points. We are, of course, allowed to call *modal* whatever operators satisfying certain formal conditions, and provided that we are only concerned with particular formal results, there is certainly no problem with this. But the issue arises as to how we should interpret these operators, and what consequences to draw about their status. Once we enter this debate, we are doing metaphysics–not logic or the foundations of physics. And at this level, as Lewis points out [29, pp.17-20], we need more than purely formal considerations. (According to Lewis: "Where we need possible worlds [...] is in applying the results of [...] metalogical investigations. Metalogical results, by themselves, answer no questions about the logic of modality. They give us conditional answers only: if modal operators can be correctly analysed in so-and-so way, then they obey so-and-so system of modal logic. We must consider whether they may indeed be so analysed; and then we are doing metaphysics, not mathematics" [29, p.17]) Now, as noted, van Fraassen surely has an account of modality to offer [20, 31]. What is not yet clear is how the modal operators introduced in his interpretation mesh with the idea that modality is a feature of our *language* (of the structures used to represent the phenomena), and do *not* refer to independently existing *possible entities*. For the necessity operator $\square$ was defined in terms of (a property of a) dynamic state, which is a theoretical construct and bears a clear relation not to any *linguistic entity*, but to a *physical* event. In this sense, we are apparently in a situation in which we are reducing modality to certain *physical* circumstances (or, at least, to events which depend on such circumstances). And this seems at odds with the idea that necessity is a *lin-guistic* feature. Moreover, the two other modal operators, $\square$ and $\lozenge$, are explicitly defined in terms of a relation of relative possibility among *situations*. But, as we saw, the notion of possible situation is primitive, and it is not clear how it relates either to physical or to linguistic events. If it is argued that these situations function only as an index set, and hence have no metaphysical import, we are back to the initial position in which no interpretation of the modal operators was offered. If these situations are to have any explanatory role *vis-à-vis* the status of the modal operators, we need an account of them. A story should be told about their status–otherwise, we will end up with no properly modal notions. (It might be argued that in order to introduce modal operators into the modal interpretation, van Fraassen can use the idea that when we accept a probabilistic theory (such as quantum mechanics) we accept it as providing a panel of expert functions to guide our personal probability, our opinion [20]. The modal operators introduced can be interpreted in terms of the expert functions: they guide our personal probability as to the observable predictions of quantum theory. But there are two problems with this suggestion. First, it does not really allow us to *interpret* the modal operators, since the latter apply to value-attributing and state-attributing propositions, which describe *unobservable* properties of a quantum system. The *empiricist* twist in introducing expert functions is to restrict them to *observable* features of the theory, but that is what the modal operators violate. Second, if empiricists adopt expert functions as an account of modal operators, they will make quantum mechanics entirely *subjective*. For quantum mechanics would then be a theory about the change of our opinion with regard to quantum states, rather than a theory about quantum phenomena themselves.)

But why does van Fraassen need to introduce modal operators in his interpretation of quantum mechanics? There are several reasons for that. These operators allow him to stress the empiricist status of his interpretation of quantum mechanics, since they cash out the claim that *at most* quantum theory tells us how the quantum world *can be*. Of course, as van Fraassen notes, any interpretation provides an account of how the world can be *if the theory is true* [13]. What is special about the modal interpretation, in van Fraassen's formulation, is that this is *the most* that the theory provides. The empiricist sentiment expressed by Descartes in the epigraph of [13], which was quoted above, underscores the empiricist nature of van Fraassen's interpretation.

Furthermore, it is via modal notions that van Fraassen talks about observables having unsharp values. One could cash out the talk of unsharp values of observables in terms of partial functions. Van Fraassen acknowledges this option but does not adopt it [13, p.307], articulating instead an account that leads more straightforwardly to the introduction of probability in quantum mechanics. Given that probability is a modality–a possibility with degrees [19, §6]–it is ultimately for reasons having to do with the introduction of modality that unsharp values are cashed out the way they are.

3.3 Truth and Modality: Theoretical Constraints

Can constructive empiricists make sense of applying truth to unobservable entities, such as dynamic states? It may be argued that they can as long as none of these claims (that refer to unobservable objects) is ever *asserted*. What this amounts to is that, although claims referring to unobservable entities are truth-apt, they are not to be evaluated in terms of their truth or falsity: some other norm has to be introduced. Which norm? Clearly empirical adequacy will not do. As formulated by van Fraassen, empirical adequacy is a property of a (class of) model(s), not a property of a particular statement [19, p.64]. Now, even if we consider the models of a proposition about dynamic states, it is not obvious that we can determine an empirical substructure corresponding to this model. As van Fraassen notes, states can be identified *in terms of* observables, but not *with* observables [13]. So empirical adequacy does not seem to be an adequate norm to evaluate interpretative claims about quantum mechanics, at least from an empiricist point of view.

This raises the broader issue: how should one evaluate interpretations of quantum mechanics (according to the empiricist)? Clearly, the interpretations cannot be taken to be true, even if they are truth-apt. For they refer to unobservable entities, which empiricists are agnostic about. Can empiricists claim that interpretations are true, but they do not believe in them? Clearly, being a kind of Moore's paradox, this is not coherent. Can empiricists claim that an interpretation, although not true, is still useful (in providing us with understanding of the quantum world)? Certainly; but an account has to be provided as to how *fictions* (since interpretations are not true) can increase our understanding. For example, how do literary fictions increase our understanding? One needs a notion of truth, or some modal notion, to cash this out. But this is precisely the problem in the first place! Let us see why either the notion of truth or some modal notion is required here.

Let us consider an example. Suppose that you assert that international aid can be problematic for an economy rather than a source of help. Let us say that I am skeptical, that I am unable to see how this can be true. To help me understand this, you construct a little, simplified economic model, in which the increase of foreign economic help generate huge economic difficulties for the country that receives help (for our purposes, we do not need to go into the details of this model). It is then true in this model that if we increase economic help, we deteriorate the economy. If I then ask you whether this model is actually true, you would then say that it is not, and that it does not have to be in order for me to understand how your claim about the downside of foreign help can be true. I then point out that something more has to be assumed. Your model does not have to be true, but it has to be consistent with accepted features of the actual situation. If there is no way in which an economy could possibly satisfy the conditions imposed by your model, the latter would simply be irrelevant to increase my understanding. In other words, a crucial claim of consistency has to be established before any assertion about understanding can be made. And surely you would grant that if your model were inconsistent with accepted information, it would not be of much use to increase my understanding of how your claim about the economy could be true. (Note that even if you were a paraconsistent theorist–who acknowledges that inconsistent theories need not be trivial, in the sense that not everything follows from them–you would still need to provide an account of why, despite its inconsistency with accepted information, the model is still coherent [32].)

What this indicates is the need for a *modal* notion–consistency–in the characterization of understanding. Can empiricists help themselves to such a notion? As usually formulated, consistency is identified with the existence of an abstract structure (the model of the relevant theory or sentence). But given that an abstract structure is not observable, it is not something an empiricist can believe in. Alternatively, consistency is understood in terms of possibility, as a claim that the relevant theory or sentence is possible, that it *can be* true. But this means that the empiricist has to assume a primitive notion of modality. It is unclear, however, how to reconcile such primitive notion of modality with the skepticism about modal notions found in constructive empiricism.

It may be argued that constructive empiricists do not have to be skeptical about *logical* consistency, since this notion is well understood on its own. I think this response makes very good sense. In fact, it is one that I will recommend below. There are, of course, several arguments to the effect that a proper characterization of modal notions requires the introduction of possible worlds or other entities, such as abstract objects, which are not part of the empiricist's ontology [29]. Constructive empiricists should resist these arguments, and offer a modalist account of logical concepts, which recognizes the need for, and the intelligibility of, a primitive logical notion of consistency [33–36].

Thus, the example about the economic model illustrates that an account of understanding requires talk of *consistency*. But it also illustrates where *truth* steps in. As noted above, it is important that *in the economic model* your claim about foreign help is *true*. You will probably say that this is only the notion of *truth in a model*; what does that have to do with *truth*?

This is, of course, an important question, and here there seems to be a tension in the use of these two notions in constructive empiricism. In some contexts, perhaps dialectical ones involving debates with the scientific realist, truth is taken as a perfect correspondence between all components of a model of a theory and the world [19, 20]; in other contexts, truth is taken in a minimalist sense [27, 28]. Truth in a structure becomes crucial in contexts in which one wishes to claim that empirical adequacy is truth about the observable domain. For empirical adequacy is a model-theoretic notion in van Fraassen's hands [19, p.64], and for it to be identified with truth about observables, truth has to be thought of in model-theoretic terms as well. After all, if truth is not conceived in this way, it cannot be a property of a *model*, as is the case with empirical adequacy. Truth *simpliciter* is not a norm of science for the constructive empiricist; but this is the notion of truth adopted by scientific realists, and van Fraassen uses it when discussing with them, in order to avoid begging the question against them. The use of a minimalist notion of truth has emerged later in constructive empiricism. The notion is employed when we are talking about our language in use, in contrast with a language that requires interpretation; in the latter case, the notion of truth in an interpretation is adopted [28]. But since the minimalist notion of truth is ontologically committing, this is not a notion that the empiricist can adopt without worry. (Given the following instantiation of the disquotational schema: "'There are electrons' is true iff there are electrons", the empiricist cannot claim that the sentence 'There are electrons' is true without being ontologically committed to the existence of electrons. That is why van Fraassen introduces a norm *different from* truth in the interpretation of science, namely, empirical adequacy.)

Suppose then that these different uses of truth are kept apart and are only explored in appropriate contexts. What is the problem? The problem is that in the context of the modal interpretation, both truth and truth in a structure are used. On the one hand, since the talk of dynamic and value states is not part of our language in use, in the sense that it is not part of our daily vocabulary, it requires an interpretation; and the truth-conditions given to state and value attributions are expressed *in terms of a model* (the one characterizing the modal interpretation). However, van Fraassen also talks about the truth of quantum mechanics plus the modal interpretation. On his view, it is in this global setting that we should consider how the world can be if quantum mechanics is true. But in this setting, he is talking about truth *simpliciter*, for he is considering the world rather than a model of it. We no longer have truth in a structure.

It might be argued that these are two different contexts: one spells out the truth-conditions for state and value attributions (and the empiricist uses truth in a structure); the other determines the truth-conditions of quantum mechanics and the modal interpretation as a whole (and the empiricist uses truth *simpliciter*). But it is not at all obvious that we have different contexts here. The modal interpretation is, of course, parasitic on quantum mechanics: it is an interpretation of that theory after all. If we were to assert its truth, we would have to assert the truth of quantum mechanics as well. (What I mean here is that, as an interpretation of quantum mechanics, and not as a rival theory, the modal interpretation has to be consistent with quantum theory.) In other words, in the *same context*–the one determined by quantum mechanics– the empiricist uses two different notions of truth, and of course the two notions are not equivalent. The problem with this is that truth in a structure is compatible with falsity of the notions under consideration, whereas truth *simpliciter* is not. (Throughout this discussion I assume with the empiricist that the underlying logic is classical.) So even if the truth-conditions for state-attributions are satisfied, the resulting propositions may actually be false. For the truth-conditions were specified in terms of truth in a structure. For example, suppose that it is true in a model that all frogs are purple provided that they live in a forest. Now, even if it is in fact true that frogs live in forests, it does not follow that frogs are purple. From a semantic viewpoint, this indicates the limitation of the proposed account.

What if the notion of truth in a structure is abandoned, and only truth *simpliciter* is used to characterize the truth-conditions of state attributions? If this is done, no longer will the empiricist be able to claim that the truth of state attributions is *not* ontologically problematic, since it was only established *with respect to a model*. (After all, the truth of a state attribution in a given model does not establish its truth.) As a result, the introduction of truth-conditions for state attributions becomes ontologically problematic for the empiricist.

But perhaps the empiricist would adopt a different approach. The idea is *not* to introduce truth in a structure as a way of alleviating ontological commitments to unobservable objects. Truth-conditions are not formulated in terms of a model, but they are expressed as truth *simpliciter*. The avoidance of ontological commitment arises from the fact that the truth of state attributions is never asserted. Rather the empiricist considers quantum mechanics and the modal interpretation as a whole package (that is, quantum mechanics plus the interpretative additions provided by the modal interpretation). In this case, the empiricist will not assert the truth of the resulting theory and interpretation–only the empirical adequacy of the compound matters. And given that empirical adequacy is weaker than truth, and does not depend on the existence of any unobservable entity, it brings no problem for the empiricist.

What this proposal amounts to is the denial that interpretations of quantum mechanics could be evaluated independently of the theory itself. In other words, we should consider quantum mechanics together with each of its interpretations, and evaluate the whole package in terms of its empirical adequacy. Since every interpretation is empirically superfluous (otherwise we would generate a rival theory), the empirical adequacy of any interpretation plus quantum mechanics is guaranteed, provided that the latter is empirically adequate. As a result, as noted above, empirical adequacy is not an informative criterion for the evaluation of interpretations of quantum mechanics. It is not an adequate norm for the interpretation of theories, since every interpretation, in virtue of being an interpretation rather than a rival theory, satisfies it.

After carefully, distinguishing the theory from its interpretations, the empiricist now seems to insist that we should evaluate both as a package. If we are to avoid ontological commitment to unobservable entities, we should only assert the empirical adequacy of the resulting compound. But why should one resist the temptation of asserting the *truth* of the latter? Except for avoiding the inflationary metaphysics associated with the interpretation, there seems to be no reason why one should not make that move. This means that only empiricists would be persuaded by this suggestion. Is there any reason that constructive empiricists can offer to those who *do not* share their view, why they should avoid making any claim beyond the empirical adequacy of the compound theory plus interpretation?

Note also that by considering quantum mechanics plus its interpretation as a package only, the constructive empiricist is in no position to assess the truth of such interpretations independently of the theory. This makes it difficult to evaluate critically rival interpretations of quantum mechanics, since any such assessment presupposes that one can demarcate the interpretation from the formalism of the theory, which clearly can be done.

We have now reached the point where the two themes of this section–truth and modality in the modal interpretation–meet. As we saw, truth was introduced in the modal interpretation to separate state-attributing propositions from value-attributing ones. But, given the empiricist constraints of van Fraassen's view, *truth* determines a norm that is *too strong* to be applicable to state-attributions. Modality has also been introduced, but here the account seems to be *too weak* to provide the required philosophical warrant for them. The question arises naturally: Is there an empiricist account that provides both a weaker notion of truth and a stronger account of modality? If so, is this account also appropriate to the particularities of the quantum domain? In order to answer these questions, da Costa and French's partial structures approach provides a useful tool.

4 Partial Structures and Quasi-Truth

The partial structures approach (as first presented in [37, 38], and then extended in [21–23]) relies on three main notions: partial relation, partial structure, and quasi-truth. (Further developments and applications of this approach can also be found in [24, 39–45].) One of the main motivations for introducing this proposal comes from the need for supplying a formal framework in which the *openness* and *incompleteness* of information dealt with in scientific practice can be accommodated in a unified way [23]. Two proposals are central to this task. First, the usual concept of structure is extended to model the partiality of information we have about a certain domain, thus leading to the concept of a *partial* structure. Second, the Tarskian characterization of the concept of truth for such *partial* contexts is advanced, and the corresponding concept of *quasi-truth* is formulated.

To characterize a partial structure, one need, first, to formulate an appropriate notion of *partial relation*. When investigating a certain domain of knowledge Δ (which is a field of the empirical sciences, such as particle physics), we formulate a conceptual framework that helps us systematize and organize the information we obtain about Δ. This domain is tentatively represented by a set D of objects, and is studied by the examination of the relations holding among D's elements. However, given a certain relation R defined over D, we often do not know whether all the objects of D (or n-tuples thereof) are related by R. This is part and parcel of the *incompleteness* of our information about Δ, and is formally accommodated by the concept of partial relation. More formally, let D be a non-empty set; an n-place partial relation R over D is a triple $\langle R_1, R_2, R_3 \rangle$, where R_1, R_2, and R_3 are mutually disjoint sets, with $R_1 \cup R_2 \cup R_3 = D^n$, and such that: R_1 is the set of n-tuples that (we know that) belong to R, R_2 is the set of n-tuples that (we know that) do not belong to R, and R_3 is the set of n-tuples for which we do not know whether they belong or not to R. (Note that if R_3 is empty, R is a usual n-place relation that can be identified with R_1.)

In order to represent the information about the domain under consideration, we need a notion of *structure*. The following characterization, formulated in terms of partial relations and based on the standard concept of structure, is meant to provide a notion that is broad enough to accommodate the partiality usually found in scientific practice. Partial relations do the main work, of course. A *partial structure S* is an ordered pair $\langle D, R_i \rangle_{i \in I}$, where D is a non-empty set and $(R_i)_{i \in I}$ is a family of partial relations defined over D. (The partiality at stake here is due to the *incompleteness* of our knowledge about the domain under investigation. Given additional information, a partial relation may become total. Hence, the partiality modeled by the partial structures approach is not understood as an ontological *partiality* in the world–an aspect about which an empiricist will be glad to remain agnostic. In other words, what is at issue is an *epistemic*, not an *ontological* partiality.)

In order to systematize our knowledge of Δ (say, again, particle physics), the domain D of the partial structure S is typically constituted by two components: (1) observable objects (in the physics of particles, configurations in a Wilson chamber, spectral lines etc.), whose set is denoted by D_1; and (2) unobservable objects (quarks, for example), whose set is denoted by D_2. It is understood that $D_1 \cap D_2 = \emptyset$, and we require that $D = D_1 \cup D_2$. In this way, the modeling of Δ involves new partial relations $R_j, j \in J$ (defined over D_2), some of which may help to extend the relations $R_i, i \in I$ (defined over D_1). As a result, if we want to be explicit, a partial structure S has the following form: $\langle D_1, D_2, R_i, R_j \rangle_{i \in I, j \in J}$. But it is usually easier to refer to it simply as $\langle D, R_k \rangle_{k \in K}$ [22, 43].

Two of the three basic notions of the partial structures approach are now defined. To formulate the last, and crucial one–quasi-truth–an auxiliary notion is required. The idea is to use, in the characterization of quasi-truth, the resources offered by Tarski's definition of truth. But since it is only defined for full structures, we have to introduce an intermediary notion of structure to connect full to partial structures. This is the first role of those structures that extend a partial structure A into a full, total structure (which are called *A-normal structures*). Their second role is purely model-theoretic, namely to put forward an interpretation of a given language and, in terms of it, to characterize basic semantic notions. The question then is: how should *A*-normal structures be defined? Here is an answer. Let $A = \langle D, R_i \rangle_{i \in I}$ be a partial structure. We say that the structure $B = \langle D', R_i' \rangle_{i \in I}$ is an *A*-normal structure if (1) $D = D'$, (2) every constant of the language in question is interpreted by the same object both in A

and in B, and (3) R_i' extends the corresponding relation R_i (in the sense that each R_i', supposed of arity n, is defined for all n-tuples of elements of D'). Although each R_i' is *defined* for all n-tuples over D', it holds for some of them (the R_{i1}'-component of R_i'), and it does not hold for others (the R_{i2}'-component).

As a result, given a partial structure A, there are several A-normal structures. Suppose that, for a given n-place partial relation R_i, we do not know whether $R_i a_1 \ldots a_n$ holds or not. One way of extending R_i into a full R_i' relation is to look for information to establish that it *does* hold, another way is to look for the contrary information. Both are *prima facie* possible ways of extending the partiality of R_i. But the same indeterminacy may be found with other objects of the domain, distinct from $a_1, \ldots, a_n$ (for instance, does $R_i b_1 \ldots b_n$ hold?), and with other relations distinct from R_i (for example, is $R_j b_1 \ldots b_n$ the case, with $j \neq i$?). In this sense, there are *too many* possible extensions of the partial relations that constitute A. Thus, we need to provide constraints to restrict the acceptable extensions of A.

In order to do that, we need first to formulate a further auxiliary notion [37]. A *pragmatic structure* is a partial structure to which a third component has been added: a set of accepted sentences P, which represents the accepted information about the structure's domain. (Depending on the interpretation of science that is adopted, different kinds of sentences are to be introduced in P: realists will typically include laws and theories, whereas empiricists will add mainly certain empirical regularities and observational statements about the domain in question.) A *pragmatic structure* is then a triple $A = \langle D, R_i, P \rangle_{i \in I}$, where D is a non-empty set, $(R_i)_{i \in I}$ is a family of partial relations defined over D, and P is a set of accepted sentences. The idea is that P introduces constraints on the ways that a partial structure can be extended.

The conditions for the existence of A-normal structures can now be spelled out [37]. Let $A = \langle D, R_i, P \rangle_{i \in I}$, be a pragmatic structure. For each partial relation R_i, we construct a set M_i of atomic sentences and negations of atomic sentences, such that the former correspond to the n-tuples that satisfy R_i, and the latter to those n-tuples that do not satisfy R_i. Let M be $\bigcup_{i \in I} M_i$. Therefore, a pragmatic structure A admits an A-normal structure iff the set $M \cup P$ is *consistent*. (Consistency here is taken as a primitive, logical notion; see [35, 36].)

Assuming that such conditions are met, we can now formulate the concept of quasi-truth. A sentence α is *quasi-true* in A according to B if (1) $A = \langle D, R_i, P \rangle_{i \in I}$ is a pragmatic structure, (2) $B = \langle D', R_i' \rangle_{i \in I}$ is an A-normal structure, and (3) α is true in B (in the Tarskian sense). If α is not quasi-true in A according to B, we say that α is *quasi-false* (in A according to B). Moreover, we say that a sentence α is *quasi-true* if there is a pragmatic structure A and a corresponding A-normal structure B such that α is true in B (according to Tarski's account). Otherwise, α is *quasi-false*.

The idea, intuitively speaking, is that a quasi-true sentence α does not describe the whole domain to which it refers, but only an aspect of it–the one modeled by the relevant partial structure A. After all, there are several different ways in which A can be extended to a full structure, and in some of these extensions α may not be true. As a result, the notion of quasi-truth is strictly weaker than truth: although every true sentence is (trivially) quasi-true, a quasi-true sentence is not necessarily true (since it may be false in certain extensions of A).

To illustrate the use of this notion, let us consider an example. As is well known, Newtonian mechanics is appropriate to explain the behavior of bodies under certain conditions, such as bodies that, roughly speaking, have *low* velocity in comparison with that of light, are not subject to strong gravitational fields etc. With the formulation of special relativity, we know that if these conditions are not satisfied, Newtonian mechanics is false. In this sense, these conditions specify a family of partial relations, which delimit the context in which the theory holds. Although Newtonian mechanics is not true (and we know under what conditions it is false), it is *quasi-true*; that is, it is true in a given context, determined by a pragmatic structure and a corresponding A-normal one [39].

Having discussed the formulation of partial structures and quasi-truth, we can now consider an application of this conceptual framework to the modal interpretation of quantum mechanics.

5 Modality, Quasi-Truth and the Modal-Structural Interpretation of Quantum Mechanics

My suggestion here is that the framework sketched above in terms of partial structures and quasi-truth provides at least a partial answer to the two questions raised at the end of section 3, namely: Is there an empiricist account that provides both a weaker notion of truth and a stronger account of modality? And is this account also appropriate to the particularities of the quantum domain? The idea is that, instead of determining the truth of state-attributions, the empiricist should change the norm of this type of theoretical discourse to quasi-truth. In this way, the commitment to unobservables can be avoided, simply because quasi-truth is weaker than truth. As we saw, in claiming that a sentence α is quasi-true, we are not committed to the claim that it captures, in full detail, all the *structure* of the domain in question: the relations defined over the

domain D_2 (which concern *unobservable* aspects of the domain under study) are left open. The only components that are *fixed* are those that refer to the *observable* aspects of the domain (which are found, roughly speaking, in D_1). So, from the fact that a theory T (about particle physics) is quasi-true, we cannot conclude that the unobservable aspects of the domain under study are fully described. Note that empirical adequacy has the same feature: if T is empirically adequate, it does not follow that it is true, i.e. that it describes, in full detail, all aspects of the world; the only aspects that are properly accommodated are the *observable* ones. But since quasi-truth is always relative to a given partial structure (which incorporates both an observable part, D_1, and an unobservable one, D_2), it readily generalizes van Fraassen's notion of empirical adequacy: roughly speaking, a theory is empirically adequate if it is quasi-true in a partial structure that describes the empirical phenomena. In this sense, quasi-truth is appropriate for an empiricist view (for details, see [24]).

Now, to say that a state-attributing proposition $[m, E]$ is quasi-true–in a partial structure A, which represents partial information about the state of the system–is to say that there is an A-normal structure B in which $[m, E]$ is true. Note that B provides one possible way of *extending* the partial information about the state; there may well be different ways of extending it. But this suggests that the concept of quasi-truth already introduces a notion of modality. In fact, this point can be explicitly made as follows (see also [43]): (a) $\Box P$ is quasi-true (in a partial structure A) iff for all A-normal structures, P is true. (b) $\Diamond P$ is quasi-true (in a partial structure A) iff there is an A-normal structure in which P is true. (The notion of relative possibility can be introduced without difficulty.) Now, as defined, these modal operators are clearly about the structures under consideration, namely those employed to represent the states of a given physical system. In order to use these operators, there is no need to go beyond the language used to describe such states. There is no need for the introduction of possible worlds, since the modal operators are explicitly formulated in terms of quasi-truth. Moreover, only a primitive notion of logical consistency is assumed in the background, as part of the partial structures approach. But there are independent reasons to favor this form of modalism, given the need for such a primitive notion in order to make sense of logical consequence and logical constants [33–36]. In the end, all we have to do is explore the representation possibilities of the relevant structures. In this sense, the proposal advanced here accommodates straightforwardly the constructive empiricist's point that modal talk is concerned with the structures we use to represent the phenomena.

But the question of the adequacy of the modal operators formulated via quasi-truth immediately arises. Can we obtain results about such operators analogous to those established in van Fraassen's interpretation? In particular, can we establish that:

> $[m, E]$ is quasi-true iff $\Box \langle m, E \rangle$ is quasi-true?

This is, of course, the crucial modal result (in terms of quasi-truth) of the modal interpretation, highlighting a consequence by which, in van Fraassen's words, the interpretation "merits at once the epithet 'modal'" [13, p.314]. The result is not difficult to establish. First, let us assume that $\Box \langle m, E \rangle$ is quasi-true. Thus, for all A-normal structures B, $\langle m, E \rangle$ is true; in other words, in each B the observable m actually has a value in E (in symbols: $\lambda(w)(m) \subseteq E$, where $\lambda(w)$, the value state, is taken as a map from observables into non-empty Borel sets). Now, since this holds *for each B*, there is an A-normal structure B' in which the state is such that the probability that the observable m has a value in E is one (in symbols: $P_x^m(E) = 1$). Hence, $[m, E]$ is true in B', and therefore it is quasi-true. Second, let us assume that $\Box \langle m, E \rangle$ is not quasi-true. In this case, there is no A-normal structure B in which $\langle m, E \rangle$ is true; that is, in no B does the observable m have a value in E. Thus, there is no B in which the state is such that the probability that m has a value in E is one. Therefore, $[m, E]$ is not quasi-true. This concludes the proof.

It is thus established that state-attributing propositions are equivalent to the necessitation of value-attributions. This is expected given the definition of state-attributions: they determine a strong, modal claim about the state of a physical system. The result helps us to see the explicit role that is played by modal notions in the interpretation of quantum mechanics since it provides a way of representing the possibilities open to the evolution of a physical system and the interrelationships between the modal operators that are formulated.

It is worth noting that, by moving to the partial structures approach, we can also represent an important feature of van Fraassen's modal interpretation. As we saw, van Fraassen's rejection of von Neumann's interpretation rule allowed the introduction of *unsharp values* to observables. Now, such unsharp values may be assigned to an observable m by a value state λ. As van Fraassen indicates, one way of spelling this out "'is to say that λ is a partial function assigning values to some observables but not to others" [13, p.307]. Although van Fraassen does not adopt this alternative, it is clear that, within the partial structures approach, this move can be represented straightforwardly. If we view a value state as a partial function (which is a particular kind of partial relation), the unsharp values are those for which λ is not defined (given our current knowledge about the state of the physical system under consideration). It then becomes clear why we

need the necessitation of value-attributions to make them equivalent to state attributions: intuitively speaking, all possibilities have to be covered.

Thus, by moving to partial structures and quasi-truth, the two problems faced by van Fraassen's modal interpretation can be overcome: we do not have to talk about the truth of state-attributing propositions (and can remain empiricist about them), and there is a straightforward way of introducing modality into the modal interpretation (via the notion of quasi-truth), which satisfies strictly empiricist constraints.

6 Conclusion

Although far more could be said, I hope to have indicated that the present framework provides a plausible setting for the empiricist to investigate the quantum domain. Given the role played by structures and modal operators in the approach, it seems adequate to call it a *modal-structural* interpretation. If developed further, it provides an interesting way of examining, in modal and empiricist terms, some of the peculiarities of the foundations of quantum mechanics.

Acknowledgements

My thanks go to Newton da Costa, Steven French, Décio Krause, and Bas van Fraassen for helpful discussions. Thanks are also due to two reviewers, whose identity has now been revealed to me, Eliahu Cohen and Danko Georgiev, for their many insightful suggestions and advice.

References

[1] von Neumann J. Mathematical Foundations of Quantum Mechanics. Beyer RT (translator), Princeton: Princeton University Press, 1955.

[2] Birkhoff G, von Neumann J. The logic of quantum mechanics. Annals of Mathematics 1936; 37 (4): 823-843. http://www.jstor.org/stable/1968621

[3] Putnam H. Is logic empirical? In: Proceedings of the Boston Colloquium for the Philosophy of Science, 1966–1968. Boston Studies in the Philosophy of Science, vol.5, Cohen RS, Wartofsky MW (editors), Dordrecht: D. Reidel, 1968, pp.216-241.

[4] Bub J. The Interpretation of Quantum Mechanics. Dordrecht: D. Reidel, 1974.

[5] Rédei M. Quantum Logic in Algebraic Approach. Dordrecht: Kluwer Academic Publishers, 1998.

[6] Manin YI. A Course in Mathematical Logic for Mathematicians, 2nd edition. Graduate Texts in Mathematics, vol.53, New York: Springer, 2010.

[7] da Costa NCA, Krause D. Schrödinger logics. Studia Logica 1994; 53 (4): 533-550. http://dx.doi.org/10.1007/BF01057649

[8] da Costa NCA, Krause D. An intensional Schrödinger logic. Notre Dame Journal of Formal Logic 1997; 38 (2): 179-194.

[9] French S, Krause D. Vague identity and quantum non-individuality. Analysis 1995; 55 (1): 20-26. http://dx.doi.org/10.1093/analys/55.1.20

[10] Krause D, French S. A formal framework for quantum non-individuality. Synthese 1995; 102 (1): 195-214. http://dx.doi.org/10.1007/BF01063905

[11] French S, Krause D. Identity in Physics: A Historical, Philosophical, and Formal Analysis. Oxford: Clarendon Press, 2006.

[12] Redhead M. Incompleteness, Nonlocality, and Realism: A Prolegomenon to the Philosophy of Quantum Mechanics. Oxford: Oxford University Press, 1987.

[13] van Fraassen BC. Quantum Mechanics: An Empiricist View. Oxford: Clarendon Press, 1991.

[14] Dummett M. Is logic empirical? In: Contemporary British Philosophy: Fourth Series. Lewis HD (editor), London: Allen and Unwin, 1976, pp.45-68. Reprinted in Dummett M, Truth and Other Enigmas, Cambridge, MA: Harvard University Press, 1978, pp.269-289.

[15] Dieks D, Vermaas PE. The Modal Interpretation of Quantum Mechanics. The Western Ontario Series in Philosophy of Science, Dordrecht: Kluwer, 1998.

[16] van Fraassen BC. A formal approach to the philosophy of science. In: Paradigms and Paradoxes: The Philosophical Challenge of the Quantum Domain. University of Pittsburgh Series in the Philosophy of Science, Colodny RG (editor), Pittsburgh: University of Pittsburgh Press, 1972, pp.303-366.

[17] van Fraassen BC. Semantic analysis of quantum logic. In: Contemporary Research in the Foundations and Philosophy of Quantum Theory, vol.2. The University of Western Ontario Series in Philosophy of Science, Hooker CA (editor), Springer Netherlands, 1973, pp.80-113. http://dx.doi.org/10.1007/978-94-010-2534-8_3

[18] van Fraassen BC. The Einstein-Podolsky-Rosen paradox. Synthese 1974; 29 (1-4): 291-309. http://dx.doi.org/10.1007/BF00484962

[19] van Fraassen BC. The Scientific Image. Oxford: Clarendon Press, 1980.

[20] van Fraassen BC. Laws and Symmetry. Oxford: Clarendon Press, 1989.

[21] da Costa NCA, French S. Pragmatic truth and the logic of induction. The British Journal for the Philosophy of Science 1989; 40 (3): 333-356. http://dx.doi.org/10.1093/bjps/40.3.333

[22] da Costa NCA, French S. The model-theoretic approach in the philosophy of science. Philosophy of Science 1990; 57 (2): 248-265. http://www.jstor.org/stable/187834

[23] da Costa NCA, French S. Science and Partial Truth: A Unitary Approach to Models and Scientific Reasoning. Oxford Studies in Philosophy of Science, New York: Oxford University Press, 2003.

[24] Bueno O. Empirical adequacy: A partial structures approach. Studies in History and Philosophy of Science A 1997; 28 (4): 585-610. http://dx.doi.org/10.1016/S0039-3681(97)00012-5

[25] van Fraassen BC. From a view of science to a new empiricism. In: Images of Empiricism: Essays on Science and Stances, with a Reply from Bas van Fraassen. Mind Association Occasional Series, Monton B (editor), Oxford: Clarendon Press, 2007, pp.337-383.

[26] van Fraassen BC. The Empirical Stance. The Terry Lectures Series, New Haven: Yale University Press, 2002.

[27] van Fraassen BC. Structure and perspective: philosophical perplexity and paradox. In: Logic and Scientific Methods: Volume One of the Tenth International Congress of Logic, Methodology and Philosophy of Science, Florence, August 1995, vol.259. Synthese Library, Dalla Chiara ML, Doets K, Mundici D, Benthem J (editors), Springer Netherlands, 1997, pp.511-530. http://dx.doi.org/10.1007/978-94-017-0487-8_29

[28] van Fraassen BC. Elgin on Lewis's Putnam's paradox. Journal of Philosophy 1997; 94 (2): 85-93. http://www.jstor.org/stable/2940777

[29] Lewis DK. On the Plurality of Worlds. Oxford: Blackwell, 1986.

[30] Monton B, van Fraassen BC. Constructive empiricism and modal nominalism. British Journal for the Philosophy of Science 2003; 54 (3): 405-422. http://dx.doi.org/10.1093/bjps/54.3.405

[31] van Fraassen BC. The only necessity is verbal necessity. Journal of Philosophy 1977; 74 (2): 71-85. http://www.jstor.org/stable/2025572

[32] Bueno O, da Costa NCA. Quasi-truth, paraconsistency, and the foundations of science. Synthese 2007; 154 (3): 383-399. http://dx.doi.org/10.1007/s11229-006-9125-x

[33] Bueno O, Shalkowski SA. A plea for a modal realist epistemology. Acta Analytica 2000; 15 (24): 175-193. http://eprints.whiterose.ac.uk/3354/

[34] Bueno O, Shalkowski SA. Modal realism and modal epistemology: a huge gap. In: Modal Epistemology. Weber E, De Mey T (editors), Brussels: Royal Flemmish Academy of Belgium, 2004, pp.93-106.

[35] Bueno O, Shalkowski SA. Modalism and logical pluralism. Mind 2009; 118 (470): 295-321. http://dx.doi.org/10.1093/mind/fzp033

[36] Bueno O, Shalkowski SA. Logical constants: a modalist approach. Noûs 2013; 47 (1): 1-24. http://dx.doi.org/10.1111/j.1468-0068.2012.00865.x

[37] Mikenberg I, da Costa NCA, Chuaqui R. Pragmatic truth and approximation to truth. Journal of Symbolic Logic 1986; 51 (1): 201-221. http://www.jstor.org/stable/2273956

[38] da Costa NCA. Pragmatic probability. Erkenntnis 1986; 25 (2): 141-162. http://dx.doi.org/10.1007/BF00167168

[39] da Costa NCA, French S. Towards an acceptable theory of acceptance: Partial structures, inconsistency and correspondence. In: Correspondence, Invariance and Heuristics, Boston Studies in the Philosophy of Science, vol.148, French S, Kamminga H (editors), Springer Netherlands, 1993, pp.137-158. http://dx.doi.org/10.1007/978-94-017-1185-2_7

[40] da Costa NCA, French S. A model theoretic approach to 'natural' reasoning. International Studies in the Philosophy of Science 1993; 7 (2): 177-190. http://dx.doi.org/10.1080/02698599308573462

[41] da Costa NCA, French S. Partial structures and the logic of Azande. American Philosophical Quarterly 1995; 32 (4): 325-339. http://www.jstor.org/stable/20009835

[42] French S. Partiality, pursuit and practice. In: Structures and Norms in Science: Volume Two of the Tenth International Congress of Logic, Methodology and Philosophy of Science, Florence, August 1995. Dalla Chiara ML, Doets K, Mundici D, van Benthem J (editors), Dordrecht: Kluwer Academic Publishers, 1997, pp.35-52.

[43] da Costa NCA, Bueno O, French S. The logic of pragmatic truth. Journal of Philosophical Logic 1998; 27 (6): 603-620. http://dx.doi.org/10.1023/A%3A1004304228785

[44] Bueno O. What is structural empiricism? Scientific change in an empiricist setting. Erkenntnis 1999; 50 (1): 55-81. http://dx.doi.org/10.1023/A%3A1005434915055

[45] Bueno O, de Souza EG. The concept of quasi-truth. Logique et Analyse 1996; 39 (153-154): 183-199.

16

Ontology and Quantum Mechanics

N. D. Hari Dass

Tata Institute of Fundamental Research (TIFR), TIFR Centre for Interdisciplinary Sciences (TCIS), Hyderabad, India
Chennai Mathematical Institute, Chennai, India
Centre for Quantum Information and Quantum Computation, Indian Institute of Science, Bangalore, India
E-mails: dass@cts.iisc.ernet.in, dass@cmi.ac.in

Editors: *Danko Georgiev & Tabish Qureshi*

The issue of ontology in quantum mechanics, or equivalently the issue of the reality of the wave function is critically examined within standard quantum theory. It is argued that though no strict ontology is possible within quantum theory, ingenious measurement schemes may still make the notion of a FAPP ontology (ontology for all practical purposes) meaningful and useful.
Quanta 2014; 3: 47–66.

1 Introduction

A cursory check as to the meaning of the word *Ontology* will turn up a bewildering response, with a wide spectrum of interpretations. So is also the case for its close relative *Epistemology*. It is not the purpose of this article to get into a general discourse on this concept. Instead, it will focus on its meaning as widely understood by physicists, more particularly the quantum physicists. Though notions of existence and of reality come frequently associated with ontology, we shall focus more on aspects of reality. In the specific context of quantum theory, this more or less concerns the so called reality of the wavefunction. Reality is in itself a heavily loaded concept were one to turn into it from general philosophical considerations. We shall therefore restrict attention to *Physicist's notion of reality*, however unsophisticated it may appear to philosophers at large!

It is fair to say that the notion of reality to most physicists is conditioned by their experience from classical physics. Many so called paradoxes in quantum theory have in fact arisen because of this. Nevertheless, a careful examination of the concept of reality in classical physics is essential as a guide to examining its counterpart in quantum theory. It is clear that even in classical physics, notions of reality are intimately tied up with aspects of observation, or of measurements. Therefore, the plan of this article is to first examine ontology in classical physics, and to identify those aspects of classical measurements and dynamics that make the notion of reality reliable and useful. We then examine the issue of ontology in quantum mechanics against the backdrop of a variety of quantum measurements all the way from the Dirac–von Neumann description to the current day explosions.

2 Ontology in Classical Physics

Reality in classical physics may be characterized by certain *robust* associations of *attributes* and *objects*. For example, when one says that a particular Rose is Red, this represents an element of reality with many important aspects, many of which appear trivial and straightforward unless carefully contemplated upon. In this case the attribute is Redness and the object is the Rose in question.

What are the mechanisms in classical physics that bring about this association, and in what sense this association is robust are questions whose answers hold the key to a finer understanding of reality in classical physics.

Before attempting to answer them, let us expand the list of attributes in this case to include, let us say, *Smell*. Classical reality says that these two attributes can peacefully coexist and that the reality of one need not interfere with the reality of the other. Now what gives an element of reality to, say, the redness, is that no matter how many times we observe the color, no matter how we observe the color, or no matter how often we interject color observations with other observations, say in this case smell, we come up with the same measure of redness for the flower. It is obvious that this is possible only if observing the color of the rose does not itself alter its color.

We can sharpen this by introducing the notion of a *state* of the object. In the above example, red is the 'value' of the attribute of color for this particular state of the rose which may be called a 'red rose'. One could have yellow roses, purple roses etc. and they would all refer to different states of the object; let us stipulate that the state of any object is specified by the *values* of its attributes. It is worth recalling a characterization of a state by Dirac [1, 2]; though it was given in the context of quantum theory, it is pertinent to any theory, and certainly to classical physics also. According to him, *a state is an embodiment of all possible measurement outcomes.*

At this point there is an important subtlety that needs to be taken care of. It appears and in reality it is indeed so, that a red rose is a different object from a yellow rose and we are not really talking of different states of an object but of different objects. In fact that may make the distinction between the object and its state artificial and unwarranted. To overcome that, we shall allow for the possibility (not altogether unrealistic) of processes that could change the color of a given rose. Then the rose, the given object, can indeed be in different states of color. If, as mentioned above, we are also considering additional attributes like smell, a characterization of the state of a rose would, in this classical context, require specifying the values of both smell and color. These values can be thought of as the outcomes of color and smell measurements thus tying up with the characterization of a state according to Dirac.

In fact, we can go to the more prosaic world of classical mechanics and consider a particle as the object, its position, velocity etc. as its attributes. The state of the particle is then specified by the values of these attributes. There is an obvious redundancy with this description. For example, one could have also considered the square of the position as an attribute, but then that would not carry any additional information from that already carried by the position on its own. So a distinction should be made between what one may call *primary* and *derived* attributes. The upshot is that it is enough to consider an optimal set of independent attributes for state description.

Let us return to the issue of reality and its robustness. The association of, say, position with the particle can be taken as an element of reality which is robust because measurement of position of the particle returns the same values within some range of errors (more on this later) no matter how often this measurement is done, how this measurement is done (as there are many means of position measurements), and in what order these measurements are done in the sense that position measurements may be interspersed with measurements of other attributes. In fact, in the classical world it would then be possible to say that this element of reality exists even if no one is actually observing the particle!

It is obvious that this is possible if and only if the measurements have no effect on the state. Such measurements can be called *non-invasive*. But not every measurement should be necessarily non-invasive even in classical physics. One could in principle adopt a measurement scheme that is deliberately invasive. For example, a position measurement of a tiny particle could be done by hitting it with a big stone. So a choice of non-invasive measurement is essential in the scheme above. In the classical world, by and large, measurements are non-invasive unless by deliberate design.

It is important to emphasize that non-invasiveness by itself is enough to guarantee a robust element of reality. Now the other crucial aspect of the classical world enters the picture and this is *determinism*. To appreciate this consider the possibility that before a measurement the particle is in a definite state (state with definite values of all its attributes). A non-invasive measurement may leave the particle in the same state, but may not necessarily yield definitive values for these attributes. This could happen when the physical processes making up the measurement are not deterministic. It could well be that a definite measured value emerges upon averaging a large number of outcomes. Such a world would exhibit both ontic and epistemic features.

But the world of classical physics is deterministic. On top of that, no separate laws have to be stipulated for measurements. Therefore in principle every classical measurement should yield definite outcomes with no errors at all. But errors do occur in classical measurements. This is for the obvious reason that even in the deterministic world of classical physics, not every source of influence in an experiment can be identified and accurately accounted for. A pragmatic approach would treat the unknowns *probabilistically* thereby introducing randomness in a perfectly deterministic world! Therefore the outcomes will have variances and actual errors can be statistically reduced through repeated measurements. Nevertheless, even this randomness introduced purely for practical purposes, governs errors that can be controlled by better experimental designs. Then, one can adopt the reasonable stand that the outcomes within such narrow and controllable errors are, for all practical purposes, making the strictly non-invasive measurements into practically non-invasive measurements.

But it is worth appreciating that any randomness, however small, does not allow for ontic descriptions, in principle. But in practice this does not pose a problem. In that sense, even the 'real' world of classical physics has a blurry edge, which we ignore all the time!

All these considerations have one profound consequence. Measurements on a single object are meaningful, and statistical errors can be meaningfully reduced arbitrarily by making sufficiently large number of repeated measurements on the same object. It should be stressed that this arose both due to the near non-invasive measurements as well as due to each measurement practically yielding full information.

Even with regard to deliberately invasive measurements, the determinism of classical physics can in principle, though tedious and heavy on resources in practice, provide a means of compensating for the invasive effects. In the example of throwing a rock to measure the position of a small particle, though the rock strongly alters the state of the particle, very careful measurements of the subsequent trajectories of both the particle and the rock can be used to accurately reconstruct the state of the particle before the collision, and restore the particle to that state. But second law can put a limit on how much invasiveness is tolerable! If for example, the invasive measurement involved setting fire and vaporizing the particle, it would be practically impossible to regain the original state!

It is of course possible that the attributes change with time. The rose of the beginning could fade. Does this mean that the element of reality that was so carefully constructed was not real at all? The physicist's answer to this is not to deny the element of reality or its robustness, but to allow for a time evolution of states and their associated attributes. This is the idea of *Dynamics*. The determinism referred to earlier then takes the form of a *Deterministic Dynamics*. These deterministic rules of dynamics not only ensure unambiguous outcomes in ideal measurements, but they also ensure that no separate rules are necessary to describe measurements, unlike in quantum theory.

3 Ontology in Quantum Mechanics

At least as per our present understanding, the standard quantum theory is *inherently random*. From our previous discussion, no strict ontology ought to be possible then. Quantum Measurements, as understood during the critical years of the development of quantum theory, and as idealized by the *Dirac–von Neumann* measurement models, are certainly invasive, and uncontrollably so. They are invasive in an unpredictable way. This too leaves no room for an ontic description. The Born probability rule has to be invoked for a consistent interpretation of quantum mechanics and that is where randomness becomes intrinsic. Paradoxically, the rules for time evolution of states, or, quantum dynamics, is completely deterministic by itself! It is only measurement that brings in indeterminacy. Nevertheless, as we shall see later, there are intriguing pointers to why there can be no ontic description of quantum mechanics, coming from purely dynamical considerations (all unitary processes are considered dynamical here).

This also makes repeated measurements of the Dirac–von Neumann meaningless when performed on a single copy. The simple reason being that the state after the first measurement bears no obvious relation to the state one started with, and the subsequent measurements can at best reveal the state after the first measurement. Therefore only *ensemble* measurements become significant. For a good account of the issues involved in getting information out of measurements on a single copy see [3].

If there are such serious obstacles to ontology in quantum mechanics, why bother to go further? There are several good reasons for it! Firstly, the extreme invasiveness of quantum measurements is certainly a distinctive feature of the Dirac–von Neumann, or more precisely, *Projective Measurements*. So the question naturally arises whether there can be other measurement schemes that are non-invasive or controllably non-invasive. It then becomes important to re-examine the ontology issue in the context of these alternate measurement schemes. As it turns out, there are so many interesting alternatives to projective measurements today [4]. It is the purpose of this article to do that examination carefully.

We set the following technical criterion for ontology: *ontology is the ability to completely determine the previously unknown state of a single copy*. Even in cases where this is not possible, we introduce the notion of *FAPP Ontology* (ontology For All Practical Purposes) as the ability to almost determine the unknown state of a single copy, or in other words, a state determination with specified amount of errors.

It may, however, be worthwhile to point out that strictly speaking, the condition of non-invasiveness may be too restrictive for ontology. Both in classical mechanics and in quantum mechanics if means are available for a *complete determination* of the state of a system albeit at the cost of a radical change of state (but not of the system itself!), one can still meaningfully ascribe reality to that state. Then one can prepare the system in this state, and as far as subsequent measurements are concerned, it would be as if the original state had not been tampered with. In the context of quantum mechanics, this entails *simultaneous and accurate* determination of the expectation values of non-commuting observables in a *single* measurement performed on a single copy. The uncertainty principle restricts the possible accuracies in such a measurement which resembles the well-known *Arthurs-Kelly Measurements*. Again, strict ontology would not be possible, but it would be interesting to investigate the best possible FAPP ontology.

Though historically it was not recognized as such, we can now trace all the essential non-classical features of quantum theory to just one principle, namely, *The Principle of Superposition of States* [1,2,5]. In fact, one can take this principle to be the defining feature of quantum theories. Other aspects like *Entanglement*, taken by many (particularly among the Quantum Information community) to be the crux of quantum mechanics, is a natural consequence of the superposition principle.

It turns out that even without a very detailed analysis, one can show the impossibility of perfectly non-invasive measurements in quantum mechanics by just invoking the superposition principle. We outline this powerful argument in subsection 3.1. Another, equally powerful argument against ontology can be given by invoking the *No Cloning Theorem*. The proof of the No Cloning Theorem involves only *Unitarity*, and makes no reference to quantum measurements at all. It is surprising that this theorem, which has nothing to do with the measurement process, could have such a strong bearing on the issue on ontology in quantum mechanics. This second argument is presented in subsection 3.2. We then analyse the projective measurements (section 4), the protective measurements (section 5), a method of cloning which we had named *Information Cloning* (section 7), the weak measurements (section 6), methods of approximate cloning (section 8) for their implications on the question of ontology in quantum mechanics.

3.1 Superposition Principle and Ontology

Let us consider a hypothetical measurement device that is perfectly non-invasive (leaves the system state undisturbed). We can consider the initial unknown system state to be $|\psi\rangle_S$. Since this does not change, we can use a state-vector representation for the system. The treatment of the apparatus will be more subtle. All that the apparatus

is required to do is produce a probability distribution of outcomes which carries complete information about the expectation value of the observable in the system state $|\psi\rangle_S$. Therefore, at least the final state of the apparatus ought to be described by a density matrix. Then one might as well describe the entire history of the apparatus by a density matrix. Because the system stays in the same state throughout, it is consistent to treat the system by a state vector, and the apparatus by a density matrix. The initial state of the system-apparatus composite can be taken to be

$$|\psi\rangle_S \otimes \rho^A(0) \tag{1}$$

Under the measurement $\mathcal{M}$, this goes to

$$|\psi\rangle_S \otimes \rho^A(0) \xrightarrow{\mathcal{M}} |\psi\rangle_S \otimes \rho^A(\langle\psi|O|\psi\rangle_S) \tag{2}$$

The measurement $\mathcal{M}$ not being a *Unitary* process, can take a pure density matrix to a mixed one. The final apparatus (reduced) density matrix is in general mixed. If such a $\mathcal{M}$ could be realized, it can be used as often as necessary to measure all the relevant observables for state tomography of $|\psi\rangle_S$ as the state is left undisturbed.

The map in Equation 2 is not consistent with the principle of linear superpositions of states. That is, if the measurement device works on $|\psi_1\rangle_S$ and $|\psi_2\rangle_S$, it will not work on an arbitrary superposition $\alpha|\psi_1\rangle + \beta|\psi_2\rangle$ i.e. the measurement does not work on an *arbitrary* unknown state. As this is a very important issue, let us state it as precisely as possible. For that, let us describe the system state also by a density matrix $\rho_S = |\psi\rangle\langle\psi|_S$ and recast Equation 2 as

$$\rho_{SA}^{\text{ini}} = \rho_S \otimes \rho^A(0) \xrightarrow{\mathcal{M}} \rho_S \otimes \rho^A(\langle O\rangle_{\rho_S}) \tag{3}$$

The map $\mathcal{M}$ can generically be expressed as

$$\rho_{SA} \xrightarrow{\mathcal{M}} \sum_i M_i \rho_{SA} M_i^\dagger \tag{4}$$

The measurement operations M_i, in the sense of positive-operator valued measures (POVMs), are independent of ρ_{SA}. Therefore, the l.h.s. of Equation 3 is bilinear in the system state $|\psi\rangle_S$. If the system state $|\psi\rangle_S = \sum_i c_i|i\rangle_S$ in some basis $|i\rangle_S$, the l.h.s. is a *quadratic form* in $\text{Re}\,c_i$, $\text{Im}\,c_i$. But the r.h.s. of the same equation is certainly not bilinear in this sense. Thus the conflict with the superposition principle.

This is a very powerful conclusion showing that the principle of linear superposition of states alone is enough to rule out ontology in quantum mechanics and one need not invoke the deep, but confusing, chain of arguments invoked by the founders like Niels Bohr. An explicit realization of this line of thinking is afforded by the measurements discussed in section 5 and section 7. In the case of Protective Measurements, the scheme requires the unknown initial states to be *non-degenerate* eigenstates of a suitable Hamiltonian. A linear superposition of such states is no longer a state of the same type. In the case of Information Cloning, the scheme requires the unknown states to be Coherent States of a Harmonic Oscillator, and again, a superposition of such coherent states is not a coherent state!

3.2 The No-cloning theorem

The No-Cloning theorem [6, 7] is one of the most striking of all results in quantum theory! Invoking nothing more than the inner-product preserving nature of unitary transformations or the superposition principle, it states that no unitary process can ever 'copy' unknown quantum states. In a lighter vein it is said that there are no quantum Xerox machines! We shall first describe the theorem, which is remarkably straightforward considering its profundity.

Consider an unknown state $|\psi\rangle_S$ of some quantum system and N identical copies of another, but *known*, state $|0\rangle_S$ of the same system (it is not really necessary that they be of the same system, though). The latter are also called 'blanks' or 'ancillaries'. A unitary transformation $\mathcal{U}$ acting on the *tensor product* Hilbert space $\mathcal{H}^{N+1}$ is said to be a *universal cloning transformation* if it satisfies

$$\mathcal{U}|\psi\rangle \otimes |0\rangle_1 \otimes \ldots \otimes |0\rangle_N = |\psi\rangle \otimes |\psi\rangle_1 \otimes \ldots \otimes |\psi\rangle_N \quad (5)$$

for every $|\psi\rangle$. The No-cloning theorem is a proof that no such universal unitary transformation can exist. For a proof based only on unitarity of $\mathcal{U}$, consider a second state $|\chi\rangle$ so chosen that $|\langle\chi|\psi\rangle| \neq 0, 1$. Then the effect of $\mathcal{U}$ on $|\chi\rangle$ has to be

$$\mathcal{U}|\chi\rangle \otimes |0\rangle_1 \otimes \ldots \otimes |0\rangle_N = |\chi\rangle \otimes |\chi\rangle_1 \otimes \ldots \otimes |\chi\rangle_N \quad (6)$$

Taking the inner product between these two equations and using unitarity of $\mathcal{U}$, one gets,

$$\langle\chi|\psi\rangle = (\langle\chi|\psi\rangle)^{N+1} \quad (7)$$

But this is possible only if $|\langle\chi|\psi\rangle| = 0, 1$ which contradicts the initial premise about $|\chi\rangle$! The same proof can also be viewed as a consequence of the superposition principle.

What is the relevance of the No-cloning theorem to our discussion of ontology? The point is, that N can be made very very large, at least in principle, either in a single application of the universal cloner or in many cascaded applications of it. Then we can set aside one out of $N + 1$ copies produced, and use the remaining N copies for an *ensemble state determination*. The accuracy of the subsequent state determination can be improved with higher and higher N. One would still be left with one

copy of the original unknown state even if the tomography with the N copies is as invasive as can be.

Thus if an universal cloner existed, one would in effect be able to make a non-invasive measurement on a single copy of an unknown state and still be able to determine its state as accurately as one wishes. It is rather remarkable that this theorem which invokes only aspects of unitary evolutions, with no explicit reference to quantum measurements, nevertheless captures the very essence of quantum measures as per the Copenhagen Interpretation! This deep connection also borders on the mystic.

However, we shall introduce a novel *Information Cloning* which bypasses the no cloning theorem in a subtle way and is a way of getting information on a single copy, albeit with errors that cannot be reduced arbitrarily.

Before proceeding, we wish to highlight some other aspects of the No-cloning theorem. As expressed in Equation 5, the cloning transformation has made additional copies of the unknown state while preserving the original. But for the purposes of ontology, the preservation of the original is quite unnecessary as long as one ends up with sufficient number of copies.

A particularly striking example where the original is totally destroyed but one copy is left behind is in *Quantum Teleportation*! In fact in that case, the copy is created in a way that it is physically separated by distance from the original. Complete destruction of the original with exactly one copy produced also happens in *swapping*, but unlike teleportation a unitary transformation can accomplish that. Though quantum teleportation is not particularly useful in determining the unknown original state, it brings to the fore another aspect of reality, namely, transport of 'real' objects. In classical mechanics for sure, an object that does not have reality, or does not exist, cannot be transported. So it would be legitimate in that context to say that anything that can be transported is also real. In that sense, quantum teleportation would accord some reality to the wavefunction (state). Being a single state there can be no strict ontology though a FAPP ontology would be possible. Since the state is not determined, there is no epistemology either.

4 Projective Measurements and Ontology

Now we analyze why the Projective or Dirac–von Neumann measurements cannot yield any ontology. Even though the arguments are simple and straightforward, we recast them in the language of joint and conditional probabilities so we can use the same framework to address the issue of ontology in other contexts like weak measurements.

In a strong or projective measurement, the state after the first measurement is changed randomly to one of the eigenstates of the observable being measured. The outcome of the apparatus is the corresponding eigenvalue. The fact that a given eigenstate-eigenvalue combination could have resulted from infinitely many unknown initial states makes their reconstruction impossible from the information available after a single such measurement. Such a reconstruction requires an *ensemble* measurements with optimally chosen observables.

If repeated strong measurements are performed on a single copy, the second and all subsequent measurements are eigenstate measurements where the eigenstate in question is the state after the first measurement. Therefore all subsequent measurements leave the system in this same eigenstate and all subsequent apparatus outcomes are exactly the same as the outcome of the first measurement. In other words, they do not generate any additional information required for the state reconstruction. The strong measurements are not only highly invasive, they do not generate any information for determining the state. These are the reasons, within standard quantum mechanics, for the impossibility of an *ontological* description.

Now let us recast these considerations in the language of conditional and joint probabilities of outcomes of repeated measurements on a single copy. Let the observable being measured is S, with the spectrum $s_i, |s_i\rangle_S$. If the initial unknown state of the system is

$$|\psi\rangle_S = \sum_i \alpha_i |s_i\rangle_S \qquad (8)$$

The probability distribution for the outcomes of the first measurement is given by

$$P(p_1) = \sum_i |\alpha_i|^2 \delta(p_1 - s_i) \qquad (9)$$

This says that the first outcome is random with the above distribution. Let the outcome of the second measurement be p_2, and as explained above, it has to be the same as p_1 because it is an eigenstate measurement. Therefore, the probability distribution for p_2 is *conditional* on the outcome p_1. In other words, the *conditional probability distribution* $P(p_2|p_1)$ for the outcome p_2, conditional on the first outcome being p_1 is

$$P(p_2|p_1) = \delta(p_2 - p_1) \qquad (10)$$

The *Joint Probability Distribution* $P(p_2, p_1)$ for the outcomes of the first two of the repeated measurements is now given by

$$
\begin{aligned}
P(p_2, p_1) &= P(p_2|p_1)P(p_1) \\
&= \sum_i |\alpha_i|^2 \delta(p_2 - p_1)\delta(p_1 - s_i) \\
&= \sum_i |\alpha_i|^2 \delta(p_2 - s_i)\delta(p_1 - s_i) \quad (11)
\end{aligned}
$$

It is straightforward to generalize these to the outcomes of N repeated measurements on a single copy:

$$P(p_N, \ldots, p_1) = \sum_i |\alpha_i|^2 \prod_{j=1}^{N} \delta(p_j - s_i) \qquad (12)$$

As usual, it is useful to introduce y_N to be the average of the first N outcomes, and consider its probability distribution $P(y)$ that is

$$y_N = \frac{\sum_i p_i}{N} \qquad (13)$$

and

$$P(y_N) = \int \cdots \int \prod_i dp_i\, P(\{p\})\, \delta(y_N - \frac{\sum_i p_i}{N}) \quad (14)$$

where the notation $\{p\}$ is used to indicate the values of the set of all p-variables. On using Equation 12 and Equation 14, it follows that

$$P(y_N) = \sum_i |\alpha_i|^2 \delta(y_N - s_i) \qquad (15)$$

The repeated measurements have not changed the nature of the distribution at all, and it remains the same as Equation 9! Though our simple reasoning had already told us this, the formalism of conditional and joint probabilities used above will prove to be useful in more complicated situations where there are no such simple reasoning available.

4.1 Sharpening the ontology criterion

The form of Equation 15, derived for Projective Measurements which are decidedly invasive and hence incapable of any ontological descriptions, suggests an even more precise technical criterion for ontology. For that, let us contrast Equation 15 with what one would expect in the case of ensemble measurements on the basis of the Central Limit Theorem:

$$P(y_N) = N\, e^{-\frac{N(y_N - \mu)^2}{\Delta^2}} \qquad (16)$$

This suggests a way to sharpen the criterion for onticity in quantum mechanics, given verbally earlier, to the following precise mathematical criterion: *exact* ontology in quantum mechanics is the ability to find non-invasive measurement schemes such that the mean of the N outcomes of repeated measurements on a *single* copy of a system in an unknown state takes the deterministic form

$$P(y_N) = \delta(y_N - \mu), \qquad \mu = \langle \psi|O|\psi\rangle \qquad (17)$$

Not surprisingly, there will be no candidates within quantum mechanics for this criterion.

The next best possibility will be the *FAPP-Ontology* discussed earlier. The following two criteria provide precise characterizations of such. The first is that the statistics of outcomes of repeated measurements on a single copy will be very similar to that obtained from measurements on an ensemble. In particular, the distribution for the average y_N will be a *single* distribution as in Equation 16, and additionally $\mu = \langle \psi | S | \psi \rangle$. The figures of merit for the FAPP ontology are (i) how close μ actually is to the expectation value, and (ii) how small the error $\epsilon = \frac{\Delta}{\sqrt{N}}$ is.

The second criterion allows for the distribution $P(y_N)$ to deviate from a single distribution but with very small deviations

$$P(y_N) = p_0\, e^{-\frac{(y_N - \langle S \rangle_\psi)^2}{\epsilon^2}} + \sum_i p_i\, e^{-\frac{(y_N - \mu_i)^2}{\epsilon_i^2}} \qquad (18)$$

In this case, the average outcome of repeated measurements will be *random*, and ensemble measurements become a necessity; measurements on a single copy will not reveal *any* information about the unknown state. In the coming sections we shall discuss explicit realizations of these criteria.

5 Protective Measurements and Ontology

Aharonov, Anandan and Vaidman [8, 9] proposed a remarkable type of experiments which they called *Protective Measurements*. They gave an explicit realization for them and showed that for a *restricted* class of states, and in a certain *ideal* limit, one could get full information about single copies of such restricted class of states *without* affecting the state. From whatever we have said so far, such a proposal would realize exact ontology in the ideal limit. Closer examination, however, shows that even these remarkable category of measurements actually provide only FAPP ontology, as the ideal limit requires measurements lasting infinitely long. Now we elaborate on the details.

They consider states that are *non-degenerate* eigenstates of some *unknown* Hamiltonian. For this reason, the states are indeed unknown. Let us briefly review the standard projective measurements to see the differences and commonalities between projective and protective measurements. For every type of measurement it is necessary to characterize the measuring apparatus. Niels Bohr was of the opinion that this necessarily had to be *classical*, whereas Dirac and von Neumann found it desirable to take this also to be a quantum system. It is also important to consider the modern picture of the Dirac–von Neumann Scheme. According to this, the final act of the measurement (the one that breaks the so called *infinite von Neumann regression*) is *environmental decoherence* which accounts for the real life situation that there is a complex environment with which both the system and the apparatus are interacting. This, technically speaking, renders the final density matrix diagonal in an apparatus Hilbert space basis which defines the *Pointer States* for the apparatus. Let R_A be the observable of the apparatus whose eigenstates are the pointer states. In the Dirac–von Neumann measurement theory formalism, one introduces an apparatus operator Q_A that is canonically conjugate to R_A that is $[R_A, Q_A] = \imath \hbar$.

For both types of measurements, the interaction between the 'apparatus' and the system is taken to be described by a Hamiltonian:

$$H_I(t) = g(t) Q_A\, S, \qquad \int g(t) dt = 1 \qquad (19)$$

Here S is the system observable that is being measured and Q_A the observable of the apparatus described above. The integral condition on $g(t)$ is a convenient normalization which can be taken without loss of generality. In addition to this interaction Hamiltonian, the time evolution of both the system and the apparatus are respectively governed by their own Hamiltonians H_A and H_S, respectively.

The projective measurements correspond to an *impulsive* $g(t)$ that is $g(t)$ is non-zero only in a very small time interval $-\frac{\epsilon}{2} < t < \frac{\epsilon}{2}$. The time-evolution unitary transformation taking pre-measurement-interaction states to post-measurement-interaction states is given by

$$U(\frac{\epsilon}{2}, -\frac{\epsilon}{2}) = e^{-\frac{\imath}{\hbar} \int_{-\frac{\epsilon}{2}}^{\frac{\epsilon}{2}} H\, dt} \qquad (20)$$

Normally this unitary transformation is given by *time ordered integral* over the *total* Hamiltonian H(t):

$$H(t) = H_A + H_S + H_I \qquad (21)$$

In the limit of the measurement interaction being extremely impulsive that is $\epsilon \to 0$, the time ordered integral is well approximated by

$$U = e^{-\frac{\imath}{\hbar} Q_A S} \qquad (22)$$

It should be noted that H_A, H_S do not contribute in this impulsive limit (it is understood that these Hamiltonians are bounded). The combined state of the system and apparatus before measurement is taken to be the disentangled state

$$|t_<\rangle = |\nu\rangle_S |\Phi(r_0)\rangle_A \qquad (23)$$

The initial apparatus state, in the Dirac–von Neumann scheme is taken to be an eigenstate of R_A with an eigenvalue, say, r_0; this corresponds to the initial reading of the apparatus. To avoid technical difficulties arising out of the use of continuous variables, the initial apparatus state $|\Phi(r_0)\rangle$ is taken to be a wavepacket *sharply* centered around the value r_0 of R_A.

Here $|v\rangle$ is the unknown system state on which a measurement of the observable S is performed. If $|s_i\rangle$ are the eigenstates of S that is $S|s_i\rangle = s_i|s_i\rangle$, and $|v\rangle = \sum_i \alpha_i |s_i\rangle$ the post-measurement interaction state is given by

$$|t_>\rangle = U|t_<\rangle = \sum_i \alpha_i \, e^{-\frac{i}{\hbar} s_i \, Q_A} \, |s_i\rangle \, |\Phi(r_0)\rangle \qquad (24)$$

As Q_A is canonically conjugate to R_A, the exponential operator shifts the value of R_A by s_i and one gets the *entangled* state

$$|t_>\rangle = \sum_i \alpha_i \, |s_i\rangle \, |\Phi(r_0 + s_i)\rangle \qquad (25)$$

This explicitly manifests the one-one correspondence between the states $|s_i\rangle$ of the system, and the states $|\Phi(r_0 + s_i)\rangle$ of the apparatus. But the state in Equation 25 is *entangled* and it hardly reflects the single outcomes expected of a good measurement! It is instructive to see how decoherence 'solves' this issue; for that, consider the *pure* density matrix corresponding to this state:

$$\rho^{S+A}(t_>) = \sum_{i,j} \alpha_i \, \alpha_j^* \, |s_i\rangle\langle s_j| \, |\Phi(r_0 + s_i)\rangle\langle\Phi(r_0 + s_j)| \quad (26)$$

Clearly this matrix is not diagonal in the pointer basis $|\Phi\rangle$. Decoherence reduces this to the mixed density matrix

$$\rho^{S+A}(t_>) = \sum_i |\alpha_i|^2 \, |s_i\rangle\langle s_i| \, |\Phi(r_0 + s_i)\rangle\langle\Phi(r_0 + s_i)| \quad (27)$$

Though this still does not explain how single outcomes come about, it has at least reduced that to a classical problem of picking from a mixture, much like picking a card out of a deck.

To pictorially contrast the projective and protective cases, we show in Figure 1a the outcomes of a standard Stern-Gerlach experiment viewed as a projective measurement. With this background, it is easy to grasp the essentials of a *Protective Measurement*. The major departure from projective measurements is that now the measurement interaction behaves oppositely to what it did in the case of projective measurements–the interaction time T is now taken to be very large, approaching infinity! Let us leave aside for now questions like the meaning of measurements that take infinitely long, and proceed. It is simplest to take $g(t)$ to be a constant. Then the normalization condition gives $g = \frac{1}{T}$, where T is the long duration of the measurement, which will tend to ∞ in the end. The total Hamiltonian becomes

$$H = H_A + H_S + \frac{1}{T} \, Q_A \, S \qquad (28)$$

which is *time independent*. Again, for simplicity we restrict analysis to the choice $[H_A, Q_A] = 0$ (for a complete discussion of the general situation see [10]). However, even in the standard Stern-Gerlach case, such a simplification does not happen. This condition allows both H_A, Q_A to be simultaneously diagonalized

$$Q_A|a_i\rangle_A = a_i|a_i\rangle_A, \quad H_A|a_i\rangle_A = E_i^A \, |a_i\rangle_A \qquad (29)$$

H_S taken to be *unknown* has the non-degenerate eigenstates $|j\rangle_S$, with eigenvalues ω_j. Because of the simplifying assumptions made, the total Hamiltonian H also commutes with Q_A and both of them can also be diagonalized simultaneously. If we take $H_A|a_i\rangle_A = E_i^A|a_i\rangle_A$, the simultaneous eigenstates of H and Q_A are of the form $|j, i\rangle_S \, |a_i\rangle_A$ with $|j, i\rangle_S$ satisfying

$$(H_S + \frac{1}{T} \, a_i \, S)|j, i\rangle_S = \Omega(j, i)|j, i\rangle_S \qquad (30)$$

It is clear that $\Omega(j, i) \xrightarrow{T\to\infty} \omega_j$ and $|j, i\rangle_S \xrightarrow{T\to\infty} |j\rangle_S$. The eigenvalues and eigenstates of the total Hamiltonian H can now be expressed as

$$\begin{aligned} H|j, i\rangle_S \, |a_i\rangle_A &= E(j, i)|j, i\rangle_S |a_i\rangle_A \\ &= (E_i^A + \Omega(j, i))|j, i\rangle_S |a_i\rangle_A \quad (31) \end{aligned}$$

For very large T, $\Omega(j, i)$ can be calculated in first order perturbation theory to get

$$\Omega(j, i) = \omega_j + \frac{1}{T} a_i \, \langle j|S|j\rangle_S \qquad (32)$$

If the unknown system state before measurement is the non-degenerate eigenstate $|k\rangle_S$ of H_S, the joint state before measurement is taken to be $|k\rangle_S \, |\Phi(r_0)\rangle_A$, with $|\Phi(r_0)\rangle_A$ being the same as what was used in projective measurements.

The joint state after time T is

$$\begin{aligned} |k, T\rangle &= U(T)|k\rangle_S|\Phi(r_0)\rangle_A \\ &= \sum_{i,j} \langle a_i|\Phi(r_0)\rangle_A \, \langle j, i|k\rangle_S \, e^{-\frac{i}{\hbar} TE(j,i)} \, |j, i\rangle_S|a_i\rangle_A \end{aligned}$$

$$\qquad\qquad (33)$$

In first order perturbation theory, $\langle j, i|k\rangle_s = \delta_{j,k}$. Putting everything together

$$|k, T\rangle \xrightarrow{T\to\infty} e^{-\frac{i}{\hbar}\omega_k T}|k\rangle_S e^{-\frac{i}{\hbar}H_A T} e^{-\frac{i}{\hbar}\langle k|S|k\rangle_S \, Q_A}|\Phi(r_0)\rangle_A$$

$$\qquad\qquad (34)$$

In other words

$$|k, T\rangle \to e^{-\frac{i}{\hbar}\omega_k T}|+\rangle e^{-\frac{i}{\hbar}H_A T}|\Phi(r_0 + \langle k|S|k\rangle))_A \qquad (35)$$

Thus under these protective measurements, *the original state is protected and the apparatus reads the expectation value $\langle k|S|k\rangle_S$!* This is modulo the $e^{-\frac{i}{\hbar}H_A T}$ factor.

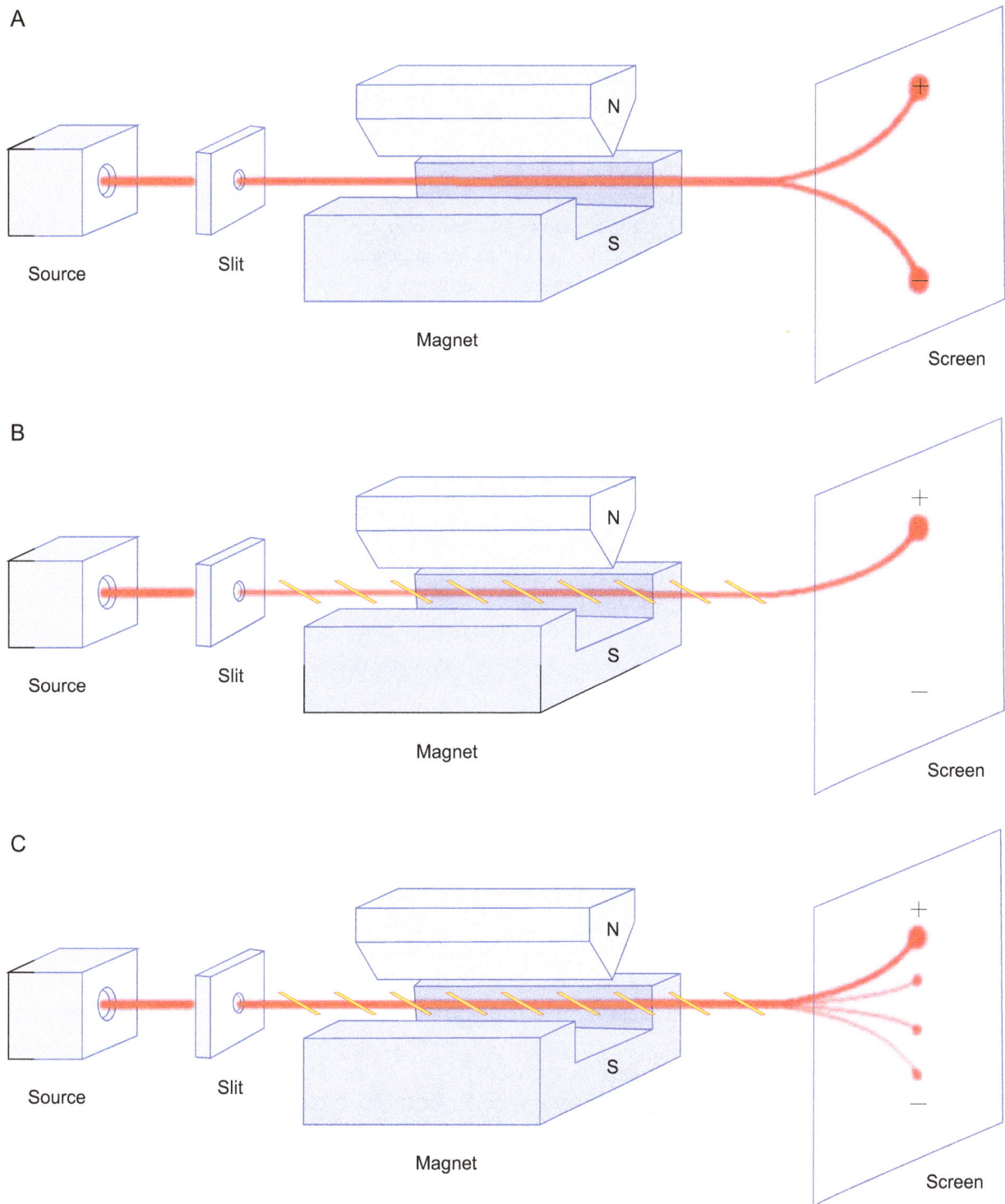

Figure 1: *A. The standard Stern-Gerlach measurement setup denoting a projective measurement on a spin-$\frac{1}{2}$ system. The source emits a collimated beam of silver atoms directed between the poles of a magnet that produces an inhomogeneous magnetic field. Since the silver atoms are neutral they do not experience a Lorentz force. However, silver has in its outer shell an unpaired 5s electron that is in a zero orbital angular momentum state. Thus the deflection by the inhomogeneous magnetic field is due to the spin of the outer electron only. The top (+) and the bottom (−) spots are hit randomly with probabilities given by the initial state. B. The situation to be expected in the case of an ideal protective measurement (that is in the extreme adiabatic limit $T \to \infty$). In this case, the beam hits the screen at only one spot which directly measures the expectation value of the observable in the initial state. The slashes denote the 'protective magnetic field' which has to be present over and above the inhomogeneous magnetic field. C. The situation in the case of the non-ideal protective measurement (that is when T is very large but finite). In this case, while the spot occurring in the ideal case has a very large probability, other cases as enumerated in subsection 5.1 occur with small but non-vanishing probabilities.*

Since the state is 'undisturbed', one can reuse it for carrying out protective measurements of all the necessary observables for complete state determination. The apparatus and the system are *disentangled*, and there is no need to take recourse to decoherence to achieve the final step in the measurement process. This is what can be called the *Ideal* protective measurements, in the sense that it is valid only in the strict $T = \infty$ limit. Figure 1b shows the situation to be expected for an ideal protective Stern-Gerlach experiment. Unlike the standard Stern-Gerlach set up, the silver atoms in an ideal protective measurement would strike the screen at only *one* spot, in between the extreme positions encountered in the standard case. Its location is a precise measure of the *expectation value* of the measured observable in the unknown initial state.

But Equation 35 is precisely the kind that had been argued to be in conflict with the superposition principle in subsection 3.1! The Aharonov–Anandan–Vaidman scheme cleverly evades this by considering the unknown initial states to be non-degenerate eigenstates of H_S; therefore, superpositions of such states can no longer be non-degenerate eigenstates of H_S!

5.1 Non ideal protective measurements

The Ideal case is obviously unphysical as it is meaningless for any measurement to last infinitely long! In real life situations T can be very very large (compared to the time scales involved) but not ∞. One may naively argue that for all practical purposes the difference between such very large T and the ideal limit should be negligible. Indeed, for ensemble measurements the difference between very large T and $T = \infty$ is negligible in the precise sense that the resulting probability distributions for outcomes differ only very slightly, and all the statistical conclusions are not affected significantly.

But for measurements on single copies, which are the only relevant measurements in the context of ontology, the situation is *dramatically* different. In Quantum Mechanics, unlike in the classical counterpart, individual outcomes of measurements are completely random and unpredictable. Even outcomes with hopelessly small probabilities can manifest. Only if their probability is *exactly* zero, will they not show up. This makes a very significant difference for protective measurements. In a nutshell, departure from $T = \infty$ causes a very small but significant entanglement between the system and the apparatus. This can cause the first protective measurement to project the unknown initial state into any state that is orthogonal to it. This way, not only is the state not protected during the first measurement, it renders meaningless the outcome of even protective measurements subsequently. No state reconstruction is possible and there is no strict

ontology. This was the criticism of protective ontology that was made by both [10] as well as by [11].

To address these issues we need to consider all sources of $\frac{1}{T}$ corrections to the ideal results. We refer the reader to [10, 12, 13] for the technical details. Here we shall list the important sources of $\frac{1}{T}$ corrections and discuss their importance. In the sum of Equation 33, one will have to take into account system states $|j \neq k, i\rangle_S$. In order to get the leading $\frac{1}{T}$ corrections, *second order* perturbation theory becomes necessary. This typically introduces corrections of the type $|k'\rangle_S Q_A^2 |\Phi(r_0)\rangle$. Schematically the effect of these corrections can be represented as

$$|T\rangle = |\text{ideal}\rangle + \frac{c}{T}|\text{non-ideal}\rangle \qquad (36)$$

It is important to note that $\langle \text{ideal}|\text{non-ideal}\rangle = 0$ because of the nature of perturbation theory. Now we can further enumerate some possibilities:

- State is protected and apparatus reads the $\langle Q_S \rangle$ in that state with $P = 1 - \frac{c^2}{T^2}$.

- State protected but pointer in *all possible* states with probability $\simeq \frac{1}{T^2}$.

- State collapses to the state *orthogonal* to it and the pointer reads the expectation value in the orthogonal state with probability $\simeq \frac{1}{T^2}$.

- State collapses to the orthogonal but pointer in all possible states with probability $\simeq \frac{1}{T^2}$

This is depicted pictorially in Figure 1c. It is worth emphasizing that in each of these cases, the system state after measurements remains correlated with the original state. This is in sharp contrast to projective measurement where the system state after the measurement is completed has no memory of the original state whatsoever.

5.2 Adiabatic two qubit interactions–another twist

As a further generalization of the protective measurement schema, Anirban Das and myself [12] considered the case where the role of the apparatus is also played by another qubit or by a quantum system with finite dimensional Hilbert space. Let us illustrate with the example of the qubit as a detector. We take the basis states to be $|d_\uparrow\rangle_A$ and $|d_\downarrow\rangle_A$. The system is also taken to be a qubit with its Hilbert space spanned by $|\uparrow\rangle_S$ and $|\downarrow\rangle_S$. The measurement interaction is taken to be represented by the unitary transformation $\mathcal{U}$:

$$|\uparrow\rangle|d_\downarrow\rangle \xrightarrow{\mathcal{U}} |\uparrow\rangle|d_\uparrow\rangle$$
$$|\downarrow\rangle|d_\downarrow\rangle \xrightarrow{\mathcal{U}} |\downarrow\rangle|d_\downarrow\rangle \qquad (37)$$

The components of spin are taken to be along the z-axis for both systems. For projective measurements where there are only two possible outcomes, it suffices to take $|d_\uparrow\rangle, |d_\downarrow\rangle$ as the pointer states. Whether there are any realistic 'environments' that can result in decoherence in this basis is not very clear. For adiabatic measurements where there can be a near-continuum of outcomes, we shall take *angular momentum coherent states* obtained by rotating, say, $|d_\uparrow\rangle$ by θ about the x-axis as the pointer states. Once again the existence of suitable decoherence mechanisms in this basis remains to be understood. More general possibilities for pointer states can also be considered. The interaction Hamiltonian that generates the unitary transformation $\mathcal{U}$ turns out to be (actually there are infinitely many such Hamiltonians!)

$$- \pi g(t) P_{z,+}^S \otimes P_{x,-}^A \qquad (38)$$

Here $P_{a,\pm}$ are the *projection operators* for spin $\pm$ along the a-direction. H_A is taken to be the rotationally invariant $\vec{S}_A \cdot \vec{S}_A$. This, being a constant, does not lead to any pointer state broadening. Let the initial unknown system state be

$$|v\rangle = \alpha |\uparrow\rangle_S + \beta |\downarrow\rangle_S \qquad (39)$$

In the ideal limit, protective measurements of this type maintain the original state and the pointer state is $\theta = \pi |\alpha|^2$. But here too, the non-ideal case is the more realistic and we enumerate the possible outcomes [12, 13].

- After accounting for the relevant $\frac{1}{T}$ corrections also, the dominant outcome is when the original state is protected and the apparatus outcome is the expectation value of the observable in the original state. But unlike the ideal case, the probability of this happening is no longer unity; instead it happens with probability $P \simeq 1 - \frac{c^2}{T^2}$.

- State collapses to its orthogonal; unlike the protective measurements considered so far, the apparatus state now is uniquely determined to be $|d_\downarrow\rangle_x$! This happens with probability $P \simeq \frac{1}{T^2}$.

- State is protected but apparatus again in $|d_\downarrow\rangle_x$, with probability $P \simeq \frac{1}{T^2}$.

- It is intriguing that the 'failed' protective measurements now always produce the same apparatus state $|d_\downarrow\rangle_x$.

We see that because of non-vanishing probabilities for deviations from the ideal case, perfect ontology is not possible. The last point mentioned above (the failed cases coming with a well defined apparatus state) might give hope that the lack of perfect ontology may somehow be overcome by exploiting this feature. Even though the apparatus state, being fixed, does not convey any information about the initial system state, the state of the system after the measurement being just orthogonal to it, carries all information about it. Unfortunately, no universal unitary transformation can transform an unknown initial state of a qubit to its orthogonal state. But what is worse, there is no way to tell, with only single copies, that the protective measurement has actually failed. The reason is that as long as the pointer states are the ones produced by rotating $|d_\downarrow\rangle$ through θ around the x-axis, $|d_\downarrow\rangle_x$ will have to be expressed as $\frac{1}{\sqrt{2}}(|d_\uparrow\rangle + |d_\downarrow\rangle)$. This being a superposition of pointer states, there are finite probabilities for different outcomes, and the failed case will behave as in a projective case. This again precludes any perfect ontological significance to these unknown states.

However, protection fails with very low probability. This means protective measurements can give practically full information about a class of unknown states in such a way as to protect the *purity* of the post-measurement ensemble to a very high degree. Further, a dramatic decrease in the *size* of the ensemble for state tomography is possible. In other words, protective measurements can provide FAPP ontology to an arbitrary degree, and this can be important and highly useful in this pragmatic sense [13, 14] though from a philosophical point of view they cannot deliver the ontological goods. Because of all these interesting aspects, it is critical that they are subjected to a proper experimental study. For some feasible suggestions, the reader is referred to [13].

6 Weak Measurements and Ontology

Now we take up another class of remarkable measurement schemes called *Weak Measurements* and *Weak Value Measurements*. These were also discovered by Aharonov and his collaborators [15, 16]. Let us first dispose off the weak value measurements as they are by design unsuited for ontology. These are also called measurements with *Post-selection*; a post-selection of the system state is made through a projective measurement, following a weak measurement on an initial, possibly unknown state. Obviously, the projective measurements involved in the post-selection stage are invasive on the system. For this reason, this class of measurements cannot have any bearing on the issues of ontology discussed here.

6.1 Weak Measurements Without Post-Selection

On the other hand, if no post-selection is made, removing thereby the invasive elements, weak measurements on their own appear to be ideal for the ontological issues. As their name suggests, they are *minimally invasive*, with this degree of invasiveness apparently under full control. Here too, it is possible to make such measurements both on ensembles and on single copies. We consider only the latter here.

As in section 4, let S be the observable of the system with $s_i, |s_i\rangle_S$ its spectrum, which we take to be *non-degenerate*. The initial states of the system and the apparatus are taken to be *pure* and as in Equation 23. The measurement interactions are also of the form of Equation 19 discussed in section 5. But there is an important difference now in that Q_A need not be as restrictive as in the Dirac–von Neumann measurement schemes.

The *Pointer States* of the apparatus denoted by $|p\rangle_A$, are taken to be eigenstates of an apparatus observable P_A. The point of view taken here is that such pointer states form the basis in which the density matrix becomes diagonal as a result of *decoherence*. They are not always labeled by the mean values of P_A in a given state of the apparatus. Therefore, the specification of an *apparatus* involves some quantum system, along with a decoherence mechanism which picks out the pointer states. The P_A, Q_A pair need not be canonically conjugate. A detailed account of many important aspects of weak measurements can be found in [17]. In what follows we shall nevertheless stick to the canonical pair for convenience.

The initial apparatus states are taken to be *Gaussian states* centered around some p_0. For $p_0 = 0$, we have

$$|\phi_0\rangle_A = N \int dp\, e^{-\frac{p^2}{2\Delta_p^2}} |p\rangle_A, \quad N^2 \sqrt{\pi \Delta_p^2} = 1 \quad (40)$$

In projective measurements, the Gaussians are taken to be very narrow that is $\Delta_p \ll 1$ so that they approximate pointer states to a high degree. In contrast, for weak measurements, $\Delta_p \gg 1$. That means that the initial apparatus state is a *very broad* superposition of pointer states with practically *equal* weight for many pointer states. Though even in the weak case, the initial apparatus state is also peaked at $p_0 = 0$, it is *not* a pointer state. This important point has led to confusing statements in literature.

The measurement interaction is still taken to be *impulsive* that is the function $g(t)$ is non-vanishing only during a very small duration, say, $-\epsilon < t < \epsilon$. We leave out the details (the reader is referred to [18] for them) and give only the essential results. The *post-measurement* density matrix turns out to be (in what follows, we shall use the notation $\{\alpha\}$ to indicate the values of the set of all α-variables):

$$\rho_{SA}^{post} = \int dp\, |N(p,\{\alpha\})|^2 |p\rangle\langle p|_A |\psi(p,\{\alpha\})\rangle\langle\psi(p,\{\alpha\})|_S \quad (41)$$

where

$$N(p,\{\alpha\}) = N \sqrt{\sum_i |\alpha_i|^2 e^{-\frac{(p-s_i)^2}{\Delta_p^2}}} \quad (42)$$

$$|\psi(p,\{\alpha\})\rangle = \frac{N}{N(p,\{\alpha\})} \sum_j \alpha_j e^{-\frac{(p-s_j)^2}{2\Delta_p^2}} |s_j\rangle_S \quad (43)$$

For an *ensemble* of weak measurements, $P(p,\{\alpha\}) = |N(p,\{\alpha\})|^2$ being the probability for outcome p, the mean outcome is

$$\langle p \rangle_\psi = \int dp\, p |N(p,\{\alpha\})|^2 = \sum_i |\alpha_i|^2 s_i \quad (44)$$

yielding the same expectation value as in strong measurements. The variance of the outcomes can be readily calculated to yield

$$(\Delta p)_\psi^2 = (\Delta p)^2 + (\Delta S)^2 \quad (45)$$

This exposes one of the major weaknesses (!) of weak measurements–the errors in individual measurements are huge. This can be reduced statistically as usual. If one considers averages over M_w measurements, the variance in the average, is $\frac{\Delta_p}{\sqrt{2M_w}}$. It makes sense to compare different measurement schemes only for a *fixed* statistical error. Therefore if averaging is done over M_s strong measurements,

$$\frac{\Delta S}{\sqrt{M_s}} = \frac{\Delta_p}{\sqrt{2M_w}} \rightarrow M_w = \left(\frac{\Delta_p}{\Delta S}\right)^2 \frac{M_s}{2} \quad (46)$$

the required resources will be super-massive!

The aspect of weak measurements that has gained great prominence is its alleged *non-invasiveness*. One possible measure of this non-invasiveness is provided by the *post-measurement reduced density matrix* of the system:

$$\rho_S^{post} = \rho^{ini} - \frac{1}{4\Delta_p^2} \sum_{i,j} (s_i - s_j)^2 \alpha_i \alpha_j^* |s_i\rangle\langle s_j| \quad (47)$$

Thus, for very large Δ_p, the reduced density matrix of the system practically equals that of the initial state.

The combination of an *exact* estimate for the expectation value, as given by Equation 44, as well as the maintenance of the state to a very high degree as per Equation 47 may give rise to the expectation that weak measurements may offer the best hopes for ontology in quantum mechanics. What would make such an expectation particularly exciting is that these measurements can be done on *any state*–they appear to offer FAPP ontology for arbitrary states! We investigate this by turning to an analysis of *repeated weak measurements on a single copy* as given in [18] with particular emphasis on ontology.

Two aspects that need to be particularly focused upon in this context are (i) how closely the averages of N outcomes approximate the exact expectation values, and, (ii) how the single state gets degraded as a result of multiple weak measurements.

The following schema defines for us repeated weak measurements of the same observable on a single copy [18]: (i) perform a weak measurement of system observable S in state $|\psi\rangle_S$ with the apparatus in the state of Equation 40 with *very large* Δ_p, (ii) let the definitive outcome, defined as above, be p_1, and the single system state be $|\psi(p_1, \{\alpha\})\rangle_S$, (iii) restore the apparatus to its initial state, (iv) repeat step (i), and so on. After N such steps, let the sequence of outcomes be denoted by $p_1, p_2 \ldots, p_N$ and the resulting system state by $|\psi(\{p\}, \{\alpha\})\rangle_S$.

The probability distribution for the first outcome $p_1, P^{(1)}(p_1)$ is given by

$$N^{(1)}(p_1, \{\alpha\})|^2 = |N(p_1, \{\alpha\})|^2 \tag{48}$$

with $N(p, \{\alpha\})$ given by Equation 42. The corresponding system state is given by $|\psi(p_1, \{\alpha\})\rangle_S$ of Equation 43. Thus the set of α for this state is given by

$$\alpha_i^{(1)} = \frac{N}{N(p_1, \{\alpha\})} e^{-\frac{(p_1 - s_i)^2}{2\Delta_p^2}} \alpha_i \tag{49}$$

Since in step (iii) the apparatus state has been restored, the probability distribution $P^{(2)}(p_2)$ for the outcome p_2 at the end of the second weak measurement, is given by

$$P^{(2)}(p_2) = |N^{(2)}(p_2, \{\alpha\})|^2 = |N^{(1)}(p_2, \{\alpha^{(1)}\})|^2 \tag{50}$$

Substituting from Equation 49, one gets

$$P^{(2)}(p_2) = \frac{(N^2)^2}{P^{(1)}(p_1)} \sum_i |\alpha_i|^2 \prod_{j=1}^{2} e^{-\frac{(p_j - s_i)^2}{\Delta_p^2}} \tag{51}$$

As stressed in [18], $P^{(2)}(p_2)$ is actually the *conditional probability* $P(p_2|p_1)$ of obtaining p_2 conditional to having already obtained p_1 (that is the reason for the explicit dependence on p_1 in Equation 51). The *joint probability distribution* $P(p_1, p_2)$ is therefore given by $P(p_2, p_1) = P(p_2|p_1)P(p_1)$ to give

$$P(p_1, p_2) = (N^2)^2 \sum_i |\alpha_i|^2 \prod_{j=1}^{2} e^{-\frac{(p_j - s_i)^2}{\Delta^2}} \tag{52}$$

The state after the second measurement is given by the exact analog of Equation 49:

$$\alpha_i^{(2)} = \frac{N}{N^{(2)}(p_2, \{\alpha^{(1)}\})} e^{-\frac{(p_2 - s_i)^2}{2\Delta_p^2}} \alpha_i^{(1)} \tag{53}$$

It is useful to explicitly write this state:

$$|\psi(p_1, p_2, \{\alpha\}) = \frac{\sum_i \prod_{j=1}^{2} e^{-\frac{(p_j - s_i)^2}{2\Delta_p^2}} \alpha_i |s_i\rangle_S}{\sqrt{\sum_i |\alpha_i|^2 \prod_{j=1}^{2} e^{-\frac{(p_j - s_i)^2}{\Delta_p^2}}}} \tag{54}$$

It is remarkable that these results are all symmetric in the outcomes p_i. Equation 52 and Equation 53 readily generalize to the case of M repeated measurements:

$$P(p_1, \ldots, p_M) = \left(N^2\right)^M \sum_i |\alpha_i|^2 \prod_{j=1}^{M} e^{-\frac{(p_j - s_i)^2}{\Delta^2}} \tag{55}$$

$$|\psi(p_1, \ldots, p_M, \{\alpha\}) = \frac{\sum_i \prod_{j=1}^{M} e^{-\frac{(p_j - s_i)^2}{2\Delta_p^2}} \alpha_i |s_i\rangle_S}{\sqrt{\sum_i |\alpha_i|^2 \prod_{j=1}^{M} e^{-\frac{(p_j - s_i)^2}{\Delta_p^2}}}} \tag{56}$$

6.2 Consequences for ontology

The *intrinsic randomness* of quantum theory makes no aspect of a *particular realization* predictable. For ensemble measurements the variables are *independently* distributed and the *Central Limit Theorem* guarantees that as long as the number of trials is large enough, averages over even particular realizations converge nicely to the true mean. To see what happens in the present context, where the outcomes are clearly not independently distributed, let us study y_M, the average of M outcomes. The expectation value of y_M in the joint probability distribution $P(p_1, \ldots, p_M)$ is

$$\bar{y}_M = \frac{1}{M} \int \cdots \int \prod_{i=1}^{M} \sum_i p_i P(\{p\}) = \sum_i |\alpha_i|^2 s_i \tag{57}$$

Which is certainly a remarkable result. With this, the repeated weak measurements on a single copy certainly pass one critical requirement for ontology. The variance in y_M can likewise be calculated and it equals $\frac{\Delta_p}{\sqrt{2M}}$. This makes it appear that in principle the errors can be reduced arbitrarily, reminding one of the situation in protective measurements, except that now no restrictions need be placed on the initial states! But such appearances turn out to be highly misleading.

As argued before the crux of the ontology issue lies in the distribution function $P(y_M)$, and not just in its mean and variance. As shown in [18], the distribution function $P(y_M)$ can itself be calculated explicitly. This is in spite of the outcomes not being independently distributed. The result is

$$P(y_M) = \int \ldots \int \prod_{i=1}^{M} dp_i \, P(\{p\}) \delta(y_M - \frac{\sum_i p_i}{M}) \quad (58)$$

Using Equation 55, this becomes

$$P(y_M) = \sqrt{\frac{M}{\pi \Delta_p^2}} \sum_i |\alpha_i|^2 e^{-\frac{(y_M - s_i)^2 M}{\Delta_p^2}} \to \sum_i |\alpha_i|^2 \delta(y_M - s_i) \quad (59)$$

where we have also displayed the limiting behavior as $M \to \infty$.

This, as per our discussions earlier, immediately negates not just ontology but even FAPP ontology! In other words, the distribution of y_M is not only not peaked at the true average, with errors decreasing as $M^{-1/2}$, it is actually a weighted sum of sharp distributions peaked around *the eigenvalues*, exactly as in the strong measurement case. This means that averages over outcomes of a particular realization will be eigenvalues, occurring randomly but with probability $|\alpha_i|^2$. It then follows that averages over outcomes of a particular realization do not give any information about the initial state, precisely as in the case of the invasive strong measurements where there can clearly be no ontology! Ensemble measurements again become inevitable.

The other issue to be settled in this context is whether the repeated measurements on single copies are invasive or not. It turns out that a very large number of repeated weak measurements on a single copy has the same invasive effect as a strong measurement. This can be seen by examining the expectation value of the system reduced density matrix, $\rho_>^{\mathrm{rep}}$:

$$\rho_>^{\mathrm{rep}} = \rho - \sum_{i,j} \alpha_i \alpha_j^* (1 - e^{-\frac{M(s_i - s_j)^2}{4\Delta_p^2}}) |s_i\rangle\langle s_j| \quad (60)$$

It is seen that as M gets larger and larger, there is significant change in the system state. In the limit $M \to \infty$, the off-diagonal parts of the density matrix get completely quenched, as in decoherence, and the density matrix takes the diagonal form in the eigenstate of S basis:

$$\rho_>^{\mathrm{rep}} \to \sum_i |\alpha_i|^2 |s_i\rangle\langle s_i| \quad (61)$$

which is exactly the post-measurement density matrix in the case of a strong measurement! The sequence of system states of Equation 56 is a *random walk* on the state space of the system (see also [19]). It follows from Equation 43 that the eigenstates of S are the *fixed points* of the probabilistic map that generates this walk. Presumably each walk terminates in one of the eigenstates but which eigenstate it terminates in is unpredictable.

6.3 Other equivalent results

Alter and Yamamoto have obtained a number of very significant results about the possibility of obtaining information about single quantum systems [3, 20, 21]. In particular they also gave an analysis based on joint and conditional probabilities applied to *repeated weak quantum non-demolition* measurements on a single state [21]. They too obtained evolutions resembling random walks in state space. They concluded that it is not possible to obtain any information on unknown single states from the statistics of repeated measurements. The degradation of the state and relation to projective measurements were not explicitly studied. In another work, they found connections between *Quantum Zeno Effect* and the problem of repeated measurements and again concluded that it is impossible to determine the quantum state of a single system. Our results on information cloning and the general results from optimal cloning discussed in the next two sections show that it may be possible to obtain partial results.

In a very interesting approach to these ontological questions, Paraoanu has investigated these issues within what he calls *partial measurements* [22, 23]. By employing a combination of repeated such measurements on a single state and the possibility of reversing such measurements, he too has concluded the impossibility of obtaining any information about single unknown states. The invasive aspects as well as the connections to strong measurements are not explored here either.

7 Information Cloning and Ontology

As we saw in subsection 3.2, a subtle inner consistency of quantum theory prevents determining the unknown state of a single copy by trying to make many clones of it. We had, however, proposed what we called *information cloning* in [24]. The main idea was to make many copies of an unknown state which are however not identical to the original state, but contain the same amount of *information* as the original. Now we discuss the implications for ontology of such a cloning scheme [25].

The details of how this type of cloning can be used to determine the state of a *single unknown coherent state of quantum harmonic oscillators* can be found in [25]. In the case of coherent states of harmonic oscillators (say, in one dimension), complete information about the state is contained in a single complex coherency parameter α. Thus by information cloning what we mean is the ability to make arbitrary number of copies of coherent states whose coherency parameter is $c(N)\alpha$ where α is the coherency parameter of the unknown coherent state and $c(N)$ is a known constant depending on the number of copies made.

To this end consider $1 + N$ systems of harmonic oscillators whose creation and annihilation operators are the set $(a, a^\dagger), (b_k, b_k^\dagger)$ (where the index k takes on values $1, .., N$). The a oscillators represent the original unknown state, and the b oscillators represent the information clones. These operators satisfy the commutation relations

$$[a, a^\dagger] = 1 \tag{62}$$
$$[b_j, b_k^\dagger] = \delta_{jk} \tag{63}$$
$$[a, b_k] = 0 \tag{64}$$
$$[a^\dagger, b_k] = 0 \tag{65}$$

Coherent states parametrized by the complex number α are given by

$$|\alpha\rangle = D(\alpha)|0\rangle \tag{66}$$

where $|0\rangle$ is the ground state and the unitary operator $D(\alpha)$ is given by

$$D(\alpha) = e^{\alpha a^\dagger - \alpha^* a} \tag{67}$$

We view the information cloning to be a unitary process. The initial composite state can be taken to be a *disentangled* state containing the unknown initial coherent state and some *known* states of the b-oscillators. It turns out to be best to take them also to be coherent states. In other words, the state before information cloning is taken as $|\alpha\rangle|\beta_1\rangle_1|\beta_2\rangle_2...|\beta_N\rangle_N$, where α is *unknown* while β_i are *known* to very high accuracy. Consider the action of the unitary transformation

$$U = e^{t(a^\dagger \otimes \sum_j r_j b_j - a \otimes \sum_j r_j b_j^\dagger)} \tag{68}$$

The most general unitary transformation of this type would involve complex r_j's. But this can be reduced to the present form through suitable redefinitions of the phases of the creation and annihilation operators [24]. Of course, such redefinitions should maintain the algebra of Equation 62. The process implemented by this unitary transformation is well known in optics and is called the *beam splitter*. But it is very important to appreciate that what we are dealing with here is when this acts on a *single photon* state, a circumstance in which the notion of a *beam* is neither meaningful nor useful. By an application of the Baker–Campbell–Hausdorff identity and the fact that $U|0\rangle|0\rangle_1..|0\rangle_N = |0\rangle|0\rangle_1..|0\rangle_N$ it follows that the resulting state is also a *disentangled* set of coherent states expressed by

$$|\alpha'\rangle|\beta_1'\rangle_1..|\beta_N'\rangle_N = U|\alpha\rangle|\beta_1\rangle_1..|\beta_N\rangle_N \tag{69}$$

In other words, the unitary transformation U acting on various coherent states induces another unitary transformation $\mathcal{U}$ among the coherency parameters. Details can

be found in [24]; we merely give the final result and discuss its physical implications. Let us define

$$a(t) = UaU^\dagger \tag{70}$$
$$b_j(t) = Ub_jU^\dagger \tag{71}$$

The explicit form of the transformation induced on the parameters (α, β_j) can be represented by the matrix $\mathcal{U}$ as

$$\alpha_a(t) = \mathcal{U}_{ab}\alpha_b. \tag{72}$$

where we have introduced the notation α_a with $a = 1, ..., N + 1$ such that

$$\alpha_1 = \alpha$$
$$\alpha_k = \beta_{k-1} \quad k \geq 2 \tag{73}$$

Then we have

$$\mathcal{U}_{1a} = \left(\cos Rt \quad \frac{r_1}{R}\sin Rt \quad .. \quad .. \quad \frac{r_N}{R}\sin Rt \right) \tag{74}$$

where $R = \sqrt{\sum_j r_j^2}$ and

$$\mathcal{U}_{ab} = -\frac{r_{a-1}}{R} \sin Rt\, \delta_{b1} + (1 - \delta_{b1})M_{a-1,b-1} \tag{75}$$

where Equation 75 is defined for $a \geq 2$. Equivalently

$$\mathcal{U} = \begin{pmatrix} \cos Rt & \frac{r_1}{R}\sin Rt & .. & .. & \frac{r_N}{R}\sin Rt \\ -\frac{r_1}{R}\sin Rt & M_{11} & .. & .. & M_{1N} \\ .. & & .. & .. & \\ .. & & .. & .. & \\ -\frac{r_N}{R}\sin Rt & M_{N1} & .. & .. & M_{NN} \end{pmatrix} \tag{76}$$

It is best to choose $\{\beta_i, r_i\}$ in such a way that all $\beta_i(t)$ become identical and we get N identical copies. This happens only when $r_i = r, \beta_i = \beta$. In that case we have

$$\beta_i(t) = -\frac{\alpha}{\sqrt{N}} \sin Rt + \beta \cos Rt \tag{77}$$

There is still the freedom to choose Rt. Let us first consider the choice of $\sin Rt = -1$ which gives N copies of the state $|\frac{\alpha}{\sqrt{N}}\rangle$. This is what we called *information cloning* in [24] as the states $|\frac{\alpha}{\sqrt{N}}\rangle$ and $|\alpha\rangle$ have the same information content. This particular choice of Rt will be seen to be optimal in the sense that it gives the least variance in the estimation of α. In this case the value of β is immaterial.

It is easily seen that Equation 7 does not pose any difficulties for information cloning, as it did for universal cloning! Now we can address the ontology issue by attempting to use the N-information clones for an ensemble determination of the information-clone state first, and then a state determination of the original unknown state subsequently, by using the fact that the information clone has the same information content as the original. More specifically, we can use the N copies of $|\frac{\alpha}{\sqrt{N}}\rangle$ to make *ensemble measurements* to estimate $\frac{\alpha}{\sqrt{N}}$ and α.

One can already sense some limitations of the method: usually, the statistical errors can be made arbitrarily small by making the ensemble size larger and larger. However, in our proposal even though the number of copies N can be made arbitrarily *large*, at least in principle, the coherency parameter given by $\frac{\alpha}{\sqrt{N}}$ becomes *arbitrarily small* while the *uncertainties* in α, being characteristic of coherent states, remain the same as in the original state. We now address the question as to how best the original state can be reconstructed.

On introducing the *Hermitian* momentum and position operators $\hat{p}, \hat{x}$ through

$$\hat{x} = \frac{(a + a^\dagger)}{\sqrt{2}} \tag{78}$$

$$\hat{p} = \frac{(a - a^\dagger)}{\sqrt{2}i} \tag{79}$$

the *probability distributions* for position and momentum in the coherent state $|\frac{\alpha}{\sqrt{N}}\rangle$ are given by

$$|\psi_{\text{clone}}(x)|^2 = \frac{1}{\sqrt{\pi}} e^{-(x - \sqrt{\frac{2}{N}}\alpha_R)^2}$$

$$|\psi_{\text{clone}}(p)|^2 = \frac{1}{\sqrt{\pi}} e^{-(p - \sqrt{\frac{2}{N}}\alpha_I)^2} \tag{80}$$

Let us distribute our N-copies into two groups of $\frac{N}{2}$ each and use one to estimate α_R through position measurements and the other to estimate α_I through momentum measurements. Let y_N denote the average value of the position obtained in $\frac{N}{2}$ measurements and let z_N denote the average value of momentum also obtained in $\frac{N}{2}$ measurements. The *central limit theorem* states that the probability distributions for y_N, z_N are given by

$$f_x(y_N) = \sqrt{\frac{N}{2\pi}} e^{-\frac{N}{2}(y_N - \sqrt{\frac{2}{N}}\alpha_R)^2}$$

$$f_p(z_N) = \sqrt{\frac{N}{2\pi}} e^{-\frac{N}{2}(z_N - \sqrt{\frac{2}{N}}\alpha_I)^2} \tag{81}$$

It is more instructive to recast these as the probability distributions for $\bar{\alpha}_{R,N}, \bar{\alpha}_{I,N}$, the average over N measurements of α_R, α_I:

$$f_R(\bar{\alpha}_{R,N}) = \frac{1}{\sqrt{\pi}} e^{-(\bar{\alpha}_{R,N} - \alpha_R)^2}$$

$$f_I(\bar{\alpha}_{I,N}) = \frac{1}{\sqrt{\pi}} e^{-(\bar{\alpha}_{I,N} - \alpha_I)^2} \tag{82}$$

Thus the original unknown α is correctly estimated, in the sense that the above distributions peak precisely at the coherency parameter α of the original state. But this is not enough and one needs to know the reliability of this estimate. For that one needs the variances. The variances for α_N are easily found out from Equation 82:

$$\Delta\alpha_{R,N} = \Delta\alpha_{I,N} = \frac{1}{\sqrt{2}} \tag{83}$$

Thus, while the statistical error in usual measurements goes as $\frac{1}{\sqrt{N}}$, and can be made arbitrarily small by making N large enough, information cloning gives an error that is fixed and equal to the variance associated with the original unknown state. For coherent states with large enough α, even these errors are quite reasonable. Another figure of merit, the so called *Fidelity* has also been adopted in [24, 26–30]. That fidelity for information cloning works out to $\frac{1}{2}$ [24], the maximum possible for *Gaussian Cloning* [26–33]. Therefore, fidelity on its own may give an unnecessarily pessimistic picture. Comparison between information cloning and optimal cloning mentioned above will again be made in section 8.

Thus we have shown that even when the coherent state is *unknown* single state, information cloning will allow its determination, but with fixed statistical errors. Nevertheless, it is a great improvement from not being able to know anything at all about the unknown state.

A comparison with our technical criteria for ontology reveals that again there is no perfect ontology but indeed there is FAPP ontology of the first kind. In contrast, protective measurements gave a FAPP ontology of the second kind. In the protective case the FAPP ontology could approach perfect ontology arbitrarily close, but never equal it. In both cases, one had to restrict the classes of states for which they would work and the restricted class did not allow linear superpositions.

8 Approximate Cloning and Ontology

Though the No-cloning theorem forbids making perfect clones of an unknown state, there seems nothing against making imperfect copies. The information cloning of the previous section was a particularly interesting variant of this theme. Then the obvious question is the closeness to perfect cloning that can be achieved. There has been an explosion of interest in this question and the reader is referred to [31] for a comprehensive review. We shall only examine the so called *optimal cloning* [26–30, 32–35] (see [31] for a review), from the ontological point of view. The details are not that critical to understanding the broad implications and chief conclusions.

In these implementations, one starts with the *original unknown state* $|\alpha\rangle$ belonging to the Hilbert space $\mathcal{H}_A$, a number of ancillary states, also known as *blank states*, $|b_0\rangle, |b_1\rangle ... |b_N\rangle$. The ancillaries are *known* states. This is the general setup for all cloning processes. The ancillaries belong to the Hilbert spaces $\mathcal{H}_{B_i}$; each of them is isomorphic to $\mathcal{H}_A$. Unlike the information cloning case, a number of additional states called machine states, also known, $|m_0\rangle, |m_1\rangle ..... |m_M\rangle$ all belonging to the Hilbert spaces isomorphic to, say, $\mathcal{H}_M$, are also considered. The combined Hilbert space has the structure $\mathcal{H}_A \otimes \mathcal{H}_M \otimes \prod_i \mathcal{H}_{B_i}$.

A *general cloning transformation* $\mathcal{T}$ has the effect

$$|\alpha\rangle \prod_0^N |b_i\rangle \prod_0^M |m_j\rangle \xrightarrow{\mathcal{T}} \sum_{i,j,k} d_{ijk} |a_i\rangle \prod_j^M |\beta_j\rangle \prod_k^N |\gamma_k\rangle \quad (84)$$

Such a general cloning is said to be *optimal* if it satisfies the two conditions: (i) *all* the reduced density matrices ρ_{i_0} obtained by tracing over the $\mathcal{H}_A$ states, the machine states and all the blank states except those belonging to $\mathcal{H}_{B_{i_0}}$, are all *identical* and (ii) each of them has *maximum* overlap with the original unknown state $|\alpha\rangle$ that is with the maximum possible value of $\langle \alpha | \rho_{i_0} | \alpha \rangle$. The reduced density matrices are in general *mixed*.

In the case of information cloning, the clones were all disentangled and one could use all of them at a time for carrying out measurements of one's choice. In the case of optimal cloning, in general the clones could be in entangled states. Depending on such details, it could even be that that at any given time it is possible to realize only a few of the reduced matrices ρ_i as different values of i require tracing over different states.

As can be gathered from [31] and the many references there, there are various types of optimal clonings. But for the ontological questions, only a part of them are of interest. Firstly, we need only look at the so called *universal* types as these can produce clones of unknown states. The so called *state dependent* cloning is not of interest. The information cloning that we discussed earlier is state dependent in one way as it can work only with coherent states, but it is also somewhat universal in the sense that the input state can be *any* coherent state. In fact, it is a particular case of Gaussian cloning [26–33]. Secondly, even among the universal optimal cloning there are results for the so called $N \rightarrow M$ type clonings. Here N is the number of copies of the unknown initial state (usually pure) and M the number of clones (usually mixed). For our ontological considerations, only $1 \rightarrow M$ types are relevant.

Let us first consider the case where the input Hilbert space is *finite dimensional*, and specifically consider only cubits. We shall only look at a few illustrative aspects. For qubits, the fidelity F, which is the overlap of the clone

with the original, is, given by

$$F(N, M) = \frac{MN + M + N}{M(N + 2)} \quad (85)$$

for the $N \rightarrow M$ case. The clone state is of the form (with $tr \rho \cdot \rho^\perp = 0$),

$$\rho^{\text{clone}} = F \rho^{\text{ini}} + (1 - F) \rho^\perp \quad (86)$$

The accuracy of the state determination with the clones requires as large a M as possible. Therefore, for $N = 1$ and $M \rightarrow \infty$, one has $F = \frac{2}{3}$. In fact, for an arbitrary M,

$$F(1, M) = \frac{2M + 1}{3M} \quad (87)$$

The largest value, for the non-trivial case $1 \rightarrow 2$ is $\frac{5}{6}$. But with only two clones the errors in the state determination are very high. But as M is increased, to get more accurate state determination, F decreases, reaching the limiting value of $\frac{2}{3}$. In that case though the errors are very small, the estimates for expectation values of observables deviates significantly from the true values. For example, for observables O with zero expectation values in $\rho^\perp$, one finds

$$\langle O \rangle_{\text{clone}} = \frac{2}{3} \langle O \rangle_{\text{true}} \quad (88)$$

failing even the first criterion for ontology rather poorly. Unlike the information cloning case, where the error was *independent* of the input state, here the error is a finite fraction ($\frac{1}{3}$) of the expectation value. The resources required even to reach this are impractically large ($M \rightarrow \infty$).

Of course, it is inappropriate to use the results obtained for optimal cloning of qubits to make a comparison with information cloning which is really a case of infinite-dimensional Hilbert space. But results are also available for optimal cloning for arbitrary, but finite dimensional Hilbert space. As an intermediary to considering continuous variable cloning, let us consider Werner's results [33] for d-dimensional Hilbert spaces. The formula for the fidelity of $N \rightarrow M$ cloning is

$$F(N, M) = \frac{N}{M} + \frac{(M - N)(N + 1)}{M(N + d)} \quad (89)$$

The clone state is given by

$$\rho^{\text{clone}} = \eta(N, M) \rho^{\text{ini}} + (1 - \eta(N, M)) \frac{I}{d} \quad (90)$$

where

$$\eta(N, M) = \frac{N}{M} \frac{M + d}{N + d} \quad (91)$$

Let us look at the continuous case by letting $d \to \infty$ first. While the fidelity approaches the limit $\frac{N}{M}$, the density matrix formula is much more tricky. Now if apply this formula for fidelity to $N = 1$, $M \to \infty$ limit relevant for our ontological concerns, we see that the fidelity vanishes!

This is because of the attempt to find a universal cloner for continuous variable case. Let us lower the expectations and consider only coherent states. It has been shown that the fidelity is *bounded* by

$$F(N, M) \leq \frac{MN}{MN + M - N} \tag{92}$$

The clone state is a mixture of coherent states centered around the unknown initial coherent state. Its explicit form is given by (see eqn.(53) of [31], but watch for a typo!):

$$\rho^{\text{clone}}(\alpha) = \frac{1}{\pi\sigma(N, M)^2} \int d^2\beta \, e^{-\frac{|\beta|^2}{\sigma(N,M)^2}} |\alpha + \beta\rangle\langle\alpha + \beta| \tag{93}$$

where $\sigma(N, M)$ stands for

$$\sigma(N, M)^2 = \frac{1}{N} - \frac{1}{M} \geq 0 \tag{94}$$

Returning to the ontology issue, we set $N = 1$. It is easy to verify that the mean values of x and p in the clone state of Equation 93 are exactly the same as in the unknown original coherent state. This was so in the case of information cloning too. But the variances in x and p for the clone state turn out to be

$$(\Delta x)^2_{\text{clone}} = \frac{1}{2} + \sigma(1, M)^2 = (\Delta p)^2_{\text{clone}} \tag{95}$$

Like the information cloning case, these variances are the same for all coherent states. But irrespective of M, the variances are worse here than there. Again there is only FAPP ontology, of a somewhat worse quality.

8.1 Probabilistic Cloning

What was described till now can be called *deterministic* cloning. There are also probabilistic cloning machines. The reader is referred to [36] to get an understanding of these. Many features and implementations are different and these cloning devices are very interesting. But from our ontological perspective, the situation is not too different; again the mean values can approach the true values and the errors cannot be completely eliminated. One can ascribe a FAPP ontology with figures of merit determined by both of these.

9 Conclusions

In this paper we have carefully examined the issue of obtaining information about the state of a single quantum system. We have equated the ability to obtain such information with the concept of *ontology* in quantum mechanics. We have given a precise technical characterization of this concept and examined the implications of a large variety of quantum measurements including projective measurements, protective measurements, weak measurements (including weak quantum non-demolition measurements) and the so called partial measurements. We have also examined the issue in the light of the no-cloning theorem on the one hand, and in the light of a variety of cloning techniques.

The impossibility of gaining information about a single quantum state is considered to be the basic tenet of quantum mechanics. Admittedly, it was based on the picture of quantum measurements that dominated during the early development of quantum theory. Central to that line of thinking were the highly invasive nature of the eigenvalue-eigenstate based projective measurements. In view of the highly invasive nature of such measurements, that thinking seemed almost obvious. But what is surprising now is that when even novel measurement schemes like weak measurements, partial measurements are around, which make such a tenet far from obvious, it still remains rock solid. Now the results that even these seemingly non-invasive measurement schemes simply cannot coax any information out of generic single states make this lack of ontology deep and perplexing, as if they are the foundational principles of quantum theory. Nevertheless, that schemes like protective measurements, information cloning in particular and optimal cloning in general exist to provide a silver lining in the form of what we have called FAPP ontology is also equally perplexing. What general principles are lurking behind these is something that all those trying to fathom the depths of quantum theory will be eagerly searching for.

Acknowledgments

I thank Tata Institute of Fundamental Research Centre for Interdisciplinary Sciences, Hyderabad for its warm hospitality during which this work was completed. I am thankful to the late Jeeva Anandan for many discussions close to the theme of this article and Tabish Qureshi for raising very interesting questions about completely invasive measurements which could reveal unknown states, and about possible implications of teleportation for ontology. I acknowledge support from the Department of Science and Technology to the project IR/S2/PU-001/2008.

References

[1] Dirac PAM. The Principles of Quantum Mechanics, 1st edition. Oxford: Oxford University Press, 1930.

[2] Dirac PAM. The Principles of Quantum Mechanics, 2nd edition. Oxford: Oxford University Press, 1934.

[3] Alter O, Yamamoto Y. Quantum Measurement of a Single System. New York: Wiley, 2001.

[4] Wiseman HM, Milburn GJ. Quantum Measurement and Control. Cambridge: Cambridge University Press, 2010.

[5] Hari Dass ND. The superposition principle in quantum mechanics–did the rock enter the foundation surreptitiously? http://arxiv.org/abs/1311.4275

[6] Wootters WK, Zurek WH. A single quantum cannot be cloned. Nature 1982; 299 (5886): 802–803. http://dx.doi.org/10.1038/299802a0

[7] Dieks D. Communication by EPR devices. Physics Letters A 1982; 92 (6): 271–272. http://dx.doi.org/10.1016/0375-9601(82)90084-6

[8] Aharonov Y, Vaidman L. Measurement of the Schrödinger wave of a single particle. Physics Letters A 1993; 178 (1-2): 38–42. http://dx.doi.org/10.1016/0375-9601(93)90724-E http://arxiv.org/abs/hep-th/9304147

[9] Aharonov Y, Anandan J, Vaidman L. Meaning of the wave function. Physical Review A 1993; 47 (6): 4616–4626. http://dx.doi.org/10.1103/PhysRevA.47.4616

[10] Hari Dass ND, Qureshi T. Critique of protective measurements. Physical Review A 1999; 59 (4): 2590–2601. http://dx.doi.org/10.1103/PhysRevA.59.2590

[11] Alter O, Yamamoto Y. Protective measurement of the wave function of a single squeezed harmonic-oscillator state. Physical Review A 1996; 53 (5): R2911–R2914. http://dx.doi.org/10.1103/PhysRevA.53.R2911

[12] Das A, Hari Dass ND. An alternate model for protective measurements of two-level systems. http://arxiv.org/abs/quant-ph/0410098

[13] Hari Dass ND. Experiments for realising pragmatic protective measurements. AIP Conference Proceedings 2011; 1384 (1): 51–58. http://dx.doi.org/10.1063/1.3635843

[14] Hari Dass ND. Cold atoms for testing quantum mechanics and parity violation in gravitation. http://arxiv.org/abs/quant-ph/9908085

[15] Aharonov Y, Albert DZ, Vaidman L. How the result of a measurement of a component of the spin of a spin-1/2 particle can turn out to be 100. Physical Review Letters 1988; 60 (14): 1351–1354. http://dx.doi.org/10.1103/PhysRevLett.60.1351

[16] Aharonov Y, Vaidman L. Properties of a quantum system during the time interval between two measurements. Physical Review A 1990; 41 (1): 11–20. http://dx.doi.org/10.1103/PhysRevA.41.11

[17] Kofman AG, Ashhab S, Nori F. Nonperturbative theory of weak pre- and post-selected measurements. Physics Reports 2012; 520 (2): 43–133. http://dx.doi.org/10.1016/j.physrep.2012.07.001 http://arxiv.org/abs/1109.6315

[18] Hari Dass ND. Repeated weak measurements on a single copy are invasive. http://arxiv.org/abs/1406.0270

[19] Korotkov AN. Continuous quantum measurement of a double dot. Physical Review B 1999; 60 (8): 5737–5742. http://dx.doi.org/10.1103/PhysRevB.60.5737

[20] Alter O, Yamamoto Y. Quantum Zeno effect and the impossibility of determining the quantum state of a single system. Physical Review A 1997; 55 (4): R2499–R2502. http://dx.doi.org/10.1103/PhysRevA.55.R2499

[21] Alter O, Yamamoto Y. Inhibition of the measurement of the wave function of a single quantum system in repeated weak quantum nondemolition measurements. Physical Review Letters 1995; 74 (21): 4106–4109. http://dx.doi.org/10.1103/PhysRevLett.74.4106

[22] Paraoanu GS. Partial measurements and the realization of quantum-mechanical counterfactuals. Foundations of Physics 2011; 41 (7): 1214–1235. http://dx.doi.org/10.1007/s10701-011-9542-7 http://arxiv.org/abs/1105.2021

[23] Paraoanu GS. Extraction of information from a single quantum. Physical Review A 2011; 83 (4): 044101. http://dx.doi.org/10.1103/PhysRevA.83.044101

[24] Hari Dass ND, Ganesh P. Information cloning of harmonic oscillator coherent states. Pramana - Journal of Physics 2002; 59 (2): 263–267. `http://dx.doi.org/10.1007/s12043-002-0116-2` `http://arxiv.org/abs/quant-ph/0202020`

[25] Hari Dass ND. Unknown single oscillator coherent states do have statistical significance. `http://arxiv.org/abs/1005.4486`

[26] Cerf NJ, Ipe A, Rottenberg X. Cloning of continuous quantum variables. Physical Review Letters 2000; 85 (8): 1754–1757. `http://dx.doi.org/10.1103/PhysRevLett.85.1754`

[27] Cerf NJ, Iblisdir S. Optimal N-to-M cloning of conjugate quantum variables. Physical Review A 2000; 62 (4): 040301. `http://dx.doi.org/10.1103/PhysRevA.62.040301`

[28] Braunstein SL, Cerf NJ, Iblisdir S, van Loock P, Massar S. Optimal cloning of coherent states with a linear amplifier and beam splitters. Physical Review Letters 2001; 86 (21): 4938–4941. `http://dx.doi.org/10.1103/PhysRevLett.86.4938`

[29] Bruß D, Macchiavello C. Optimal state estimation for d-dimensional quantum systems. Physics Letters A 1999; 253 (5-6): 249–251. `http://dx.doi.org/10.1016/S0375-9601(99)00099-7` `http://arxiv.org/abs/quant-ph/9812016`

[30] Acín A, Latorre JI, Pascual P. Optimal generalized quantum measurements for arbitrary spin systems. Physical Review A 2000; 61 (2): 022113. `http://dx.doi.org/10.1103/PhysRevA.61.022113`

[31] Scarani V, Iblisdir S, Gisin N, Acín A. Quantum cloning. Reviews of Modern Physics 2005; 77 (4): 1225–1256. `http://dx.doi.org/10.1103/RevModPhys.77.1225`

[32] Lindblad G. Cloning the quantum oscillator. Journal of Physics A: Mathematical and General 2000; 33 (28): 5059–5076. `http://dx.doi.org/10.1088/0305-4470/33/28/310`

[33] Keyl M, Werner RF. Optimal cloning of pure states, testing single clones. Journal of Mathematical Physics 1999; 40 (7): 3283–3299. `http://dx.doi.org/10.1063/1.532887` `http://arxiv.org/abs/quant-ph/9807010`

[34] Fiurášek J. Optical implementation of continuous-variable quantum cloning machines. Physical Review Letters 2001; 86 (21): 4942–4945. `http://dx.doi.org/10.1103/PhysRevLett.86.4942`

[35] Grosshans F, Grangier P. Quantum cloning and teleportation criteria for continuous quantum variables. Physical Review A 2001; 64 (1): 010301. `http://dx.doi.org/10.1103/PhysRevA.64.010301` `http://arxiv.org/abs/quant-ph/0012121`

[36] Duan L-M, Guo G-C. Probabilistic cloning and identification of linearly independent quantum states. Physical Review Letters 1998; 80 (22): 4999–5002. `http://dx.doi.org/10.1103/PhysRevLett.80.4999` `http://arxiv.org/abs/quant-ph/9804064`

17

Morlet Wavelets in Quantum Mechanics

John Ashmead

School of Engineering and Applied Science, University of Pennsylvania, Philadelphia, Pennsylvania, United States.
E-mail: john.ashmead@timeandquantummechanics.com

Editors: *Boaz Tamir, José R. Croca & Danko Georgiev*

Wavelets offer significant advantages for the analysis of problems in quantum mechanics. Because wavelets are localized in both time and frequency they avoid certain subtle but potentially fatal conceptual errors that can result from the use of plane wave or δ function decomposition. Morlet wavelets in particular are well-suited for this work: as Gaussians, they have a simple analytic form and they work well with Feynman path integrals. But to take full advantage of Morlet wavelets we need to supply an explicit form for the inverse Morlet transform and a manifestly covariant form for the four-dimensional Morlet wavelet. We construct both here. Quanta 2012; 1: 58–70.

1 Introduction

Wavelet transforms represent a natural development of Fourier transforms and may be used for similar purposes. Where the Fourier transform lets us decompose a wave function into its component plane waves, a wavelet transform lets us decompose a wave function into its component wavelets. If we think of the plane waves as corresponding to pure tones, we may think of the wavelets as corresponding to the notes produced by physical instruments: of finite duration and spanning a finite range of tones. Wavelets have two advantages over plane waves. First, they are localized in time and frequency. This can make them a better fit to the wave forms found in nature, which are always localized in both time and frequency. As a result, wavelet series will often converge faster than corresponding Fourier series. Second, there are many different wavelets to choose from. Therefore, we can tailor our wavelets to our problem. These advantages have resulted in their application to a wide variety of practical problems in acoustics, astronomy, medical imaging, computer graphics, meteorology, and so on. Morlet's original references are [1, 2]. Wavelets are discussed in Chui, Meyer, and other texts [3–9]. Wavelets also have significant if less numerous applications on the theory side: canonical quantization of the electromagnetic field using a discrete wavelet basis [10], analysis of localization properties of photons using windowed wavelets [11], regularization of Euclidean field theories [12], and use of wavelets to provide Lorentz covariant, singularity-free, finite energy, zero action, localized solutions to the wave equation [13].

Wavelets offer significant benefits for the study of foundational questions in quantum mechanics as well. We will focus here specifically on Morlet wavelets. These are Gaussians, so are both easy to work with and a natural fit to path integrals [14–21], which typically consist of long series of Gaussian integrations. Use of Morlet wavelets can let us (1) avoid any need to invoke the problematic collapse of the wave function in the analysis of the Stern-Gerlach experiment, (2) avoid the use of artificial convergence factors or Wick rotation in computing path integrals, and (3) compute path integrals in a time symmetric way. But to prepare Morlet wavelets for their new responsibilities we need to (1) supply an explicit form for the 'admissibility constant' that is needed to define the inverse Morlet transform, and (2) provide a manifestly covariant extension of Morlet wavelets to four dimensions.

2 Three applications of Morlet wavelets

2.1 Analyzing the Stern-Gerlach experiment

In the original Stern-Gerlach experiment [22–24] a beam of silver atoms is sent through an inhomogeneous magnetic field. The beam is split into two: those atoms with spin up getting a kick in one direction; those with spin down in the opposite. This was striking first because it demonstrated the existence of spin and secondly because a classical system would have shown a continuous range of values for the spin, not just up and down. This split is regarded as a classic demonstration of the measurement problem, explained in the Copenhagen interpretation [25] as a collapse of the wave function into up and down components.

In the Stern-Gerlach experimental the initial wave function is typically modeled as the product of a plane wave and a spin vector. Replacing the plane waves with Gaussian test functions provides a more physically realistic model. Gondran and Gondran [26] have looked at the time evolution of such Gaussian wave functions in a Stern-Gerlach apparatus. They show that when the wave function is modeled with Gaussian test functions, the spin up and spin down components split without any need to invoke a collapse. It works a bit like a diffraction experiment: there is coherent interference at two spots, incoherent at the rest. One may think of this as an internal diffraction effect.

Gondran and Gondran intended their work at least partly in support of the Bohm interpretation; however the math is independent of the interpretation. The implication is that – at least in this case – there is no need to invoke the highly problematic [27–30] collapse of the wave function.

The use of a single Gaussian test function is not of itself general. But with the use of the Morlet wavelet transform we can write an arbitrary square-integrable wave function as a sum over Gaussian test functions, making the Gondran and Gondran approach completely general.

To be sure, we could attempt to restore the honor of the plane wave by arguing that we could build up a Gaussian test function as a sum over such. But then why not eliminate the middleman and start with Gaussian test functions?

There are several related analyses of the Stern-Gerlach effect. Cruz-Barrios and Gómez-Camacho [31, 32] argue that we can explain the effect by modeling the atom with coherent internal states, whereas Venugopalan and coauthors [33–35] argue the effect is a result of decoherence. The Gondran and Gondran result is simpler in that it posits no additional structure (coherent internal states) or additional interaction (decoherence); standard quantum mechanics of its own suffices.

2.2 Ensuring convergence of path integrals

Morlet wavelets can assist in establishing convergence of Feynman path integrals without recourse to convergence factors as used in [14,15] or Wick rotation as in [20]; convergence of the slice-by-slice integrals in the path integral is a side-effect of the initial wave function being composed of Gaussians, for which convergence is automatic. It is sufficient to examine the free case.

We start with the free Schrödinger equation:

$$\imath \frac{d}{d\tau} \psi_\tau (\vec{x}) = -\frac{1}{2m} \nabla^2 \psi_\tau (\vec{x}) \tag{1}$$

The path integral expression for the kernel is given by:

$$K_\tau (\vec{x}; \vec{x}') = \lim_{N \to \infty} \left(\frac{m}{2\pi\imath\hbar\varepsilon} \right)^{\frac{3N}{2}}$$
$$\times \int d\vec{x}_1 \ldots d\vec{x}_{N-1} e^{\frac{\imath\varepsilon}{\hbar} \frac{m}{2} \sum_{j=0}^{N-1} \left(\frac{\vec{x}_{j+1} - \vec{x}_j}{\varepsilon} \right)^2} \tag{2}$$

A typical integral is:

$$K_j \left(\vec{x}_{j+1}; \vec{x}_{j-1} \right) = \left(\frac{m}{2\pi\imath\hbar\varepsilon} \right)^{\frac{3}{2}} \int d\vec{x}_j e^{\frac{\imath\varepsilon}{\hbar} \frac{m}{2} \left[\left(\frac{\vec{x}_{j+1} - \vec{x}_j}{\varepsilon} \right)^2 + \left(\frac{\vec{x}_j - \vec{x}_{j-1}}{\varepsilon} \right)^2 \right]} \tag{3}$$

where $\varepsilon \equiv \frac{\tau}{N}$ is the width of a time slice.

The typical integral does not converge. We can force convergence by adding a small imaginary part $\imath\sigma$ to the mass: $m \to m + \imath\sigma$. Equivalently we could add a small imaginary part to the time step: $\varepsilon \to \varepsilon - \imath\sigma$. Or we could rotate time in the complex plane: $t \to \imath t$.

Now, focus attention on the first step:

$$\psi_1 (\vec{x}_1) = \int d\vec{x}_0 K_1 (\vec{x}_1; \vec{x}_0) \psi_0 (\vec{x}_0) \tag{4}$$

Assume the initial wave function is a Gaussian:

$$\psi_1 (\vec{x}_1) = \left(\frac{m}{2\pi\imath\hbar\varepsilon} \right)^{\frac{3}{2}} \left(\pi\sigma^2 \right)^{-\frac{3}{4}} \int d\vec{x}_0 e^{\frac{\imath\varepsilon}{\hbar} \frac{m}{2} \left(\frac{\vec{x}_1 - \vec{x}_0}{\varepsilon} \right)^2 - \frac{(\vec{x}_0 - \langle \vec{x}_0 \rangle)^2}{2\sigma^2}} \tag{5}$$

This integral is convergent of itself. The result is a (slightly wider) Gaussian. We can do an infinite series of these, with the initial wave function showing an increasing amount of middle-aged spread but with all of the integrals converging.

As an arbitrary wave function may be written, via Morlet wavelets, as a sum over Gaussian test functions, we have convergence in the general case, without the introduction of artificial convergence factors.

2.3 Summing path integrals in a time symmetric way

One immediate benefit of not needing convergence factors or Wick rotation is that we can treat time in a more symmetric way. One case where we might want to do this is in setting up a path integral analysis of the Stückelberg-Schrödinger equation:

$$\imath \frac{d\psi_u(x)}{du} = H\psi_u(x) \tag{6}$$

Here u is a formal parameter, a scalar of some kind – perhaps the particle's proper time – and H is a Lorentz invariant Hamiltonian. There are examples in Feynman [36,37] and more recently in work by Land, Horwitz, and Seidewitz [21, 38, 39]. This approach has been sufficiently interesting that there are regular conferences held by the International Association for Relativistic Dynamics (IARD) devoted to this and related questions.

In the free case H might be given by:

$$H = -\frac{1}{2m}\left(\imath\frac{\partial}{\partial x^\mu}\right)\left(\imath\frac{\partial}{\partial x_\mu}\right)$$

$$= \frac{1}{2m}\left(\frac{\partial^2}{\partial t^2} - \frac{\partial^2}{\partial x^2} - \frac{\partial^2}{\partial y^2} - \frac{\partial^2}{\partial z^2}\right) \tag{7}$$

Note that because of the Lorentz invariance the time and space parts enter into H with opposite sign, so in the path integral will have a problem converging in a Lorentz covariant way.

Path integral form for the kernel:

$$K_\tau(x;x') = \lim_{N\to\infty}\left(\frac{m}{2\pi\sqrt{\imath\hbar\varepsilon}}\right)^{2N}\int dt_1 d\vec{x}_1\ldots dt_{N-1}d\vec{x}_{N-1}$$

$$\times e^{-\frac{\imath\varepsilon}{\hbar}\frac{m}{2}\sum_{j=0}^{N-1}\left[\left(\frac{t_{j+1}-t_j}{\varepsilon}\right)^2 - \left(\frac{\vec{x}_{j+1}-\vec{x}_j}{\varepsilon}\right)^2\right]} \tag{8}$$

The pre-factor for the time part is the complex conjugate of the pre-factor for the space part:

$$\left(\frac{m}{2\pi\imath\hbar\varepsilon}\right)^{\frac{1}{2}} \to \left(\frac{\imath m}{2\pi\hbar\varepsilon}\right)^{\frac{1}{2}} \tag{9}$$

A typical slice:

$$K_j\left(\vec{x}_{j+1};\vec{x}_{j-1}\right) = \left(\frac{m}{2\pi\sqrt{\imath\hbar\varepsilon}}\right)^2\int d\vec{x}_j$$

$$\times e^{-\frac{\imath\varepsilon}{\hbar}\frac{m}{2}\left[\left(\frac{t_{j+1}-t_j}{\varepsilon}\right)^2 - \left(\frac{\vec{x}_{j+1}-\vec{x}_j}{\varepsilon}\right)^2 + \left(\frac{t_j-t_{j-1}}{\varepsilon}\right)^2 - \left(\frac{\vec{x}_j-\vec{x}_{j-1}}{\varepsilon}\right)^2\right]} \tag{10}$$

But now the addition of a small imaginary part to mass or time fails; any change that causes the time integrals to converge will cause the space integrals to diverge and vice versa. If we use different signs for time and space, then we break covariance.

Wick rotation fails for the same reason. Here we rotate time in the complex plane: $t \to \pm \imath t$. But if we pick one sign, the integral over the past will diverge; the other, the integral over the future.

Again, look at the first step:

$$\psi_1(t_1,\vec{x}_1) = \int dt_0 d\vec{x}_0 K_1(t_1,\vec{x}_1;t_0,\vec{x}_0)\psi_0(t_0,\vec{x}_0) \tag{11}$$

Assume our initial wave function is given by a Gaussian test function:

$$\psi_1(t_1,\vec{x}_1) = \left(\frac{m}{2\pi\sqrt{\imath\hbar\varepsilon}}\right)^2\frac{1}{\pi\sigma^2}\int dt_0 d\vec{x}_0$$

$$\times e^{-\frac{\imath\varepsilon}{\hbar}\frac{m}{2}\left[\left(\frac{t_1-t_0}{\varepsilon}\right)^2 - \left(\frac{\vec{x}_1-\vec{x}_0}{\varepsilon}\right)^2\right] - \frac{t_0^2 + (\vec{x}_0 - \langle\vec{x}_0\rangle)^2}{2\sigma^2}} \tag{12}$$

Now the integrals converge step by step. As any square-integrable wave function may be written as a sum over such (see below) we have convergence. Of course to do this, we need covariant Morlet wavelets (see further below).

In many cases, an asymmetric treatment of time is harmless. But if we are analyzing time itself, then we do not want to wire the assumption that it is asymmetric into the maths. To do so would result in circular reasoning. The use of small imaginary factors or Wick rotation will not work for an analysis that is of time itself, as such approaches implicitly prejudge the conclusion.

3 Morlet wavelets in one dimension

We will first review the Morlet wavelet transform, then show how to compute the inverse Morlet wavelet transform explicitly.

To generate a set of wavelets we start with a mother wavelet $\phi(t)$. We get the general wavelet $\phi_{sl}(t)$ by scaling the mother wavelet by a scale factor s and displacing her by a displacement l:

$$\phi_{sl}(t) \equiv |s|^{-\frac{1}{2}}\phi\left(\frac{t-l}{s}\right) \tag{13}$$

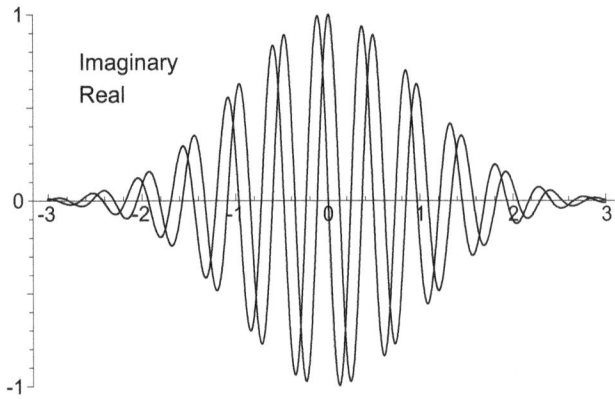

Figure 1: *Real and imaginary parts of a mother Morlet wavelet* $\phi(t)$ *given by Equation 14 with* $f = 13$.

For Morlet wavelets the mother wavelet is given by:

$$\phi(t) = \left(e^{-\imath f t} - e^{-\frac{1}{2}f^2}\right)e^{-\frac{1}{2}t^2} \tag{14}$$

The second term is needed to satisfy the admissibility condition, discussed below. The parameter t is often the time and f may then be thought of as a reference frequency.

In some practical applications (i.e. [40]) the second term is dropped. However it is needed in general to ensure convergence of the inverse Morlet wavelet transform. The exact value of f does not matter in principle, provided it is non-zero. If f is zero, the mother wavelet is zero (and useless). We keep f a variable to help in calculating the value of the admissibility constant C_f.

The general Morlet wavelet is created from the mother Morlet wavelet by scaling by s and displacing by l:

$$\phi_{sl}(t) \equiv |s|^{-\frac{1}{2}} \left(e^{-\imath f\left(\frac{t-l}{s}\right)} - e^{-\frac{1}{2}f^2}\right)e^{-\frac{1}{2}\left(\frac{t-l}{s}\right)^2} \tag{15}$$

Both scale s and displacement l run from $-\infty$ to ∞. A negative scale s gives the complex conjugate of the Morlet wavelet with positive scale:

$$\phi_{-s,l}(t) = \phi_{s,l}^*(t) \tag{16}$$

The mother Morlet wavelet herself is given in this notation as the Morlet wavelet with scale factor one, displacement zero:

$$\phi(t) = \phi_{1,0}(t) \tag{17}$$

Any square integrable function ψ may be expressed as a sum over Morlet wavelets. In principle this excludes δ functions and plane waves. We will see below they are handled correctly however. The Morlet wavelet transform of a wave function ψ is given by:

$$\tilde{\psi}_{sl} = \int_{-\infty}^{\infty} dt \phi_{sl}^*(t)\,\psi(t) \tag{18}$$

We get the original ψ back by integrating over the displacement and scale:

$$\psi(t) = \frac{1}{C_f} \int_{-\infty}^{\infty} \frac{ds}{s^2} \int_{-\infty}^{\infty} dl \phi_{sl}(t)\,\tilde{\psi}_{sl} \tag{19}$$

The admissibility constant is given by an integral over the square of the Fourier transform of the mother wavelet:

$$C_f \equiv 2\pi \int_{-\infty}^{\infty} \frac{d\omega}{|\omega|}\left|\hat{\phi}(\omega)\right|^2 \tag{20}$$

In the general case we could use a different set of wavelets for the forward and the inverse transforms; it is one of the attractions of Morlet wavelets that we do not need to do this.

The wavelet decomposition fails if C_f is not finite. For C_f to be finite, we see we need the zero frequency component of the Fourier transform of the Morlet wavelet mother to be zero:

$$\hat{\phi}(0) = 0 \Rightarrow \int_{-\infty}^{\infty} dt \phi(t) = 0 \tag{21}$$

The Fourier transform of the Morlet mother is:

$$\hat{\phi}(\omega) = \left(e^{f\omega} - 1\right)e^{-\frac{1}{2}(f^2+\omega^2)} \tag{22}$$

By inspection, we see the zero frequency component is zero. As noted above, the second term of the Morlet mother wavelet was included precisely to ensure this.

The Fourier transform of the general Morlet wavelet is:

$$\hat{\phi}_{sl}(\omega) = |s|^{\frac{1}{2}} e^{\imath l\omega}\left(e^{fs\omega} - 1\right)e^{-\frac{1}{2}(f^2+s^2\omega^2)} \tag{23}$$

It may be written in terms of the Fourier transform of the mother:

$$\hat{\phi}_{sl}(\omega) = |s|^{\frac{1}{2}} e^{\imath l\omega}\hat{\phi}(s\omega) \tag{24}$$

3.1 Normalization

Morlet wavelets are not wave functions, but do not object to being treated as such. Their normalization is independent of their scale and displacement:

$$\int dt \phi_{sl}^*(t)\,\phi_{sl}(t) = \sqrt{\pi}\left(e^{-f^2} - 2e^{-\frac{3}{4}f^2} + 1\right) \tag{25}$$

We can therefore write normalized Morlet wavelets as:

$$\phi_{sl}^{(norm)}(t) = \left[\sqrt{\pi} \left(e^{-f^2} - 2e^{-\frac{3}{4}f^2} + 1 \right) \right]^{-\frac{1}{2}} \phi_{sl}(t) \quad (26)$$

3.2 Resolution of unity

We can establish the completeness of the wavelet transform by very general methods, see [5].

But if we are only concerned with Morlet wavelets, we can take advantage of their specific character to give a less general but more immediate proof.

If we substitute the integral for $\tilde{\psi}_{sl}$ (Equation 18) in the integral for the inverse Morlet wavelet transform (Equation 19) we get:

$$\psi(t) = \frac{1}{C_f} \int_{-\infty}^{\infty} \frac{dsdldt'}{s^2} \phi_{sl}(t) \phi_{sl}^*(t') \psi(t') \quad (27)$$

This will be true if we have:

$$\delta(t - t') = \frac{1}{C_f} \int_{-\infty}^{\infty} \frac{dsdl}{s^2} \phi_{sl}(t) \phi_{sl}^*(t') \quad (28)$$

This looks like a familiar decomposition in terms of a set of states weighted by s^{-2}. If we can show this directly, we will have a resolution of unity. To do this, we define the integral:

$$I(t,t') \equiv \frac{1}{C_f} \int_{-\infty}^{\infty} \frac{dsdl}{s^2} \phi_{sl}(t) \phi_{sl}^*(t') \quad (29)$$

We wish to show that this integral gives the δ function. We write the Morlet wavelets in terms of their Fourier transforms to get:

$$I(t,t') = \frac{1}{C_f} \int_{-\infty}^{\infty} \frac{dsdl}{s^2} \int \frac{d\omega}{\sqrt{2\pi}} e^{-\iota\omega t} \hat{\phi}_{sl}(\omega)$$
$$\times \int \frac{d\omega'}{\sqrt{2\pi}} e^{\iota\omega' t'} \hat{\phi}_{sl}^*(\omega') \quad (30)$$

Then we write the Fourier transforms of the wavelets in terms of the Fourier transform of the mother wavelet (Equation 24):

$$\frac{1}{C_f} \int_{-\infty}^{\infty} \frac{dsdl}{|s|} \int \frac{d\omega}{\sqrt{2\pi}} e^{-\iota\omega(t-l)} \hat{\phi}(s\omega)$$
$$\times \int \frac{d\omega'}{\sqrt{2\pi}} e^{\iota\omega'(t'-l)} \hat{\phi}^*(s\omega') \quad (31)$$

We recognize the integral over l as a δ function in ω and ω'.

$$\int \frac{dl}{2\pi} e^{\iota(\omega-\omega')l} = \delta(\omega - \omega') \quad (32)$$

We use this hitherto disguised δ function to do the integral over ω'.

$$\frac{1}{C_f} \int \int \frac{dsd\omega}{|s|} e^{-\iota\omega t} e^{\iota\omega t'} \hat{\phi}(s\omega) \hat{\phi}^*(s\omega) \quad (33)$$

We break the integral up into positive and negative s parts:

$$\frac{1}{C_f} \int_0^{\infty} \frac{ds}{s} \int_{-\infty}^{\infty} d\omega e^{-\iota\omega t} e^{\iota\omega t'} \hat{\phi}(s\omega) \hat{\phi}^*(s\omega)$$
$$+ \frac{1}{C_f} \int_{-\infty}^0 \frac{ds}{|s|} \int_{-\infty}^{\infty} d\omega e^{-\iota\omega t} e^{\iota\omega t'} \hat{\phi}(s\omega) \hat{\phi}^*(s\omega) \quad (34)$$

In the second term, replace s by $-s$ and flip the sense of the integration:

$$\frac{1}{C_f} \int_0^{\infty} \frac{ds}{s} \int_{-\infty}^{\infty} d\omega e^{-\iota\omega t} e^{\iota\omega t'} \hat{\phi}(s\omega) \hat{\phi}^*(s\omega)$$
$$+ \frac{1}{C_f} \int_0^{\infty} \frac{ds}{s} \int_{-\infty}^{\infty} d\omega e^{-\iota\omega t} e^{\iota\omega t'} \hat{\phi}(-s\omega) \hat{\phi}^*(-s\omega) \quad (35)$$

We change the variable of integration to $s' = s\omega$, then combine the two terms:

$$\frac{1}{C_f} \int_0^{\infty} \frac{ds'}{s'} \int_{-\infty}^{\infty} d\omega e^{-\iota\omega t} e^{\iota\omega t'} \left[\hat{\phi}(s') \hat{\phi}^*(s') + \hat{\phi}(-s') \hat{\phi}^*(-s') \right]$$
$$\quad (36)$$

We identify the ω integration as still another δ function, one which can come outside of the integrals:

$$2\pi\delta(t - t') \frac{1}{C_f} \int_0^{\infty} \frac{ds'}{s'} \left[\hat{\phi}(s') \hat{\phi}^*(s') + \hat{\phi}(-s') \hat{\phi}^*(-s') \right]$$
$$\quad (37)$$

We replace s' by $-s'$ in the second term:

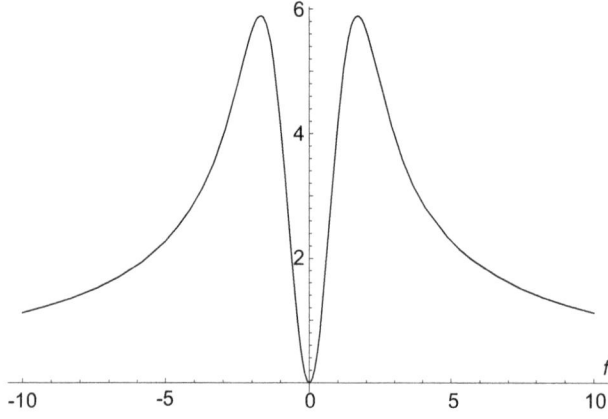

Figure 2: *The admissibility constant C_f as a function of f.*

when f is zero. As ω goes to zero, the integrand goes as $f^2\omega$ so is well-behaved in the small ω limit. And as ω goes to ∞ the integrand goes as $\frac{e^{2f\omega - \omega^2}}{|\omega|}$ so is also well-behaved in the large ω limit. Therefore we can write $I(f)$ as:

$$I(f) = \int_0^f df \frac{dI(f)}{df} \tag{45}$$

The advantage of taking the derivative with respect to f is that it gets rid of the troubling factor of $|\omega|$ in the denominator. We break up the integral over ω into negative and positive parts:

$$I(f) = \int_{-\infty}^0 \frac{d\omega}{|\omega|} \left(e^{\omega f} - 1\right)^2 e^{-\omega^2} + \int_0^\infty \frac{d\omega}{|\omega|} \left(e^{\omega f} - 1\right)^2 e^{-\omega^2} \tag{46}$$

then change variables from $\omega \rightarrow -\omega$ in the negative part to get:

$$I(f) = \int_0^\infty \frac{d\omega}{|\omega|} \left[\left(e^{-\omega f} - 1\right)^2 + \left(e^{\omega f} - 1\right)^2\right] e^{-\omega^2} \tag{47}$$

The derivative of I with respect to f is:

$$\frac{dI(f)}{df} = 2 \int_0^\infty d\omega \left[\left(e^{2\omega f} - e^{-2\omega f}\right) - \left(e^{\omega f} - e^{-\omega f}\right)\right] e^{-\omega^2} \tag{48}$$

which can be re-written as:

$$\frac{dI(f)}{df} = 4 \int_0^\infty d\omega \left[\sinh(2\omega f) - \sinh(\omega f)\right] e^{-\omega^2} \tag{49}$$

After integration with respect to ω we have:

$$\frac{dI(f)}{df} = 2 \sqrt{\pi} \left[e^{f^2} \operatorname{erf}(f) - e^{\frac{1}{4}f^2} \operatorname{erf}\left(\frac{f}{2}\right)\right] \tag{50}$$

We integrate this with respect to f to get a pair of generalized hypergeometric functions $_2F_2$:

$$I(f) = f^2 \left[_2F_2\left(1, 1; \frac{3}{2}, 2; f^2\right) - _2F_2\left(1, 1; \frac{3}{2}, 2; \frac{f^2}{4}\right)\right] \tag{51}$$

Therefore, we have for C_f:

$$\int_0^\infty \frac{ds'}{s'} \left[\hat{\phi}(-s') \hat{\phi}^*(-s')\right] \tag{38}$$

$$= -\int_{-\infty}^0 \frac{ds'}{s'} \left[\hat{\phi}(s') \hat{\phi}^*(s')\right] \tag{39}$$

$$= \int_{-\infty}^0 \frac{ds'}{|s'|} \left[\hat{\phi}(s') \hat{\phi}^*(s')\right] \tag{40}$$

Giving for the integral:

$$\int_{-\infty}^\infty \frac{ds'}{|s'|} \hat{\phi}(s') \hat{\phi}^*(s') \tag{41}$$

Which is $\frac{C_f}{2\pi}$ (see Equation 20) so we have, as required:

$$I(t, t') = \delta(t - t') \tag{42}$$

3.3 Calculation of admissibility constant

To actually use the inverse Morlet transform we need an explicit expression for the value of the admissibility constant. By substituting the Fourier transform of the mother Morlet wavelet (Equation 22) in the formula for the admissibility constant (Equation 20) we get:

$$C_f = 2\pi e^{-f^2} \int_{-\infty}^\infty \frac{d\omega}{|\omega|} \left(e^{\omega f} - 1\right)^2 e^{-\omega^2} \tag{43}$$

For convenience, we define a new integral:

$$I(f) \equiv \int_{-\infty}^\infty \frac{d\omega}{|\omega|} \left(e^{\omega f} - 1\right)^2 e^{-\omega^2} \tag{44}$$

For $f = 0$, we find $I(f) = 0$ by inspection. This is expected given that the original mother wavelet is zero

$$C_f = 2\pi e^{-f^2} f^2 \left[{}_2F_2\left(1, 1; \frac{3}{2}, 2; f^2\right) \right.$$
$$\left. - {}_2F_2\left(1, 1; \frac{3}{2}, 2; \frac{f^2}{4}\right) \right] \tag{52}$$

For $f = 1$ we obtain $C_1 \approx 4.1636$. This can be checked by doing the original integral numerically, see subsection 6.2.

For small f, C_f goes as:

$$\lim_{f \to 0} C_f \to 2\pi f^2 \tag{53}$$

For large f, C_f goes as:

$$\lim_{f \to \infty} C_f \to e^{-f^2} \tag{54}$$

At this point we have an explicit form for the inverse Morlet transform, so have reached our objective. We now apply the Morlet wavelet transform to some interesting cases.

3.4 Gaussian test functions

Gaussian test functions (squeezed states) are the most important case:

$$\psi_{\sigma E\tau}(t) = \left(\pi\sigma^2\right)^{-\frac{1}{4}} e^{-\imath E(t-\tau) - \frac{(t-\tau)^2}{2\sigma^2}} \tag{55}$$

The Fourier transform of this Gaussian test function is:

$$\hat{\psi}_{\sigma E\tau}(\omega) = \left(\frac{\sigma^2}{\pi}\right)^{\frac{1}{4}} e^{\imath\omega\tau - \frac{(E-\omega)^2\sigma^2}{2}} \tag{56}$$

3.4.1 Analysis

Per (Equation 18), the Morlet wavelet transform of a Gaussian test function is:

$$\tilde{\psi}_{sl}^{(\sigma E\tau)} = \int_{-\infty}^{\infty} dt\, \phi_{sl}^*(t)\, \psi_{\sigma E\tau}(t) \tag{57}$$

By inspection, we can write the Morlet wavelet in the transform as the sum of two Gaussians:

$$\phi_{sl}^*(t) = \pi^{\frac{1}{4}} \left(\psi_{s\frac{f}{s}l}^*(t) - e^{-\frac{f^2}{2}} \psi_{s0l}^*(t) \right) \tag{58}$$

This means the transform reduces to a pair of Gaussian integrations:

$$\tilde{\psi}_{sl}^{(\sigma E\tau)} = \int_{-\infty}^{\infty} dt'\, |s|^{-\frac{1}{2}} \left[e^{\imath f\frac{(t'-l)}{s}} - e^{-\frac{f^2}{2}} \right] e^{-\frac{1}{2}\left(\frac{t'-l}{s}\right)^2}$$
$$\times \left(\pi\sigma^2\right)^{-\frac{1}{4}} e^{-\imath E(t'-\tau) - \frac{(t'-\tau)^2}{2\sigma^2}} \tag{59}$$

The integral is elementary, giving:

$$\tilde{\psi}_{sl}^{(\sigma E\tau)} = \left(\frac{2\sqrt{\pi}\sigma|s|}{s^2+\sigma^2}\right)^{\frac{1}{2}} \left\{ e^{-\imath\frac{fs(l-\tau)}{s^2+\sigma^2} - \frac{1}{2}\frac{(Es-f)^2\sigma^2}{s^2+\sigma^2}} - e^{-\frac{1}{2}\left[f^2 + \frac{(Es\sigma)^2}{s^2+\sigma^2}\right]} \right\}$$
$$\times e^{-\imath E\frac{\sigma^2(l-\tau)}{s^2+\sigma^2} - \frac{1}{2}\frac{(l-\tau)^2}{s^2+\sigma^2}} \tag{60}$$

The Morlet transform looks like a sum of inner products of Gaussians:

$$\tilde{\psi}_{sl}^{(\sigma E\tau)} = \pi^{\frac{1}{4}} \left(\left\langle \psi_{s\frac{f}{s}l} \middle| \psi_{\sigma E\tau} \right\rangle - e^{-\frac{f^2}{2}} \left\langle \psi_{s0l} \middle| \psi_{\sigma E\tau} \right\rangle \right) \tag{61}$$

This suggests (looking just at the leading term) that the greatest contributions to the transform will come when $s \sim \sigma$, $\frac{f}{s} \sim E$, and $l \sim \tau$.

3.4.2 Inverse Morlet wavelet transform

We expect the original Gaussian function will be recovered by (Equation 19):

$$\psi_{\sigma E\tau}(t) = \frac{1}{C_f} \int_{-\infty}^{\infty} \frac{ds\,dl}{s^2} \phi_{sl}(t)\, \tilde{\psi}_{sl}^{(\sigma E\tau)} \tag{62}$$

Without loss of generality, we simplify by assuming that $\tau = 0$ in the original Gaussian test function and write the Morlet wavelet as the sum of a pair of Gaussians:

$$\psi_{\sigma E}(t) = \frac{1}{C_f} \int_{-\infty}^{\infty} \frac{ds\,dl}{s^2} \left(\frac{2\sqrt{\pi}\sigma}{s^2+\sigma^2}\right)^{\frac{1}{2}} \left[e^{-\imath f\left(\frac{t-l}{s}\right)} - e^{-\frac{f^2}{2}} \right] e^{-\frac{1}{2}\left(\frac{t-l}{s}\right)^2}$$
$$\times \left[e^{-\imath\frac{fsl}{s^2+\sigma^2} - \frac{1}{2}\frac{(Es-f)^2\sigma^2}{s^2+\sigma^2}} - e^{-\frac{f^2}{2} - \frac{1}{2}\frac{(Es\sigma)^2}{s^2+\sigma^2}} \right] e^{-\imath E\frac{\sigma^2 l}{s^2+\sigma^2} - \frac{1}{2}\frac{l^2}{s^2+\sigma^2}} \tag{63}$$

The integral over l is straightforward, as all the terms are Gaussians in l:

$$\psi_{\sigma E}(t) = \frac{1}{C_f} \int\limits_{-\infty}^{\infty} \frac{ds}{|s|} 2\pi^{\frac{3}{4}} \left(\frac{\sigma}{2s^2 + \sigma^2} \right)^{\frac{1}{2}}$$

$$\times \left[e^{-\frac{2E^2 s^2 \sigma^2 + 2f^2\left(2s^2+\sigma^2\right) + 2\iota E\sigma^2 t + t^2}{2\left(2s^2+\sigma^2\right)}} \right.$$

$$-2e^{-\frac{3f^2 s^2 + 2\sigma^2\left(f^2 - fEs + E^2 s^2\right) + 2\iota t\left(fs + E\sigma^2\right) + t^2}{2\left(2s^2+\sigma^2\right)}}$$

$$\left. +e^{-\frac{2\sigma^2\left(f - Es\right)^2 + 2\iota t\left(2fs + E\sigma^2\right) + t^2}{2\left(2s^2+\sigma^2\right)}} \right] \tag{64}$$

The limit of the integrand as s goes to zero is:

$$2e^{-f^2 - \iota Et - \frac{t^2}{2\sigma^2}} f^2 \pi^{\frac{3}{4}} \sigma^{-\frac{9}{2}} |s|$$

$$\times \left(\sigma^2 + E^2\sigma^4 - 2\iota E\sigma^2 t - t^2 \right) + O\left[s^2\right] \tag{65}$$

The limit as s goes to $\pm\infty$ is:

$$\pm \frac{\sqrt{2}\sigma\pi^{\frac{3}{4}} e^{-f^2 - \frac{E^2\sigma^2}{2}} \left(1 - 2e^{\frac{1}{4}f^2} + e^{f^2} \right)}{s^2} + O\left[\frac{1}{s}\right]^3 \tag{66}$$

Our integral is therefore neither singular at the origin nor divergent at infinity. Of course, we expect this since we are guaranteed by the decomposition theorem that this integral will give the original Gaussian. To show explicitly we get the original Gaussian we take the Fourier transform of both sides, with respect to t. The simplification is dramatic – most of the factors come outside of the integral over s. On the left we have (Equation 24):

$$\hat{\psi}_{\sigma E}(\omega) = \left(\frac{\sigma^2}{\pi} \right)^{\frac{1}{4}} e^{-\frac{(E-\omega)^2\sigma^2}{2}} \tag{67}$$

On the right we get:

$$\frac{1}{C_f} 2\pi e^{-f^2} \left(\frac{\sigma^2}{\pi} \right)^{\frac{1}{4}} e^{-\frac{(E-\omega)^2\sigma^2}{2}} \int\limits_{-\infty}^{\infty} \frac{ds}{|s|} e^{-s^2\omega^2} \left(e^{fs\omega} - 1 \right)^2 \tag{68}$$

We change variables in the integral $s' \equiv s\omega$:

$$\left(\frac{\sigma^2}{\pi} \right)^{\frac{1}{4}} e^{-\frac{(E-\omega)^2\sigma^2}{2}} \frac{1}{C_f} 2\pi e^{-f^2} \int\limits_{-\infty}^{\infty} \frac{ds'}{|s'|} 2e^{-s'^2} \left(e^{fs'} - 1 \right)^2 \tag{69}$$

We note the integral is essentially the admissibility constant (Equation 43). Factors cancel yielding:

$$\left(\frac{\sigma^2}{\pi} \right)^{\frac{1}{4}} e^{-\frac{(E-\omega)^2\sigma^2}{2}} \tag{70}$$

Which is identical to the left hand side, as was to be shown.

3.5 Other Applications

We will compute the Morlet wavelet transforms of δ functions, plane waves, and – to achieve maximum self-referentiality – a Morlet wavelet itself.

3.5.1 δ functions

Since the δ function is not a square-integrable function, we are not guaranteed the wavelet transform will work. We therefore write the δ function as a limit of Gaussian test functions:

$$\delta(x) = \lim_{\sigma \to 0^+} \frac{1}{\sqrt{2\pi}\sigma} e^{-\frac{x^2}{2\sigma^2}} \tag{71}$$

This lets us use the result for a Gaussian test function (Equation 60):

$$\tilde{\delta}_{sl}(\tau) = \lim_{\sigma \to 0^+} \left(s^2 + \sigma^2 \right)^{-\frac{1}{2}} |s|^{\frac{1}{2}}$$

$$\times \left[e^{-\iota\frac{fs}{s^2+\sigma^2}(l-\tau) - \frac{1}{2}\frac{f^2\sigma^2}{s^2+\sigma^2}} - e^{-\frac{f^2}{2}} \right] e^{-\frac{1}{2}\frac{(l-\tau)^2}{s^2+\sigma^2}} \tag{72}$$

Taking the limit as σ goes to zero:

$$\tilde{\delta}_{sl}(\tau) = |s|^{-\frac{1}{2}} \left[e^{-\iota f\left(\frac{l-\tau}{s}\right)} - e^{-\frac{f^2}{2}} \right] e^{-\frac{1}{2}\frac{(l-\tau)^2}{s^2}} \tag{73}$$

This is itself a Morlet wavelet:

$$\tilde{\delta}_{sl}(\tau) = \phi_{sl}^*(\tau) \tag{74}$$

We get the same result by computing the Morlet wavelet transform directly:

$$\tilde{\delta}_{sl}(\tau) = \int\limits_{-\infty}^{\infty} dt' \phi_{sl}^*(t') \delta(t' - \tau)$$

$$= |s|^{-\frac{1}{2}} \left(e^{\iota f\frac{\tau-l}{s}} - e^{-\frac{f^2}{2}} \right) e^{-\frac{1}{2}\left(\frac{\tau-l}{s}\right)^2} \tag{75}$$

Since the demonstration of the resolution of unity only applies to square-integrable functions, we verify the inverse transform. We want to show:

$$\delta(t - \tau) = \frac{1}{C_f} \int\limits_{-\infty}^{\infty} \frac{dsdl}{s^2} \phi_{sl}(t) \phi_{sl}^*(\tau) \tag{76}$$

However this is just what we showed when we computed the admissibility constant (Equation 20), so we are done.

3.5.2 Plane waves

The Morlet wave transform of a plane wave:

$$\chi_E(t) \equiv \frac{1}{\sqrt{2\pi}} e^{-\imath E t} \tag{77}$$

is given by:

$$\tilde{\chi}_{sl}(E) = \int dt \, |s|^{-\frac{1}{2}} \left[e^{\imath f \frac{t-l}{s}} - e^{-\frac{f^2}{2}} \right]$$

$$\times e^{-\frac{1}{2}\left(\frac{t-l}{s}\right)^2} \frac{1}{\sqrt{2\pi}} e^{-\imath E t} \tag{78}$$

The integral is essentially the Fourier transform of a Morlet wavelet:

$$\tilde{\chi}_{sl}(E) = \hat{\phi}_{sl}^*(E) \tag{79}$$

For the inverse transform to be valid we require:

$$\chi_E(t) = \frac{1}{C_f} \int \frac{ds\,dl}{s^2} \phi_{sl}(t) \tilde{\chi}_{sl}(E)$$

$$= \frac{1}{C_f} \int \frac{ds\,dl}{s^2} \phi_{sl}(t) \hat{\phi}_{sl}^*(E) \tag{80}$$

To show this, we take the Fourier transform (Equation 113) of each side. On the left side we get:

$$\delta(\omega - E) \tag{81}$$

On the right side we write the Fourier transforms of the Morlet wavelets (Equation 24) in terms of the Fourier transforms of their mothers (Equation 22):

$$\frac{1}{C_f} \int \frac{ds\,dl}{s^2} |s|^{\frac{1}{2}} e^{\imath l\omega} \hat{\phi}(s\omega) |s|^{\frac{1}{2}} e^{-\imath l E} \hat{\phi}^*(sE) \tag{82}$$

The integral over l is a δ function, which we pull out of the integral, leaving the now familiar admissibility constant (Equation 20) behind:

$$\frac{2\pi}{C_f} \delta(\omega - E) \int \frac{ds}{|s|} \hat{\phi}(s\omega) \hat{\phi}^*(s\omega) = \delta(\omega - E) \tag{83}$$

3.5.3 Morlet wavelet transform of a Morlet wavelet

We look at the Morlet wavelet transform of a Morlet wavelet (Equation 15) with σE replacing f and σ replacing s:

$$\Phi_{\sigma E\tau}(t) \equiv |\sigma|^{-\frac{1}{2}} \left[e^{-\imath E(t-\tau)} - e^{-\frac{\sigma^2 E^2}{2}} \right] e^{-\frac{1}{2}\left(\frac{t-\tau}{\sigma}\right)^2} \tag{84}$$

Per (Equation 18), the Morlet wavelet transform is given by:

$$\tilde{\Phi}_{sl}^{(\sigma E\tau)} = \int_{-\infty}^{\infty} dt' \phi_{sl}^*(t) \Phi_{\sigma E\tau}(t) \tag{85}$$

To apply the results for Gaussian test functions we split the incoming Morlet wavelet $\Phi_{\sigma E\tau}(t)$ into its two Gaussians then use the results for Gaussian test functions (Equation 60) to read off the results:

$$\tilde{\Phi}_{sl}^{(\sigma E\tau)} = \left(2\pi \frac{\sigma}{s^2 + \sigma^2} |s| \right)^{\frac{1}{2}}$$

$$\times \left\{ \left[e^{-\imath \frac{fs}{s^2+\sigma^2}(l-\tau) - \frac{1}{2}\frac{(Es-f)^2\sigma^2}{s^2+\sigma^2}} - e^{-\frac{f^2}{2} - \frac{1}{2}\frac{(Es\sigma)^2}{s^2+\sigma^2}} \right] e^{-\imath E \frac{\sigma^2}{s^2+\sigma^2}(l-\tau)} \right.$$

$$\left. - \left[e^{-\imath \frac{fs}{s^2+\sigma^2}(l-\tau) - \frac{1}{2}\frac{f^2\sigma^2}{s^2+\sigma^2} - \frac{\sigma^2 E^2}{2}} - e^{-\frac{f^2}{2} - \frac{\sigma^2 E^2}{2}} \right] \right\} e^{-\frac{1}{2}\frac{(l-\tau)^2}{s^2+\sigma^2}} \tag{86}$$

4 Covariant Morlet wavelets

4.1 Strategy

We would like to generalize Morlet wavelets to four dimensions (one time, three space) in a way that is manifestly covariant. We will do this by taking the direct product of Morlet wavelets in time and the three space dimensions. The natural generalization of the Gaussian part of the one-dimensional Morlet wavelet is:

$$e^{-\frac{1}{2}x^2} \rightarrow e^{-\frac{1}{2}(x_\mu x^\mu)} = e^{\frac{1}{2}(t^2 - x^2 - y^2 - z^2)} \tag{87}$$

This clearly diverges in t. We have to fix this without losing manifest covariance.

We will assume we start in a specific frame M, possibly the center-of-mass frame. We will define the four-dimensional Morlet wavelet as the product of four one-dimensional Morlet wavelets, then write our results in a way that is Lorentz-invariant.

4.2 Construction

We take the four-dimensional mother Morlet wavelet as the direct product of four one-dimensional mother Morlet wavelets, one for each coordinate:

$$\phi(t) \rightarrow \phi(t)\phi(x)\phi(y)\phi(z) \tag{88}$$

We write the four dimensional mother Morlet wavelet as the product of four one-dimensional mother Morlet wavelets (Equation 14):

$$\phi(t, x, y, z) = \left(e^{-\imath f_0 t} - e^{-\frac{f_0^2}{2}} \right)\left(e^{\imath f_1 x} - e^{-\frac{f_1^2}{2}} \right)$$

$$\times \left(e^{\imath f_2 y} - e^{-\frac{f_2^2}{2}} \right)\left(e^{\imath f_3 z} - e^{-\frac{f_3^2}{2}} \right) e^{-\frac{t^2 + x^2 + y^2 + z^2}{2}} \tag{89}$$

By scaling and displacing each component separately we get:

$$\phi_{sl}(t, x, y, z) = |s_0 s_1 s_2 s_3|^{-\frac{1}{2}} \left(e^{-\iota f_0 \frac{t-l_0}{s_0}} - e^{-\frac{f_0^2}{2}} \right)$$

$$\times \left(e^{\iota f_1 \frac{x-l_1}{s_1}} - e^{-\frac{f_1^2}{2}} \right) \left(e^{\iota f_2 \frac{y-l_2}{s_2}} - e^{-\frac{f_2^2}{2}} \right) \left(e^{\iota f_3 \frac{z-l_3}{s_3}} - e^{-\frac{f_3^2}{2}} \right)$$

$$\times e^{-\frac{1}{2}\left[\left(\frac{t-l_0}{s_0}\right)^2 + \left(\frac{x-l_1}{s_1}\right)^2 + \left(\frac{y-l_2}{s_2}\right)^2 + \left(\frac{z-l_3}{s_3}\right)^2 \right]} \tag{90}$$

Using (Equation 22), the Fourier transform of the mother Morlet wavelet is:

$$\hat{\phi}(E, p_x, p_y, p_z) = \left(e^{f_0 E} - 1 \right) \left(e^{f_1 p_x} - 1 \right) \left(e^{f_2 p_y} - 1 \right)$$

$$\times \left(e^{f_3 p_z} - 1 \right) e^{-\frac{f_0^2 + f_1^2 + f_2^2 + f_3^2}{2} - \frac{E^2 + p_x^2 + p_y^2 + p_z^2}{2}} \tag{91}$$

The Fourier transform of the general Morlet wavelet is:

$$\hat{\phi}_{sl}(E, p_x, p_y, p_z) = |s_0 s_1 s_2 s_3|^{\frac{1}{2}} e^{\iota(l_0 E - l_1 p_x - l_2 p_y - l_3 p_z)}$$

$$\times \left(e^{s_0 f_0 E} - 1 \right) \left(e^{s_1 f_1 p_x} - 1 \right) \left(e^{s_2 f_2 p_y} - 1 \right) \left(e^{s_3 f_3 p_z} - 1 \right)$$

$$\times e^{-\frac{f_0^2 + f_1^2 + f_2^2 + f_3^2}{2} - \frac{s_0^2 E^2 + s_1^2 p_x^2 + s_2^2 p_y^2 + s_3^2 p_z^2}{2}} \tag{92}$$

Now we have to promote various non-covariant bits to covariant bits.

The scale factors enter into the inverse Morlet integral in a slightly awkward way:

$$\int \frac{ds_0}{s_0^2} \frac{ds_1}{s_1^2} \frac{ds_2}{s_2^2} \frac{ds_3}{s_3^2} \tag{93}$$

The simplest approach to this is to treat the four scale factors as so many scalars.

The obvious choices for the displacement l and the reference frequency f are to treat them as four vectors. For the displacement a single four vector will suffice:

$$l = (l_0, l_1, l_2, l_3) \tag{94}$$

We will need one four vector for each reference frequency:

$$F^{(0)} \equiv (f_0, 0, 0, 0) \tag{95}$$

$$F^{(1)} \equiv (0, f_1, 0, 0) \tag{96}$$

$$F^{(2)} \equiv (0, 0, f_2, 0) \tag{97}$$

$$F^{(3)} \equiv (0, 0, 0, f_3) \tag{98}$$

For convenience, we define the sum over all four F's as:

$$F \equiv \sum_{n=0}^{3} F^{(n)} = (f_0, f_1, f_2, f_3) \tag{99}$$

This is also a four vector. Note that the raw frequencies f_0, f_1, f_2, f_3 are themselves scalars since they are defined with respect to the specific frame M.

To represent the sums as Lorentz invariants we define a set of second rank tensors (with their inverses):

$$\Sigma_\mu^{(n)\nu} \equiv \begin{pmatrix} s_0^{-n} & 0 & 0 & 0 \\ 0 & -s_1^{-n} & 0 & 0 \\ 0 & 0 & -s_2^{-n} & 0 \\ 0 & 0 & 0 & -s_3^{-n} \end{pmatrix} \tag{100}$$

$$\left(\frac{1}{\Sigma^{(n)}} \right)_\mu^\nu \equiv \begin{pmatrix} s_0^n & 0 & 0 & 0 \\ 0 & -s_1^n & 0 & 0 \\ 0 & 0 & -s_2^n & 0 \\ 0 & 0 & 0 & -s_3^n \end{pmatrix} \tag{101}$$

We need the explicit forms for n from 0 to 2:

$$\Sigma_\mu^{(0)\nu} \equiv \begin{pmatrix} 1 & 0 & 0 & 0 \\ 0 & -1 & 0 & 0 \\ 0 & 0 & -1 & 0 \\ 0 & 0 & 0 & -1 \end{pmatrix} \tag{102}$$

$$\Sigma_\mu^{(1)\nu} \equiv \begin{pmatrix} s_0^{-1} & 0 & 0 & 0 \\ 0 & -s_1^{-1} & 0 & 0 \\ 0 & 0 & -s_2^{-1} & 0 \\ 0 & 0 & 0 & -s_3^{-1} \end{pmatrix} \tag{103}$$

$$\Sigma_\mu^{(2)\nu} \equiv \begin{pmatrix} s_0^{-2} & 0 & 0 & 0 \\ 0 & -s_1^{-2} & 0 & 0 \\ 0 & 0 & -s_2^{-2} & 0 \\ 0 & 0 & 0 & -s_3^{-2} \end{pmatrix} \tag{104}$$

The choice of signature $(1, -1, -1, -1)$ ensures convergence.

With these definitions the mother Morlet wavelet is:

$$\phi(x_\mu) = \left\{ \prod_{n=0}^{3} \left[e^{-\iota F^{(n)\mu} x_\mu} - e^{-\frac{1}{2} F^{(n)\mu} F_\mu} \right] \right\} e^{-\frac{1}{2} x^\mu \Sigma_\mu^{(0)\nu} x_\nu} \tag{105}$$

and the general Morlet wavelet is:

$$\phi_{\Sigma l}(x_\mu) = \sqrt{\det(\Sigma^{(1)})} e^{-\frac{1}{2}(x^\mu - l^\mu)\Sigma_\mu^{(2)\nu}(x_\nu - l_\nu)}$$

$$\times \left[\prod_{n=0}^{3} e^{-\iota F^{(n)\mu} \Sigma_\mu^{(0)\varpi} \Sigma_\varpi^{(1)\nu}(x_\nu - l_\nu)} - e^{-\frac{1}{2} F^{(n)\mu} \Sigma_\mu^{(0)\nu} F_\nu^{(n)}} \right] \tag{106}$$

While we have worked this out in frame M, as it is written in terms of covariant quantities it is valid in all frames. We have therefore guaranteed Lorentz covariance of the Morlet wavelets.

Note that the choice of frame defines a set of Morlet wavelets; with each frame there is a distinct set of Morlet wavelets. If we have multiple frames we wish to work with we will need to tag each Morlet wavelet with the frame it comes from. Usually there is an obvious choice of frame, i.e. the center-of-mass frame.

With these definitions, the Fourier transform of the mother Morlet wavelet is:

$$
\hat{\phi}(p) = \left\{ \prod_{n=0}^{3} \left[e^{F^{(n)} \frac{1}{\Sigma^{(0)}} p} - 1 \right] \right\} e^{-F \frac{1}{2\Sigma^{(0)}} F - \frac{1}{2} p \frac{1}{2\Sigma^{(0)}} p} \quad (107)
$$

The Fourier transform of the general Morlet wavelet is:

$$
\hat{\phi}_{\Sigma l}(p) = \sqrt{\frac{1}{\det\left(\Sigma^{(1)}\right)}} e^{\iota p l - F \frac{1}{2\Sigma^{(0)}} F - \frac{1}{2} p \frac{1}{2\Sigma^{(2)}} p}
$$

$$
\times \left\{ \prod_{n=0}^{3} \left[e^{F^{(n)} \frac{1}{\Sigma^{(1)}} p} - 1 \right] \right\} \quad (108)
$$

4.3 Resolution of unity

Any square integrable function $\psi(t, x, y, z)$ may be expressed as a sum over these Morlet wavelets. The covariant Morlet wavelet transform is given by:

$$
\tilde{\psi}_{\Sigma l} = \int_{-\infty}^{\infty} d^4 x \phi_{\Sigma l}^*(x) \psi(x) \quad (109)
$$

And the inverse is given by:

$$
\psi(x) = \frac{1}{C_{f_0} C_{f_1} C_{f_2} C_{f_3}} \int_{-\infty}^{\infty} \left(\prod_{n=0}^{3} \frac{ds_n}{|s_n|^2} \right) \int_{-\infty}^{\infty} d^4 l \phi_{\Sigma l}(x) \tilde{\psi}_{\Sigma l} \quad (110)
$$

The resolution of unity and the values of the constants of admissibility follow directly from the results for one dimension.

The solutions for Gaussian test functions, δ functions, and plane waves are merely the direct products of the corresponding one-dimensional wave functions.

We have therefore reached our second and final objective: to generalize the Morlet wavelet transform to four dimensions in a way which is manifestly covariant.

4.4 Alternative approaches

Alternative (and more sophisticated) lines of attack are possible. For instance in [9] or in [41] two dimensional wavelets are generated from the mother wavelet by using displacements, rotations (in the xy plane), and a single scale factor:

$$
\phi_{Rls}(x, y) = \frac{1}{s} \phi \left[\overleftrightarrow{R} \cdot \frac{\left(\vec{r} - \vec{l} \right)}{s} \right] \quad (111)
$$

where R is a rotation matrix (in two dimensions).

By analogy, we could generalize one-dimensional wavelets to four dimensions by using displacements l, Lorentz transformations:

$$
\phi_{\Lambda l s}(x_\mu) = \frac{1}{s^2} \phi \left[\frac{1}{s} \Lambda_\mu^\nu (x_\nu - l_\nu) \right] \quad (112)
$$

But establishing convergence, verifying the resolution of unity, and computing the admissibility constant for these wavelets would be a new project. Our immediate requirement is merely to establish that there is at least one set of covariant Morlet wavelets.

5 Summary

The naive use of plane wave or δ function decomposition can create artificial difficulties in the analysis of foundational questions of quantum mechanics. The use of Morlet wavelet decomposition avoids these difficulties. With the explicit calculation of the admissibility constant and the demonstration of covariant Morlet wavelets, we have eliminated two of the barriers to full use of this powerful technology for the analysis of foundational questions in quantum mechanics.

6 Appendix

6.1 Conventions for the Fourier transform

For the Fourier transform from time t to frequency ω we are using:

$$
\hat{f}(\omega) \equiv \frac{1}{\sqrt{2\pi}} \int_{-\infty}^{\infty} dt e^{\iota \omega t} f(t) \quad (113)
$$

with inverse Fourier transform:

$$
f(t) = \frac{1}{\sqrt{2\pi}} \int_{-\infty}^{\infty} d\omega e^{-\iota \omega t} \hat{f}(\omega) \quad (114)
$$

In the Fourier transform in four dimensions time and space enter with opposite signs:

$$
\hat{f}\left(\omega, \vec{k}\right) \equiv \frac{1}{4\pi^2} \int_{-\infty}^{\infty} dt d\vec{x} e^{\iota \omega t - \iota \vec{k} \cdot \vec{x}} f(t, \vec{x}) \quad (115)
$$

with inverse Fourier transform:

$$
f(t, \vec{x}) = \frac{1}{4\pi^2} \int_{-\infty}^{\infty} d\omega d\vec{k} e^{-\iota \omega t + \iota \vec{k} \cdot \vec{x}} \hat{f}\left(\omega, \vec{k}\right) \quad (116)
$$

6.2 Direct calculation of the admissibility constant

To check the results for the admissibility constant (Equation 43), we ran a numeric calculation of the values of f from $-3\pi \to 3\pi$. The integrand: $|\omega|^{-1}\left(e^{f\omega}-1\right)^2 e^{-\omega^2}$ is real, smooth, and positive definite, making it a perfect candidate for numerical integration. Numerical integration produced results visually indistinguishable from the formula for C_f given by Equation 52. On a test of 100 evenly spaced points from $-3\pi \to 3\pi$, with $f = 1$, the maximum error was 2.72×10^{-8} and the average absolute error was 8.2×10^{-10}. These residuals are easily explained in terms of rounding.

Much of the work here was done using the program Wolfram's Mathematica (version 7). However a certain amount of care is need when using this tool. An attempt to have it compute directly the integral for $I(f)$ given by Equation 44 produces the expression given in Equation 51 plus five additional terms: $2\pi\text{erfi}\left(\frac{1}{2}|f|\right) - \pi\text{erfi}(|f|) + \ln(-f) + \ln(f) - \ln(f^2)$. This is incorrect by inspection: the three ln terms add a value of $\ln(-1) = \iota\pi$ which is nonsense, given that the original integral is real. The other terms are real so offer no escape. The usual moral in working with Mathematica or any math software applies: as Walter Donovan, the villain in *Indiana Jones and the Last Crusade*, said to Indiana Jones: *trust no one*.

References

[1] Morlet J, Arens G, Fourgeau E, Glard D. Wave propagation and sampling theory. Part I: Complex signal and scattering in multilayered media. Geophysics 1982; 47 (2): 203-221. http://dx.doi.org/10. 1190/1.1441328

[2] Morlet J, Arens G, Fourgeau E, Giard D. Wave propagation and sampling theory. Part II: Sampling theory and complex waves. Geophysics 1982; 47 (2): 222-236. http://dx.doi.org/10.1190/1. 1441329

[3] Chui CK. An Introduction to Wavelets. Wavelet Analysis and Its Applications, vol.1, Boston: Academic Press, 1992.

[4] Meyer Y. Wavelets and Operators. Cambridge Studies in Advanced Mathematics, Cambridge: Cambridge University Press, 1992.

[5] Kaiser G. A Friendly Guide to Wavelets. Boston: Birkhäuser, 1994.

[6] van den Berg JC. Wavelets in Physics. Cambridge: Cambridge University Press, 1999.

[7] Addison PS. The Illustrated Wavelet Transform Handbook: Introductory Theory and Applications in Science, Engineering, Medicine and Finance. Bristol: Institute of Physics Publishing, 2002.

[8] Bratteli O, Jørgensen PET. Wavelets through a Looking Glass: The World of the Spectrum. Applied and Numerical Harmonic Analysis, Boston: Birkhäuser, 2002.

[9] Antoine J-P, Murenzi R, Vandergheynst P, Ali ST. Two-Dimensional Wavelets and their Relatives. Cambridge: Cambridge University Press, 2004. http://dx.doi.org/10.1017/ CBO9780511543395

[10] Havukainen M. Wavelets as basis functions in canonical quantization, 2000. http://arxiv. org/abs/quant-ph/0006083

[11] Kim YS. Wavelets and information-preserving transformations, 1996. http://arxiv.org/abs/ quant-ph/9610018

[12] Altaisky MV. Wavelet based regularization for Euclidean field theory, 2003. http://arxiv.org/ abs/hep-th/0305167

[13] Visser M. Physical wavelets: Lorentz covariant, singularity-free, finite energy, zero action, localized solutions to the wave equation. Physics Letters A 2003; 315 (3-4): 219-224. http://arxiv.org/ abs/hep-th/0304081

[14] Feynman RP, Hibbs AR. Quantum Mechanics and Path Integrals. New York: McGraw-Hill Companies, 1965.

[15] Schulman LS. Techniques and Applications of Path Integration. New York: John Wiley & Sons, 1981.

[16] Swanson MS. Path Integrals and Quantum Processes. New York: Academic Press, 1992.

[17] Khandekar DC, Lawande SV, Bhagwat KV. Path-Integral Methods and Their Applications. Singapore: World Scientific Publishing Company, 1993.

[18] Marchewka A, Schuss Z. Path-integral approach to the Schrödinger current. Physical Review A 2000; 61 (5): 052107. http://arxiv.org/abs/ quant-ph/9903076

[19] Kleinert H. Path Integrals in Quantum Mechanics, Statistics, and Polymer Physics, and Financial Markets. Singapore: World Scientific Publishing Company, 2004.

[20] Zinn-Justin J. Path Integrals in Quantum Mechanics. Oxford Graduate Texts, Oxford: Oxford University Press, 2005.

[21] Seidewitz E. Foundations of a spacetime path formalism for relativistic quantum mechanics. Journal of Mathematical Physics 2006; 47 (11): 112302-112329. http://arxiv.org/abs/quant-ph/0507115

[22] Gerlach W, Stern O. Der experimentelle Nachweis des magnetischen Moments des Silberatoms. Zeitschrift für Physik 1922; 8 (1): 110-111. http://dx.doi.org/10.1007/BF01329580

[23] Gerlach W, Stern O. Der experimentelle Nachweis der Richtungsquantelung im Magnetfeld. Zeitschrift für Physik A Hadrons and Nuclei 1922; 9 (1): 349-352. http://dx.doi.org/10.1007/BF01326983

[24] Gerlach W, Stern O. Das magnetische Moment des Silberatoms. Zeitschrift für Physik 1922; 9 (1): 353-355. http://dx.doi.org/10.1007/BF01326984

[25] von Neumann J. Mathematical Foundations of Quantum Mechanics. Investigations In Physics, Princeton: Princeton University Press, 1955.

[26] Gondran M, Gondran A. A complete analysis of the Stern-Gerlach experiment using Pauli spinors, 2005. http://arxiv.org/abs/quant-ph/0511276

[27] Schrödinger E. Die gegenwärtige Situation in der Quantenmechanik. Naturwissenschaften 1935; 23 (48): 807-812. http://dx.doi.org/10.1007/BF01491891

[28] Schrödinger E. Die gegenwärtige Situation in der Quantenmechanik. Naturwissenschaften 1935; 23 (49): 823-828. http://dx.doi.org/10.1007/BF01491914

[29] Schrödinger E. Die gegenwärtige Situation in der Quantenmechanik. Naturwissenschaften 1935; 23 (50): 844-849. http://dx.doi.org/10.1007/BF01491987

[30] Schrödinger E. The present situation in quantum mechanics: a translation of Schrödinger's 'cat paradox' paper. In: Quantum Theory and Measurement, Wheeler JA, Zurek WH (editors), New Jersey: Princeton University Press, 1983, pp.152-167.

[31] Cruz-Barrios S, Gómez-Camacho J. Semiclassical description of scattering with internal degrees of freedom. Nuclear Physics A 1998; 636 (1): 70-84. http://dx.doi.org/10.1016/S0375-9474(98)00176-6

[32] Cruz-Barrios S, Gómez-Camacho J. Semiclassical description of Stern-Gerlach experiments. Physical Review A 2000; 63 (1): 012101. http://arxiv.org/abs/quant-ph/0010079

[33] Venugopalan A, Kumar D, Ghosh R. Analysis of the Stern-Gerlach measurement, 1995. http://arxiv.org/abs/quant-ph/9501022

[34] Venugopalan A. Decoherence and Schrödinger-cat states in a Stern-Gerlach-type experiment. Physical Review A 1997; 56 (5): 4307-4310. http://dx.doi.org/10.1103/PhysRevA.56.4307

[35] Venugopalan A. Pointer states via decoherence in a quantum measurement. Physical Review A 1999; 61 (1): 012102. http://arxiv.org/abs/quant-ph/9909005

[36] Feynman RP. Mathematical formulation of the quantum theory of electromagnetic interaction. Physical Review 1950; 80 (3): 440-457. http://dx.doi.org/10.1103/PhysRev.80.440

[37] Feynman RP. An operator calculus having applications in quantum electrodynamics. Physical Review 1951; 84 (1): 108-128. http://dx.doi.org/10.1103/PhysRev.84.108

[38] Land MC, Horwitz LP. Off-shell quantum electrodynamics, 1996. http://arxiv.org/abs/hep-th/9601021

[39] Horwitz LP. Second quantization of the Stueckelberg relativistic quantum theory and associated gauge fields, 1998. http://arxiv.org/abs/hep-th/9804155

[40] Johnson RW. Symmetrization and enhancement of the continuous Morlet transform. International Journal of Wavelets, Multiresolution and Information Processing 2012; 10 (1): 1250009. http://arxiv.org/abs/0912.1126

[41] Perel MV, Sidorenko MS. New physical wavelet 'Gaussian wave packet'. Journal of Physics A: Mathematical and Theoretical 2007; 40 (13): 3441-3461. http://arxiv.org/abs/math-ph/0701051

18

Popper and Bohr on Realism in Quantum Mechanics

Don Howard

Department of Philosophy, University of Notre Dame, Notre Dame, Indiana, United States. E-mail: dhoward1@nd.edu

Editors: *Danko Georgiev & Peter J. Lewis*

Popper's program in the foundations of quantum mechanics defending objectivity and realism developed out of a profound dissatisfaction with the point of view associated with Bohr, which is usually designated the Copenhagen interpretation. Here I will argue that while Popper's aim is a noble one, his program does not succeed on two counts: he does not succeed in showing that Bohr's philosophy must be rejected as a variety of subjectivism, and his alternative interpretation of indeterminacy rests on a highly questionable assumption according to which simultaneously precise conjugate parameters are possible. Nevertheless I like Popper's propensity interpretation of probability and think that the propensity idea deserves further research.
Quanta 2012; 1: 33–57.

1 Introduction

For years I have been puzzled, by the attitudes of my colleagues in philosophy and physics toward Popper's work in the foundations of quantum mechanics. His program developed out of a profound dissatisfaction with the point of view associated with Bohr, which is usually designated the *Copenhagen interpretation*. He has severely criticized Bohr's philosophy, calling it the "ruling dogma" [1, p. 7], and he presents his own interpretation of the quantum theory as an explicit alternative. And yet, with the exception of one essay by Paul Feyerabend [2, 3], no one who professes sympathy with Bohr's interpretation has troubled to respond, in a systematic way, to Popper's criticism, nor have they undertaken a comprehensive critique of Popper's views. It may be that Bohr's sympathizers view Popper's work as unworthy of a reply. But my guess is that their reticence is to be explained not so much by arrogance, as by a disheartening sense of their own failure to have understood fully what Bohr intended. In any case, the silence is embarrassing; a reckoning is in order.

Popper's program in the foundations of quantum mechanics is continuous with his efforts elsewhere to defend *objectivity* and *realism*. It is an essential part of his larger philosophical enterprise because the apparent novelties of the quantum theoretical description of nature have, on occasion, been called upon to give scientific authority to philosophical attacks on realism and objectivity. The defense is two-pronged. It includes, on the one hand, a critical analysis of the subjectivism which allegedly infects the Copenhagen interpretation, and, on the other hand, provision of objective interpretations of both indeterminacy and quantum mechanical probabilities. I will argue that while Popper's aim is a noble one, his program does not succeed on two counts: he does not succeed in showing that Bohr's philosophy must be rejected as a variety of subjectivism, and his alternative interpretation of indeterminacy rests on a highly questionable assumption. But I like his interpretation of probability, the *propensity* interpretation. I think the propensity idea is Popper's one genuine contribution to the foundations of quantum mechanics, and its further development ought to be one of the principal aims of current research. But I also believe that when the propensity interpretation is divorced from the remainder of Popper's program, it no longer looks like an alternative to Bohr's position; it appears, instead, to confirm some of the latter's basic insights.

2 Realism, objectivity and the interpretation of quantum mechanics

In 1927, Popper was finishing his Ph.D. at the University of Vienna when two important ideas made their appearance. The first was Heisenberg's uncertainty or indeterminacy principle [4]; the second was Bohr's complementarity interpretation of quantum mechanics [5–7]. The exact meaning of each is to this day the subject of considerable disagreement, but to a first degree of approximation they can be characterized as follows.

From a formal point of view, the indeterminacy principle asserts the existence of an upper bound on the product of the standard deviations of any two conjugate parameters of a physical system. In the familiar case of position and momentum, this limitation is expressed by the well-known equation: $\Delta q_x \Delta p_x \geq \frac{\hbar}{2}$, where Δq_x and Δp_x represent, respectively, the indeterminacies of the x-components of the position and momentum of a system. But while the derivation of this equation and its analogues for other pairs of conjugate coordinates is a fairly straightforward matter (in the case of the energy-time indeterminacy relation, even the derivation is a matter of dispute; the basic difficulty concerns the possibility of representing the time by a Hermitian operator) [8, pp. 136-156], their interpretation has been a topic of controversy, various thinkers having argued (and this is by no means an exhaustive catalogue of the different points of view) that the indeterminacies represent (1) measures of an intrinsic indefiniteness of the coordinates themselves, (2) measures of the precision with which individual measurements of the coordinates can be carried out, or (3) measures of the spread of results obtained in a series of measurements, each of which may, itself, be as precise as desired.

Bohr's interpretation of quantum mechanics shares with the indeterminacy principle a concern with the apparently novel relationship between conjugate coordinates. Bohr argues that the kinds of experimental arrangements suited to the measurement of conjugate coordinates are invariably mutually exclusive, in the sense that, for example, a position measurement can never be performed with exactly the same apparatus, nor even in the presence of the same apparatus, which we use for a momentum measurement [9, p. 699]. Since he also believes that,

in addition to their role in measurement, experimental arrangements are somehow crucial in the very definition of the properties of quantum objects (this is not merely a variety of *operationism*; for Bohr the experimental conditions are necessary, but not sufficient, conditions for definition) [9, p. 700], Bohr concludes:

> Consequently, evidence obtained under different experimental conditions cannot be comprehended within a single picture, but must be regarded as complementary in the sense that only the totality of the phenomena exhausts the possible information about the objects. [10, p. 210]

Philosophers, and even many physicists, were quick to attempt to extract philosophical lessons from quantum mechanics. In the very lecture in which he introduced complementarity, Bohr comments:

> Now, the quantum postulate implies that any observation of atomic phenomena will involve an interaction with the agency of observation not to be neglected. Accordingly, an independent reality in the ordinary physical sense can neither be ascribed to the phenomena nor to the agencies of observation. [7, p. 54]

Reading this and many similar pronouncements, Popper thought he detected a common theme, which he took to be characteristic of the views thought to be shared by Bohr and his associates at Copenhagen:

> ... the *Copenhagen interpretation of quantum mechanics* ... says that *"objective reality has evaporated"* and that *quantum mechanics does not represent particles, but rather our knowledge, our observations, or our consciousness, of particles.* [1, p. 7]

Later I will question the assumption that there is a single Copenhagen interpretation, for there are important differences between Bohr and even his close colleagues, like Heisenberg. But the contrary impression is widespread, so it is not surprising that Popper adopts it uncritically (Heisenberg, himself, is largely responsible for the confusion [11]. He freely uses the label *Copenhagen interpretation* to denote both his views and Bohrs [12]). In any case, the threat which such a viewpoint poses to Popper's realism is substantial, for it suggests that realism is contradicted by a physical theory with impeccable empirical credentials. Moreover, a number of thinkers who allied themselves with the Vienna positivists were quick to draw just such a conclusion. The confluence of both philosophical and scientific doubts about the existence of a real external world surely made the case against realism look formidable.

Popper's arguments against positivism are familiar enough not to need repeating here. But his arguments against drawing idealist or subjectivist conclusions on the basis of quantum mechanics are far less well known, in spite of the importance which Popper himself attaches to them.

According to Popper, realism and objectivity are not contradicted by the formulae of quantum mechanics, the correctness by of which he does not question, but only by a mistaken interpretation of these formulae [13, §75]. And the primary reason for this mistaken interpretation, so he claims, is a failure to understand correctly the nature of quantum mechanical probability statements [14, p. 73] [1, p. 28]:

> ... the problems of the interpretation of quantum mechanics can all be traced to problems of the interpretation of the calculus of probability. [15, p. 104]

The quantum mechanical formalism describes a physical system by means of a mathematical device called a *state function* (commonly referred to as the ψ function), a different state function being assigned to the system for every different state that it can occupy (hence one typically speaks indifferently of *states* and *state functions*). Various purely mathematical manipulations of the state function enable one to derive predictions concerning the outcome of measurements that can be performed on the system. The Heisenberg indeterminacy formulae are themselves the result of such manipulations. The fact that the ψ function arises as a solution of the Schrödinger equation (the fundamental dynamical equation of quantum mechanics), which bears a formal analogy to the classical wave equation, led Schrödinger and other non-Copenhagen quantum theorists to believe that electrons and other elementary *particles* in some sense *really* are waves [8, pp. 24-33] [16] [17, pp. 43-80]. But ever since Born's pioneering statistical interpretation of the ψ function, there has been widespread agreement that the ψ function is best viewed as providing a measure of the probability of a system's being found, upon measurement, to have a specific value of a given variable [18] [8, pp. 38-44].

While agreeing that quantum mechanics is a probabilistic theory, Popper at first denied that this fact alone proves that nature itself is not deterministic. In the 1930s he argued that the determinism issue was essentially a metaphysical question, and hence not scientific, but he betrayed a certain sympathy for determinism, at least in the form of a heuristic maxim to the effect that scientists should always search for strict laws [13, §78]. Later,

however, he came to regard the determinism thesis as a testable hypothesis, and expressed his preference for a type of objective indeterminism [19, 20]. In this respect, Popper stands alone among the realist critics of Bohr and Heisenberg, many of whom, including Einstein have held the opinion that realism and determinism go hand in hand. In a letter to Born, Einstein says:

> I admit, of course, that there is a considerable amount of validity in the statistical approach ... I cannot seriously believe in it, because the theory cannot be reconciled with the idea that physics should represent a reality in time and space, free from spooky actions at a distance. [21, p. 158]

Instead of focusing on the determinism issue, Popper's dispute with Bohr and Heisenberg concerns the meaning of the probability statements which quantum mechanics necessarily employs. In Popper's account, the Copenhagen view traces the need for probabilities to quantum mechanical limitations on our *knowledge* of the properties of atomic systems, and thus ascribes a *subjective* interpretation to quantum mechanical probabilities. Popper, on the other hand, traces the probabilistic character of quantum mechanics to the *statistical* character of its problems, and would thus give an objective interpretation to its probability statements. In the 1930s this meant for him a classical *frequency* interpretation, later it meant a *propensity* interpretation. What exactly is meant by the various kinds of interpretations will be considered in some detail below.

The distinction between objective and subjective interpretations is crucial, for Popper holds that all of the objectionable consequences of the Copenhagen interpretation, including its denial of realism, stem from its predilection for a subjective interpretation of probability statements. In his own words:

> ... it is this mistaken belief that we have to explain the probabilistic character of quantum theory by our (allegedly necessary) lack of knowledge, rather than by the statistical character of our problems, which has led to the intrusion of the observer, or the subject into quantum theory. [1, p. 17]

If this is the source of the problem, then to defend realism and objectivity from the quantum mechanical challenge, one need only establish the tenability of an objective interpretation of quantum me chanical probabilities. This is precisely Popper's strategy.

3 Indeterminacy and subjectivism

The topic of Popper's first disagreement with the Copenhagen theorists was the interpretation of the Heisenberg indeterminacy formulae; it was here that Popper thought he detected most clearly the Copenhagen penchant for subjectivism. But while indeterminacy was the immediate focus of Popper's initial critique, the implications of that critique were widespread, touching already upon the central problem of the meaning of the quantum mechanical state function. Since the supposed need to combat the malaise of subjectivism is what gives meaning and purpose to Popper's whole program in the philosophy of quantum mechanics, we would do well to examine this part of his argument in considerable detail.

From the start, the discussion of indeterminacy, even among those identified as the members of the Copenhagen school, was characterized by a tension between at least two different interpretations. In some cases one even finds a vacillation between the two in the work of an individual physicist. The first takes indeterminacy to be merely a *limitation on measurement*, while the second takes it to be a reflection of some *intrinsic indefiniteness* of the coordinates of atomic systems. Both views entail a limitation on how precisely we can *know* a system's coordinates.

Heisenberg, himself, tended toward the former interpretation, though he was by no means wholly consistent in his published pronouncements on the subject. Typically, he illustrated the indeterminacy relations with thought experiments which purport to show how a measurement of an electron's position, following close up on a momentum measurement, would necessarily interfere with or disturb the electron's momentum in an inherently unpredictable manner, so that the prior momentum measurement would be rendered useless for the purposes of prediction [22, pp. 21-30]. This means that the physical disturbance caused by the position measurement necessarily renders imprecise our previously obtained *knowledge* of the electron's momentum, and, because it thus focuses our attention on limitations on knowledge, Poper labels this reading of indeterminacy *subjective* [13, p. 215].

Following the lead of Schrödinger's wave mechanics, a number of other physicists opted for the intrinsic indefiniteness interpretation of indeterminacy. According to this view, the Heisenberg formulae imply that it is physically impossible for a system like an electron even to possess simultaneously definite values of any two conjugate variables The unavoidable imprecision in our knowledge of the value of these coordinates is thus explained as the reflection of a deeper-lying indefiniteness of the coordinates themselves. Because it regards indeterminacy as a property of the system itself, and not directly as a

function of our limited knowledge of what might actually be a well-defined system, Popper terms this approach *objective* [13, p. 215].

At the time he published *Logik der Forschung* (1935), Popper had had no formal advanced training in physics – he says of his abilities in this area: "I felt in the end that I was not really good enough" [15, p. 57] and yet he already sensed the physicists' confusion over indeterminacy more clearly than did many physicists themselves. In an interesting analysis, Popper traces the vacillation between the *subjective* and *objective* approaches to indeterminacy to the physicists' desire to reconcile what, at the time, Popper took to be ultimately inconsistent features of the quantum theory as it was then understood [13, pp. 210-211]. On the one hand, the apparent success of quantum mechanics in explaining such fundamental phenomena as the stimulated emission of radiation from atoms, the photoelectric effect, and the specific heats of solids (all of which were embarrassing problems for classical theories of radiation and matter), together with the resolution of doubts about the strict conservation of energy in individual atomic processes, encouraged physicists to view quantum mechanics as a fundamental theory of the mechanics of individual systems. Such a view also accorded with a traditional conception of physics as the basic science of nature, the science which concerns itself with the deepest level of structure. According to Popper, it was this attitude which initially inclined physicists toward the *objective* approach. At the same time, however, the emerging consensus, especially among those termed the Copenhagen theorists, was that quantum mechanics is an irreducibly probabilistic theory, and this belief had just received impressive reinforcement at the time *Logik der Forschung* was being written, in the form of von Neumann's famous, though, as we now know, mistaken, proof of the impossibility of deterministic, hidden variable interpretations of quantum mechanics [23, pp. 313-325] [24]. The problem, Popper suggests, was in understanding how quantum mechanics could be *both* a theory of individual systems *and* a probabilistic theory. In Popper's account, it was the physicists' striving to resolve this difficulty that led them down the garden path to subjectivism.

On the face of it, there is no difficulty; the *objective* approach to indeterminacy seems to hold the key. One could simply argue that theory's probabilistic character stemmed from an intrinsic indefiniteness in the states of individual systems, or, more specifically, from an indefiniteness in some or all of the parameters, such as position and momentum, which go to make up the state; the uncertainties mentioned in the Heisenberg formulae would be the measures of this indefiniteness. Under these circumstances, probabilistic predictions would be the norm, for if, say, the position of a system is intrinsically indefinite at a given time, then no matter how *deterministic* the theory's dynamical laws are, they can at best yield only a range of possible positions for the system at any subsequent time. But however attractive this analysis might initially appear, certain considerations raise serious doubts about its cogency, doubts which go far beyond the mere offense to our classical intuitions, according to which indeterminacy is always but an appearance, masking a still deeper deterministic reality.

From a physical point of view, the most vexing difficulty concerns the famous problem of the *collapse* or *reduction* of the wave packet [13, pp. 231-232] [1, pp. 34-38]. If we are to regard indefiniteness as an intrinsic property of an individual system, then we must regard the state function as a description of the intrinsic state of that system (as opposed to regarding it as a representation of our knowledge of the system). In the most general case, which is typical of a system *prior* to the performance of a measurement on it, the state function is of a type known as a superposition, a weighted, complex sum of a number of different state functions, each of which, by itself, corresponds to a state wherein the system has a definite value of the observable to be measured. At the very least, a system described by a superposition cannot be *known*, with certainty, to be in any one of the component states, called the *eigenstates* of the observable; but one can calculate the *probability* of the system's being found, upon measurement, to have any specific value of the observable. Repeated measurements on identically prepared systems will yield different results, but the distribution of those results will conform to the predicted probabilities. Whether the system can be said actually to *be* in one of the eigenstates prior to the measurement is an important question of interpretation. (It is the presence of superpositions in the formalism that gives rise to non-zero indeterminacies. The indeterminacy principle's restriction on conjugate parameters is related to the formalism's not admitting simultaneous eigenstates of conjugate observables. If a system is in an eigenstate of one observable, it will be in a superposition with respect to any conjugate observable.)

Being committed to the view that the state function describes the state of the system itself, the proponents of intrinsic indefiniteness must conceive of the state corresponding to a superposition as somehow "blurred" or "smeared". They would have to deny that a system thus described *is in any definite eigenstate*, be it known or unknown. The problem with this view is that measurement always yields a definite result, and if the measurement is a non-destructive one (for example, a filter that only allows systems with a definite value of the observable to pass), then immediately *after* the measurement a system is always describable by a state function corresponding to

one of the eigenstates represented in the original superposition. Such a description of the post-measurement state of the system is correct, in the sense that it leads to the proper predictions of the system's subsequent behavior. It would appear then, from the *objective* point of view, as though a system originally in a "smeared", indefinite state *collapses* upon measurement to some definite state.

But – and this is why the collapse is a problem – quantum mechanics has no way of explaining the collapse. If the state function is taken to describe a system's intrinsic properties, then it is hard to avoid the conclusion that the collapse is a real physical phenomenon, but the dynamics of the quantum theory make no provision for such a quasi-instantaneous, discontinuous transition; certainly the Schrödinger equation, the basic dynamical equation of quantum mechanics, cannot account for such a process. The best the proponents of the *objective* approach can do in the face of this difficulty is to follow von Neumann in taking the collapse as an independent postulate (the so-called *projection postulate*) [23, pp. 347-358] [8, pp. 226-230]. This strategy is not objectionable from an empirical point of view, but the *ad hoc* character of the postulate makes its presence an aesthetic defect in an otherwise elegant theory.

Popper was neither the first nor the only commentator to draw attention to the troubling implications of the problem of wave packet collapse. But over the years he voiced other criticisms of the idea that a superposition describes the intrinsic state of an individual system. One of the most important is his contention that from the point of view of the *objective* indefiniteness interpretation, the quantum formalism contains *metaphysical elements* [13, p. 215]. The details of the pertinent arguments will be reviewed and criticized in section 4, but briefly, what Popper maintains is that while precise *predictions* of all aspects of the future behavior of quantum systems may be impossible, completely precise calculations of all aspects of the *past* behavior of quantum systems are possible according to the formalism. If, as the *objective* indefiniteness interpretation asserts, quantum systems never posses simultaneously precise values of all coordinates, then the results of these calculations are, to say the least, anomalous.

Among the alternatives open to those who are dissatisfied with the *objective* indefiniteness interpretation of the state function and the Heisenberg formulae, two are prominent in Popper's deliberations. The one favored by Popper continues to regard quantum mechanics as a probabilistic theory, but denies that it is itself a fundamental theory of individual systems, viewing the state function instead as a description of the statistical properties of an *ensemble* of systems, and the Heisenberg formulae as what Popper calls *statistical scatter relations*. This view assumes that individual systems do always possess simultaneously definite values of all their coordinates, an ensemble of systems displaying the appropriate scatter in these values. The other alternative is to continue to regard quantum mechanics as a probabilistic theory of individual systems, but to construe the state function as describing not the intrinsic physical properties of microsystems, but, in some way, our *knowledge* of these properties. This view is of a piece with the limitations-on-measurement interpretation of indeterminacy, if the latter is construed, following Popper, as a statement about limitations on the possible extent of our knowledge of atomic systems; and like the limitations-on-measurement idea, it earns from Popper the designation *subjective* [13, p. 215]. However, to avoid begging any questions, I will refer to it as the epistemic interpretation of the state function.

This interpretation has its advantages. For instance, to say that a superposition correctly describes a system is merely to say, on this approach, that our knowledge of the system is limited in such a way that we know only the likelihood of one of its variables being found to have a certain value; no claim about the intrinsic state of the system is entailed, and the resulting difficulties of the *objective* interpretation are avoided. In particular, the collapse of the state function is regarded, from this perspective, simply as a sudden, discontinuous change in the degree of precision of our knowledge of the system, something which is not at all surprising, given that the aim of measurement is to enhance our knowledge of the object of measurement. Another advantage is that the epistemic interpretation enables us to continue regarding quantum mechanics as a theory of individual systems, at least in the sense that the knowledge which it takes as fundamental can be viewed as knowledge, however imprecise, of individual systems.

Problems such as the collapse of the wave packet were a common topic of discussion among the early developers of the quantum theory, and undoubtedly were crucial in leading some to favor the epistemic interpretation of the state function. A general sympathy with positivist scruples regarding talk of unobservable entities should also not be discounted as an influence on their thinking. But Popper has suggested that the situation in the foundations of probability was among the most important factors favoring this choice. He notes that in the late 1920s and early 1930s, crucial years in the development of the *Copenhagen interpretation*, the only *objective* theory of probability under serious discussion was the relative frequency interpretation, which defines the probability of one kind of event, called an *outcome*, as its frequency, relative to other outcomes, in an infinite sequence of other events, called *trials*. According to Popper, this definition presents a problem for those who attempt to construe indeterminacy and quantum mechanical probabilities generally as *objective* properties of individual quantum systems. The difficulty is that, as it stands, the frequency theory does not permit the unambiguous ascription of probabilities to individual events, such as getting a six on a *specific* throw of a die; one could, in principle, have different probabilities for that event, depending on which infinite sequence of trials it is included in. But this means that the attribution of an *objective* probability, in the relative frequency sense, to a specific event is meaningful only *relative* to a specification of an infinite sequence of trials, which is really to say that such a probability is a property not of the individual event, but of the infinite sequence itself. Since the interpretation of quantum mechanics as a fundamental theory of individual systems seems to require the attribution of probabilities to individual events (such as the passage of a specific photon through a filter), these probabilities cannot, therefore, be understood as relative frequencies. And if the relative frequency interpretation is the only *objective* interpretation of probability, it follows that individual quantum mechanical probabilities *are not objective* [13, §71] [15, §34].

In the period under discussion, the principal alternative to the frequency interpretation was one which derives the probability function from a measure of a person's *degree of confidence* in the truth of a proposition. This is commonly referred to as the *subjective* interpretation of probability. It poses no obstacle to our regarding probabilities as properties of individual events and systems, as long as it is understood that the connection with individual events is indirect, in the sense that probabilities, *subjectively* construed, are, strictly speaking, measures of one's faith in the truth of propositions describing the events, and only secondarily properties of the events and systems themselves. In particular, *subjective* probabilities are not infected with the same kind of ambiguity which plagues probabilities regarded as relative frequencies; different individuals might assign different probabilities to one and the same proposition (this is, of course, a large part of what is meant by calling such probabilities *subjective*, and a major reason why Popper opposes this interpretation), but there is no systematic obstacle to any one individual's assigning a unique probability to a proposition describing a single event. (For Popper's objections to the subjective interpretation of probability, see [25] [13, §48 and §62]). It is precisely because the *subjective* interpretation gives us a way to speak of probabilities as properties of individual events and systems that Popper identifies it as the primary motivation for many physicists' having interpreted quantum mechanical probabilities, and thus the state function, in an epistemic, or, in Popper's words, *subjective*, fashion. He says:

Now frequency theorists hold that there are objective questions concerning mass phenomena, and corresponding objective answers. But they have to admit that whenever we speak of the probability of a *single* event, *qua* element of a mass phenomenon, the objectivity becomes problematic; so that it may well be asserted that with respect to single events, such as the emission of one photon, probabilities merely evaluate our ignorance. For the objective probability tells us only what happens on the average if this sort of event is repeated many times: about the single, event itself the objective statistical probability says nothing. It was here that subjectivism entered quantum mechanics ... [15, p. 178]

Popper marshals evidence, in the form of quotations from a variety of authors, in order to convince us that, from the time of the consolidation of the modern quantum theory in the late 1920s,

> physics had become a stronghold of subjectivist philosophy, and it has remained so ever since. [15, p. 177]

We read from Moritz Schlick, one of the fathers of Vienna positivism:

> Of natural events themselves it is impossible to assert meaningfully any such things as "haziness" or "inaccuracy". It is only to our own thoughts that anything of this sort can apply... [13, pp. 215-216] [26]

Sir James Jeans is found commending a similar view:

> In brief, the particle picture tells us that our knowledge of an electron is indeterminate; the wave picture that the electron itself is indeterminate, ... Yet the content of the uncertainty principle must be exactly the same in the two cases. There is only one way of making it so: we must suppose that the wave picture [the ψ function is intended here] provides a representation not of objective nature, but only of our knowledge of nature ... [13, p. 229] [27, p. 235]

Popper comments on the preceding quotation:

> Schrödinger's waves are thus for Jeans *subjective probability waves*, waves of our knowledge. And with this, the whole subjectivist probability theory invades the realm of physics. [13, p. 229]

The list of citations could easily be extended; there is no denying that quantum mechanics has frequently been construed as referring to our knowledge, rather than to the world, especially by thinkers either belonging to or influenced by the tradition of Vienna positivism. To give just one more example, consider the following more recent remark by Heisenberg:

> The conception of the objective reality of the elementary particles has thus evaporated in a curious way, not into the fog of some new, obscure, or not yet understood reality concept, but into the transparent clarity of a mathematics that represents no longer the behavior of the elementary particles but rather our knowledge of this behavior. [28, p. 100]

But Popper wants to claim much more than this. He wants to claim, first, that the *dominant* interpretive tradition since the 1930s – such is the stature, in his estimation, of the Copenhagen interpretation – is *subjectivistic* in import and intent, second, that the allegedly subjectivistic Copenhagen interpretation is the result of a mistaken *subjective* interpretation of quantum mechanical probabilities; and third, that this "intrusion of the subject" into quantum theory leads the Copenhagen theorists to conclude that "objective reality has evaporated". About each of these claims serious questions must be raised. In particular, as I mentioned earlier, I want to criticize the assumption that the opinions of Bohr and Heisenberg coalesce into a single interpretation. A subjectivist epidemic may have raged in Copenhagen, but some people might not have succumbed.

First we must pause to consider carefully what Popper means when he labels a point of view *subjective*. That an unflattering contrast with *objective* is intended is clear. He remarks, for example:

> ... if one interprets the Heisenberg formulae (directly) in a subjective sense, then the position of physics as an objective science is imperilled. [13, p. 229]

But the meaning of *subjective* is not exhausted by this contrast. We have found that what earns an interpretation the designation *subjective* from Popper is its being somehow concerned with our knowledge of physical systems, rather than with the systems themselves. But if *subjective* interpretations of quantum mechanics are to be opposed because they represent a threat to *objectivity*, there is something puzzling about simply making reference to knowledge the criterion of *subjectivity*, for as Popper himself has stressed in other contexts, there are both *subjective* and *objective* types of knowledge:

Now I wish to distinguish between two kinds of 'knowledge': subjective knowledge (which should better be called organismic knowledge, since it consists of the dispositions of organisms); and objective knowledge, or knowledge in the objective sense which consists of the logical content of our theories, conjectures, guesses ... [29, p. 73]

In terms of Popper's later "three worlds" ontology, *subjective* knowledge is knowledge in its *World 2* aspect, where *World 2* is "the mental world, or world of mental states" [29, p. 154] (in Popper's philosophy, there is no inconsistency in regarding subjective knowledge as both a mental state and a disposition of an organism. He rejects a Cartesian dualism of mental and physical *substances*, but advocates a dualism of mental and physical *states*. A mental state is, thus, for Popper, just a special kind of dispositional state of an organism [29, p. 231]), and *objective* knowledge is knowledge in its *World 3* aspect, this world being characterized as:

> ... the world of intelligibles, or of ideas in the objective sense; it is the world of possible objects of thought: the world of theories in themselves, and their logical relations; of arguments in themselves; and of problem situations in themselves. [29, p. 154]

It is the existence of just this ambiguity in our talk of *knowledge* which led me to speak above of *epistemic* interpretations, rather than *subjective* ones, the point being that talk of *knowledge* may not be, automatically, a sign of *subjectivism*.

In view of the prominent role which Popper assigns to the objective/subjective knowledge distinction, one would assume that when he indicts an interpretation as *subjectivist* because it makes essential reference to knowledge, he means to indict it for referring to knowledge in the subjective sense, but while his bill of particulars usually includes the expression *subjective knowledge* or one of its cognates [29, p. 141], he does not always specify the nature of the crime this carefully. Popper no doubt wants to criticize reference to either kind of knowledge in any interpretation of a physical theory which makes our knowledge, rather than the world, the focus of the theory's attention, for the claim that theories tell us about our knowledge, rather than the world, does not harmonize with a realist attitude. Moreover, one of the traditional (though, I would argue, misleading) uses of the terms *subjective* and *objective* is precisely to mark the knowledge/world distinction. So while it may be confusing, the distinction between subjective and objective knowledge might simply be irrelevant to Popper's discussion of epistemic interpretations of quantum mechanics.

But I do not think that this is the case. Popper was too systematic a philosopher not to have noted such a confusion and sought to correct it. My suggestion is that he believed that *subjectivism* is the same sin, wherever it might be encountered, and, thus, that the *subjective* character of an interpretation of quantum mechanics consists in nothing more and nothing less than its taking the theory to tell us about our *subjective* knowledge of events, knowledge in the *World 2* sense, rather than the events themselves. In a later essay he said:

> ...it is not surprising that neglect of the third world – and consequently a subjectivist epistemology – should be still widespread in contemporary thought. Even where there is no connection with Brouwerian mathematics there are often subjectivist tendencies to be found within the various specialisms, I will here refer to some such tendencies in logic, probability theory, and physical science. [29, p. 140]

The story which he then sketches about *subjectivism* in probability theory and physics is the same one we have been reviewing in this section. It is only in his early discussion of subjectivism in quantum mechanics, in *The Logic of Scientific Discovery*, that Popper fails to heed the distinction between subjective and objective knowledge, but in that discussion the same work is performed by a distinction which prefigures the later one, namely, a distinction between "our knowledge" *simpliciter* and "objective science", the former being defined, just as subjective knowledge is later defined, as a "system of *dispositions*". And about "knowledge", thus defined, Popper adds:

> But all this interests only the psychologist. It does not even touch upon problems like those of the logical connections between scientific statements, which alone interest the epistemologist. [13, p. 80]

Thus, whether Popper speaks just of "knowledge", in his early work, or more pointedly of *subjective knowledge*, in his later work, his charging an interpretation with subjectivism means simply that it reads quantum mechanics as telling us about a kind of knowledge which is the private possession of individual knowing subjects, rather than about the physical world.

There is a good reason for being careful to show that what motivates Popper's charge of subjectivism is a worry about reference to knowledge of the specifically *subjective* type. For this criticism, directed against Bohr, Heisenberg and the other Copenhagen theorists, is a crucial part of Popper's argument; it is what first establishes the need for an alternative, objective interpretation of quantum mechanics, which Popper then provides. But I want to argue that the charge, thus understood, is unfounded.

As noted above, it is easy to accumulate a list of references to knowledge in the writings of commentators on quantum mechanics who are loosely associated with Bohr. Some of them, surely, can plausibly be read as references to subjective knowledge (Schlick's mention of "our own thoughts" is an example), but many, if not most, cannot (I would be reluctant to read Jeans talk of "our knowledge of nature" in this way). The fact is that many, often ill-considered things have been said about quantum mechanics by philosophers and, especially, by physicists. And it is pointless to press these remarks too hard with fine, philosophical distinctions. I would guess that if we explained the subjective/objective knowledge distinction to Jeans, he would say that, of course, he meant knowledge of the objective sort, but it is in the nature of the case that such conjectures cannot easily be established. It is a far better strategy for critic and defender simply and straight-forwardly to consider the facts. Thus, we must address the question as to whether or not there is anything about the quantum, theory that would lead someone to assign subjective knowledge a special place in its interpretation.

The hallmark of objective knowledge, according to Popper, is its susceptibility to inter-subjective testing and criticism [13, p. 25]. Is there any reason to suspect that our knowledge of quantum objects and events falls short of this standard? The original source of the worry over subjectivism in quantum mechanics is doubtless Heisenberg's unfortunate explanation of indeterminacy as the inevitable consequence of the observer's disturbance of the object of observation. Popper is quite explicit about this. He describes the Copenhagen interpretation as:

> ... the claim that, in atomic theory, we have to regard *"the observer"* or *"the subject"* as particularly important, because atomic theory takes its peculiar character largely from *the interference of the subject or the observer (and his "measuring agencies") with the physical object under investigation.* [1, p. 10]

We seem to be confronted here with a clear-cut failure of objectivity, since the magnitude of the disturbance allegedly occasioned by the observer's intervention presumably varies from observer to observer, and from observation to observation, owing to its inherent unpredictability (or "uncontrollability", in Bohr's quaint, early terminology) [22, p. 3] [7, pp. 57-68].

But it would be seriously misleading to describe this situation simply as a failure of objectivity, and to make it the basis for the charge that physics has become the stronghold of subjectivism. For one thing, we must be careful about just who or what does the disturbing. While Heisenberg himself sometimes speaks of the observer as the "subject", [22, p. 2] conjuring up the image of the denizens of *World 2*, the proper object of our concern is the *purely physical interaction* between the quantum system under investigation and the observer, the latter being considered now as nothing more than an especially complicated physical system. Indeed, it is better to speak simply of an interaction between the object and a *measuring instrument*, and to leave conscious human observers out of the picture entirely, for whatever the novelties of observation in quantum mechanics, they remain the same if human observers are replaced by instrumentation attached to automatic recording devices. Thus, even if it were true that every observation disturbs the observed object in a different way, the difference is not due to the involvement of different human *subjects*. It is, instead, a mundane physical fact of a purely objective, *World 3* sort. Furthermore, factoring out the human observer makes it clear that the knowledge derived even from "disturbing" observations is totally objective, for the result preserved by the recording device needs only to be formulated as a statement in order to become a publicly debatable, *World 3* entity, something that all researchers can agree upon. Some physicists would question the claim that the recorded results are objective. The most straightforward description of the measurement process implies that an instrument interacting with a system described by a superposition will itself be in a superposition at the conclusion of the measurement interaction, whereas I am assuming that the instrument winds up in a definite state. If it is in a superposition, then different observers might, conceivably, find it in different states when they look at it. Of course, I question this simple description of measurement, but, unfortunately, there is no consensus among physicists on an adequate quantum theory of measurement [8, pp. 471-521] [30, pp. 159-226].

Another reason to be wary of drawing substantive philosophical conclusions from Heisenberg's disturbance analysis of indeterminacy is that talk about "disturbance" or "interference" seems to beg one of the crucial questions at issue in debates over objectivity and realism in quantum mechanics. Specifically, it presupposes that the observed system has a definite value of the parameter which is the conjugate of the measured parameter (for example, a definite momentum in the case of a position measurement), for otherwise it would be unclear what, if anything, is being disturbed.

It is indeed true that, from a philosophical point of view, one of the most important implications claimed for the quantum theory is that it mandates a radical revision in our understanding of the observer-observed relationship. Bohr, Heisenberg and others have argued that the picture of the detached observer, which underlay the edifice of classical scientific realism, must be abandoned in favor of a model in which there is a far more intimate connection between observer and object. In fact, a case can be made for the claim that the two lose their separate identities during, and for some time following the observation interaction. This is a consequence of what is known as the *nonseparability* of the quantum mechanical description of interacting systems [30, pp. 75-156]. Interestingly, Wigner proposes an extremely subjectivist interpretation of quantum mechanics [31]. He adopts the account of the measurement process described above and argues that the superposition representing the instrument's state is reduced in the consciousness of the observer. Still, in all of this, and Wigner's musings about the reduction of superpositions in consciousness notwithstanding [31], it should be stressed that Bohr's position is not at all like Wigner's, and there need be no mention of observing *subjects*. If a conscious human observer plays any role at all, his subjective consciousness is of no consequence; for an understanding of observation in the quantum domain, we need only consider the observer as a physical system.

No doubt some confusion about this matter has been caused by the fact that Bohr occasionally speaks as if what were at issue here *is* the "subject-object" relationship, as when he talks of the urgent necessity, in the consideration of quantum mechanical observations, of paying "proper attention to the placing of the object-subject separation" [32, p. 79]. But a careful reading of these passages reveals that this manner of speaking is nothing more than Bohr's concession to standard philosophical parlance when he is attempting to explain to a broad audience the "epistemological implications of the lesson regarding our observational position, which the development of physical science has impressed upon us," [32, p. 78] and, in particular, when he is attempting to point up the structural analogy between the observer-observed relationship in quantum mechanics and the subject-object relationship in psychology [33]. Ordinarily Bohr speaks simply of the relationship between the observed object and the *measuring instruments* we employ in observation, with no mention of an observing subject. In this cleaner idiom, the need to attend to the placing of the object-subject separation becomes the

> necessity of discriminating in each experimental arrangement between those parts of the physical system considered which are to be treated as measuring instruments and those which constitute the objects under investigation. [9, p. 701]

Ironically, the very passage which Popper cites from Bohr to back up his claim that the Copenhagen interpretation stresses the interference of *the subject* with the object under investigation, refers not to the subject, but to the measuring instruments:

> Indeed the *finite interaction between object and measuring agencies* . . . entails . . . the necessity of a final renunciation of the classical ideal . . . and a radical revision of our attitude towards the problem of physical reality. [9, p. 697]

The fact of the matter is that Bohr rejects, unequivocally, the suggestion that his interpretation of quantum mechanics assigns *the subject* any special role:

> . . .the decisive point is that in neither case [neither relativity nor complementarity] does the appropriate widening of our conceptual framework imply any appeal to the observing subject, which would hinder unambiguous communication of experience . . . in complementary description all subjectivity is avoided by proper attention to the circumstances required for the well-defined use of elementary physical concepts. [34, p. 394]

The expression "unambiguous communication" alludes to Bohr's own conception of objectivity. In his more careful moments, Heisenberg is equally clear about there being no special role for the subject in quantum mechanics:

> Of course the introduction of the observer must not be misunderstood to imply that some kind of subjective features are to be brought into the description of nature. The observer has, rather, only the function of registering decisions, . . . and it does not matter whether the observer is an apparatus or a human being. [35, p. 121]

Thus, though it is perhaps warranted to speak, as Popper does, of Bohr's having sanctioned an "intrusion of the observer" into quantum theory, there is no warrant for speaking, as Popper also does, of an "intrusion of the subject", as if the two kinds of intrusion came to the same thing. It should be pointed out here that Bohr was, from the start, sceptical of Heisenberg's disturbance analysis of indeterminacy [10] [32, p. 73]. As early as 1927 he was exhorting readers who wished to grasp the full significance of Heisenberg's formulae to consider not only the "possibilities of observation", but also the "possibilities of definition", meaning that we have to go beyond an analysis of the observer-observed interaction to examine the way concepts such as *position* and *momentum* are employed in the description of quantum phenomena [7, p. 73].

We earlier saw that Popper's account of the etiology of subjectivism in the Copenhagen interpretation of quantum mechanics assigns a major role to the way Bohr and colleagues are supposed to have interpreted the probability calculus, the allure of the subjective interpretation of probability being cited as a principal reason for their alleged choice of a *subjective* epistemic interpretation of the state function. Do we have here the makings of a better argument for the charge of subjectivism? Recall that the subjective interpretation of probability amounts, roughly, to the claim that the probability function is a measure of an individual's degree of confidence in the truth of a proposition describing the event whose probability we want to know. It is *subjective* because different individuals could assign different probabilities to the same event. "Degree of confidence", especially when understood as something like "strength of belief", has a pronounced *World 2* character. What is the evidence that Bohr and the other Copenhagen theorists subscribed to this view of probability?

Popper bases his case on the fact that many of the thinkers he associates with the Copenhagen school portray the lesson of indeterminacy as the existence of limitations on our knowledge of the properties of quantum systems. For example, Heisenberg, the source of this interpretive tradition, comments:

> The uncertainty principle refers to the degree of indeterminateness in the possible present knowledge of the simultaneous values of various quantities with which the quantum theory deals. [22, p. 20]

As Popper tells it, subjectivism infected the thinking of the Copenhagen school because its members give the wrong reasons for the probabilistic character of the quantum theory, and

> Foremost among these reasons is the argument that it is our (necessary) lack of knowledge – especially the limitations to our knowledge discovered by Heisenberg and formulated in his *"principle of indeterminacy"* or *"principle of uncertainty"* – which forces us to adopt a probabilistic, and consequently a statistical, theory. [1, p. 17]

As to why this leads to subjectivism, he continues:

> ... it is this mistaken belief that we have to explain the probabilistic character of quantum theory by our (allegedly necessary) *lack of knowledge*, rather than by the statistical character of our problems, which has led to *the intrusion of*

the observer, or the subject into quantum theory. It has led to this intrusion because the view that a probabilistic theory is the result of lack of knowledge leads inescapably to the *subjectivist interpretation of probability theory*; that is, to the view that the probability of an event measures the degree of somebody's (incomplete) knowledge of that event, or of his "belief" in it. [1, pp. 17-18]

There can be no denying that adoption of a subjective interpretation of probability would incline one to interpret the quantum mechanical state function in a subjectively epistemic manner. But several objections can be raised to Popper's analysis of the Copenhagen attitude toward probability. To begin with, it is by no means obvious that a limitations-on-knowledge interpretation of indeterminacy leads "inescapably" to the subjective interpretation of probability. It would depend, in part, on whether the knowledge in question is of the subjective, *World 2* variety or the objective, *World 3* variety. If it is the latter, then, *at best* one might speak of an *epistemic* interpretation of probability, but not of a *subjective* interpretation, at least not in the special pejorative sense that *subjective* carries in Popper's vocabulary. What I am here calling an *epistemic* interpretation has much in common with the "logical" theory of probability, or "probability$_1$" in Carnap's terminology, which regards probability statements, such as $p(a, e) = r$, as logical truths expressing something like the degree of evidential support which a body of evidence, *e*, provides for an hypothesis, *a*. Popper considers the logical theory to be merely a "variant" of the subjective interpretation [13, §48, p. 136]. But this is a mistake, encouraged partly by some of the logical theory's proponents' having spoken of logical probability as a measure of degree of *rational belief*, and partly by Popper's systematic neglect of the distinction that one might, in any case, draw between belief, *simpliciter*, and rational belief, for talk of *rational belief* presumably alludes to objective norms of rationality [25] [36, pp. 43-47, 238-241]. Indeterminacy would entail limitations on *subjective* knowledge only if it were a consequence of an observing *subject's* special relationship with the observed object. But, as I have argued, Popper's attribution of this view of indeterminacy to Bohr and Heisenberg is a misreading of their positions. When indeterminacy is correctly viewed as a feature of the physical relationship between the observed object and the instruments used to measure its various properties, then whatever limitations it entails on our knowledge of quantum systems are properly seen as purely objective limitations, the same for all observers. Popper's failure to distinguish subjective and objective knowledge in the context of the interpretation of probability is evident in the following passage, where it is unclear which kind of knowledge is intended:

> the various subjective interpretations have all one thing in common: probability theory is regarded as a means of dealing with the incompleteness of our knowledge. [37, p. 25]

On a deeper level, it is not clear why *tracing the need* for probabilities in quantum mechanics to our necessarily limited knowledge of quantum systems (if this is in fact what Bohr and Heisenberg do) commits one to *interpreting* those probabilities in an epistemic fashion. To take an example from a less controversial domain, there would be no inconsistency in my claiming that I have to describe the behavior of a tossed coin probabilistically because there are no practical means by which I could know all of the relevant initial conditions of a toss, while nevertheless insisting that what I *mean* when I make a probabilistic prediction about the outcome of a toss is that in a hypothetical infinitely long run of similar tosses a certain relative frequency of heads and tails would be found.

But the most serious objection to Popper's analysis is simply that, aside from Heisenberg's characterization of the uncertainty principle as referring to an "indeterminateness" in our knowledge (which, one should note, says nothing about probability, *per se*), there is little basis for the claim that the Copenhagen interpretation views limitations on knowledge as the explanation of the probabilistic character of quantum mechanics. In fact, Bohr *explicitly rejects* this account. In explaining why quantum mechanics differs essentially from statistical mechanics, where resort to a probabilistic description *is* generally conceded to be the result merely of a lack of complete, knowledge of all the relevant parameters of the systems under investigation (and a science in which Popper also finds significant traces of subjectivism), Bohr remarks:

> ...we have in each experimental arrangement suited for the study of proper quantum phenomena not merely to do with an ignorance of the value of certain physical quantities, but with the impossibility of defining these quantities in any unambiguous way. [9, p. 699]

Bohr's position, in other words, is that we have to resort to a probabilistic description in quantum mechanics because of limitations more basic than any limits on our knowledge of the parameters of quantum systems. In another place, Bohr comments:

> ... the statistical character of the uncertainty relations in no way originates from any failure of measurements to discriminate within a certain latitude between classically describable states of the object, but rather expresses an essential limitation of the applicability of classical physical ideas to the analysis of quantum phenomena. [33, p. 311]

Bohr's interpretation is not a variety of the intrinsic indefiniteness interpretation. In his view, the parameters conjugate to a measured parameter are not indefinite. Instead, he would say that the concepts corresponding to these parameters do not even apply in such a context (e.g. the concept of *momentum* does not apply in the context of a *position* measurement) [38, 39].

One must be cautious about imputing views on the interpretation of probability to Bohr and Heisenberg. As far as I know, neither ever made any explicit pronouncements on the interpretation question, and what little each does say about the role of probabilities in quantum mechanics can just as easily be read as suggesting an objective interpretation as a subjective one. Ultimately, the question is one of the possibilities for testing the quantum theory's probabilistic predictions, for, as Popper correctly observes, the mark of an objective interpretation of probability is its securing the objective testability of probability statements [37, p. 25]. Popper argues that the logical theory is not an objective interpretation precisely because logical probability statements are "untestable tautologies" [25, p. 357]. It is true that such statements are not subject to empirical testing, but Popper has noted elsewhere that the real test of the objectivity of scientific statements is their susceptibility to inter-subjective criticism [13, §8, p. 22] and logical probability statements fare quite well by this criterion, for we can question whether the degree of evidential support has been properly assessed, just as we can criticize a deduction in mathematics. Furthermore, Popper offers no evidence of Bohr's or Heisenberg's having said anything that suggests doubt about the testability of quantum mechanical predictions. My guess is that if we had explained the options to them, Bohr, at least, would have declared, without hesitation, that he had always understood quantum mechanical probabilities in an objective sense, as referring to the relative frequencies of results one would expect to find in repeated measurements of a given parameter on similarly prepared systems. Such an exercise being impossible, however, the best course of action is to suspend judgment. A case, of sorts, can even be made for attributing something like a propensity interpretation to Bohr; see section 5.

We must conclude that Popper has not established his claim that physics has become "a stronghold of subjectivist philosophy" under the influence of the interpretation of quantum mechanics championed by Bohr. The allegation that the Copenhagen interpretation accords a special role in quantum mechanics to the observing "subject" and the charge that it incorporates a subjective interpretation of probability are both unsubstantiated. Quotation of a few equivocal remarks by a peripheral figure like Jeans, a non-physicist like Schlick and even by a major figure like Heisenberg, who, though responsible for some of the most important formal developments in quantum mechanics, was indirectly criticized by Bohr for the superficiality of his views on interpretation, does not suffice to show that Bohr's position – the most sophisticated, subtle and systematic of the views lumped together under the Copenhagen designation – is subjectivist. This is not to say that Bohr never made mistakes, is never obscure and never inconsistent. He is guilty of all these failings. But he is not a subjectivist.

A more sympathetic reading of Bohr, and one that attends to the many differences between Bohr and the other Copenhagen thinkers, reveals that he is just as anxious as Popper to preserve the objectivity of quantum mechanics, even if his approach is somewhat different. In a discussion of the aim of science, Bohr remarks:

> ... our task must be to account for [human] experience in a manner independent of individual subjective judgement and therefore objective in the sense that it can be unambiguously communicated in the common human language. [40, pp. 63-64]

And of his own complementarity interpretation of quantum mechanics he says:

> The notion of complementarity does not imply any renunciation of detailed analysis limiting the scope of our inquiry, but simply stresses the character of objective description, independent of subjective judgment, in any field of experience where unambiguous communication essentially involves regard to the circumstances in which evidence is obtained. [41, p. 1105]

Unambiguous communicability is not the same thing as intersubjective testability (though it might turn out that both are secured by the same conditions), but a disagreement over the definition of *objectivity* does not, by itself, entitle Popper to conclude that Bohr's interpretation is not objective.

Finally, what about Popper's claim that the Copenhagen interpretation asserts that "objective reality has evaporated"? He offers no independent argument for this claim, apparently on the assumption that it is implied by the subjectivism charge. Is this claim any more warranted than the latter? That Heisenberg believed that objective

reality has evaporated is certain, for he is the source of the quoted remark [28, p. 100]. But this is another place where we must be careful to distinguish Heisenberg's views from Bohr's, for while Bohr is clearly critical of the presuppositions of the kind of realism associated with the world-view of classical physics, his criticism does not extend as far as Popper suggests. Bohr's position is not that there is no quantum mechanical reality, but rather that quantum mechanics forces us to reexamine and redefine the circumstances under which we can speak of the reality of the properties of quantum systems, the fundamental claim of his complementarity interpretation being that we can regard complementary properties as real only under incompatible experimental conditions, and never simultaneously [42], because

> these conditions constitute an inherent element of the description of any phenomenon to which the term 'physical reality' can be properly attached [9, p. 700]

We will return to Bohr's attitude toward realism in section 4, but it should already be clear that on this issue, too, Popper's attack on Bohr is, if nothing else, a bit hasty. In his *Autobiography*, Popper says the following: "About Bohr I said little in *Logik der Forschung* because he was less explicit than Heisenberg, and because I was reluctant to saddle Bohr with views which he might not hold" [15, p. 111]. One wishes that Popper had continued to display such reserve.

There *is* a deep dispute between Bohr and Popper, for, as the previous quotation suggests, Bohr denies the possibility of the simultaneous definiteness of conjugate variables, while Popper asserts that they are always simultaneously definite, and this assertion is a central part of Popper's own interpretation of quantum mechanics. Moreover, to someone who believes in simultaneous definiteness, the claim that we cannot simultaneously know both the position and momentum of a system with arbitrary precision might appear to say more about our knowledge than about the world, and hence, might appear as a kind of subjectivism. But not only would such an inference run the risk of the confusion, discussed above, between the objective and subjective senses of *knowledge*, it would also be objectionable on the ground that, since it is based on the assumption of simultaneous definiteness, it thus ignores the possibility that the apparent limitations on our knowledge are a genuine reflection of some important features of objective, quantum mechanical reality.

4 Simultaneous definiteness and the statistical scatter interpretation of indeterminacy

As we saw above, Popper believes that physicists were led down the garden path of subjectivism by their desire to reconcile the theory's probabilistic character with their belief that it is also a fundamental theory, a theory of individual quantum systems. At the time he wrote *Logik der Forschung*, Popper believed that the way to avoid subjectivism is to give up the latter belief. Accordingly, he proposed an objective *statistical* interpretation of quantum mechanics, which holds that the theory is true only of *ensembles* or *aggregates* of systems [13, p. 234], and the centerpiece of this alternative is the proposal that we view the Heisenberg indeterminacy formulae as neither the consequence of limitations on measurement nor an expression of intrinsic indefiniteness, but as *statistical scatter relations* [13, §75]. Popper later believed that we have an objective probabilistic theory of individual systems, which is embodied in his later *propensity* interpretation of quantum mechanics, and thus he gave up the claim that quantum mechanics only refers to ensembles; but he still defended the statistical scatter view of indeterminacy and all that it presupposes [1, pp. 20-21].

To interpret the indeterminacy formulae as statistical scatter relations is, first, to assume that atomic systems can possess simultaneously sharp values of their conjugate parameters, and, then, to construe the formulae as expressing lower limits on the scatter of the results of measurements of these parameters on any ensemble of such systems. For example, imagine that we have a device, such as a screen with a slit in it, which selects from a beam of incoming particles only those whose (definite) positions are confined to the region $q_x \pm \frac{1}{2}\Delta q_x$. This selection defines an ensemble of particles according to their positions. The scatter interpretation would then imply that the (definite) results of measurements of the momenta of the particles of this ensemble must be scattered over a range no smaller than $\Delta p_x \geq \frac{\hbar}{2\Delta q_x}$. In Popper's words, the Heisenberg formulae, construed thus, "mean only that there are limits to the *statistical homogeneity*" of the results [1, p. 21].

What is really controversial is not the scatter interpretation *per se*, but the assumption upon which it rests, that, contrary to the standard interpretations of indeterminacy, simultaneously precise conjugate parameters are possible. In a bold gesture, Popper claims not only that this assumption is *compatible* with quantum mechanics, but also that simultaneously precise *measurements* of these parameters are possible and, in fact, *necessary* for testing the statistical predictions of the theory [1, p. 20]. All of these assertions are questionable.

Popper's argument regarding simultaneously precise measurements starts with a critique of some early thought experiments which Heisenberg used to illustrate his own interpretation of indeterminacy. The experiments use standard methods of measuring either the position or momentum of an atomic system, and they all have this much in common: when the physical details of the measurement procedure are examined, say in the case of a position measurement, it is allegedly found that the measurement necessarily produces a mechanical disturbance of the observed system's momentum of precisely the magnitude required to satisfy the indeterminacy relations [22, pp. 21-30].

A typical way to measure momentum (the product of mass and velocity) is to combine two position measurements. Knowing a particle's mass, we measure its position at two different times and divide the resulting distance by the time of flight to get the velocity, which we multiply by the mass in order to get the momentum. If we allow enough time to elapse between the position measurements, momentum measurements of theoretically unlimited accuracy are possible. However, as Heisenberg points out, the accuracy thus obtained pertains only to the particle's momentum *before* the second measurement, because the momentum uncertainty necessarily engendered by this measurement renders the result computed from the two measurements useless for predictions of the particle's momentum *after* the second position measurement. Heisenberg concludes that while *retrodictive* momentum measurements of unlimited accuracy are possible, *predictive* measurements must always conform to the restrictions of the indeterminacy formulae [22, p. 25]. (The terminology – *predictive* and *retrodictive* measurements – is Popper's, not Heisenberg's [1, p. 25]).

Popper's criticism of Heisenberg first took the form of a head-on attempt to show, by means of another thought experiment, that predictive measurements violating the indeterminacy relations were also possible [43] [13, §77]. But shortly after the publication of the experiment Einstein detected an error in it [13, §*xii]. With the failure of this approach, Popper shifted the focus of his criticism to Heisenberg's concession that retrospective violations of the indeterminacy principle are possible. Heisenberg explained his attitude toward such violations as follows:

> This formulation makes it clear that the uncertainty relation does not refer to the past; if the velocity of the electron is at first known and the position then exactly measured, the position for times previous to the measurement may be calculated. Then for these past times $\Delta p\Delta q$ is smaller than the usual limiting value, but this

knowledge of the past is of a purely speculative character, since it can never (because of the
unknown change in momentum caused by the
position measurement) be used as an initial condition in any calculation of the future progress
of the electron and thus cannot be subjected
to experimental verification. It is a matter of
personal belief whether such a calculation concerning the past history of the electron can be
ascribed any physical reality or not. [22, p. 20]

Popper welcomes the concession, but quite rightly attacks
Heisenberg's crudely positivistic interpretation of the significance of retrodictions. Referring to the last sentence
of the previous quotation. Popper remarks:

But it is not a matter of personal belief: the measurements in question are needed for testing the
statistical laws (1) and (2) [the indeterminacy
formulae] that is, the scatter relations ... To
question whether the so ascertained "past history of the electron can be ascribed any physical
reality or not" is to question the significance of
an indispensible standard method of measurement (retrodictive, of course); indispensible,
especially, for quantum mechanics. But once
we ascribe physical reality to measurements for
which, as Heisenberg admits, $\Delta p \Delta q \ll h$, the
whole situation changes completely: for now
there can be no question whether, according
to the quantum theory, an electron can *have* a
precise position and momentum. It can. [1, p.
27]

Up to a point, Popper is right. If simultaneously precise
retrodictive measurements are possible, we have little
choice but to ascribe reality to their results. But he should
not have been so quick to trust Heisenberg's authority
regarding the existence of such measurements. Popper's
argument for simultaneous definiteness rests largely on
Heisenberg's analysis. But maybe Heisenberg was wrong.

In fact, I think Heisenberg and, thus, Popper are both
wrong, and the error in their reasoning is simple to detect.
Consider the case of a retrodictive momentum measurement, one consisting of two successive position measurements. The putative violation of indeterminacy arises
because we can presumably infer the particle's momentum at any time prior to the second position measurement,
and thus, in particular, at the time of the first position
measurement, with arbitrary precision. Whatever the uncertainty attaching to the first position measurement, Δq_x,
it would follow that we could infer a momentum at the
time of that measurement more precise than permitted by
the indeterminacy relations, that is, with $\Delta p_x \ll \frac{\hbar}{2\Delta q_x}$ (the

reasoning, and the criticism, are the same if we look at the
particle's position and momentum immediately prior to the
second position measurement). But this conclusion is
incorrect. What we infer from the two position measurements is not the instantaneous momentum of the particle,
but only its *average* momentum during the interval. If we
assume that the momentum was constant during that interval, then the average and instantaneous momenta would
be equal, but otherwise, no conclusion about the momentum at *any* specific time is licensed. Of course, one can
typically advance theoretical arguments for assuming the
approximate constancy of the momentum, such as the absence of any external forces. But notice that the assertion,
for whatever reason, of a precisely constant momentum
– the necessary hidden assumption in the argument for
retrospective violations of indeterminacy – again begs the
very question at issue, in as much as it would be the assertion of constant *definite* momentum, which is exactly
what the indeterminacy relations are believed by Bohr
and others to disallow (unless, of course, the *position* is
completely indeterminate).

Popper, on his own, adduced other examples of putative
simultaneously precise measurements of conjugate variables. I think they all fail, but for now, I will only examine
the most interesting one. It draws upon a thought experiment first proposed by Einstein, Podolsky and Rosen
(EPR) in 1935 as part of their famous attempt to prove the
incompleteness of quantum mechanics [44], but I will discuss Popper's analysis as it applies to another experiment,
suggested by David Bohm, which is simpler than the EPR
experiment, though formally identical to it in all relevant
respects [45, pp. 614-622]. The simplicity of the Bohm
version of the experiment is largely due to its employing
discrete variables, rather than the continuous position and
momentum variables of the EPR experiment. Consider
the following situation. A spin-zero particle decays into
two spin-$\frac{1}{2}$ particles, A and B. Assume that we have filters
which allow us to select pairs of decay products traveling
in opposite directions without affecting their spins, and
that we have devices (called Stern-Gerlach apparatuses)
which allow us eventually to measure the spins of the
decay particles along any spin axis orthogonal to the particles' paths. The quantum mechanical description of the
decay process entails that total spin is conserved. Thus,
the spin of the original particle being zero, the sum of the
spins of the decay products must also be zero along all
spin axes. Since the decay products are spin-$\frac{1}{2}$ particles,
the spin conservation principle implies that if we measure
the spin of particle A along the z-axis and get a result
of $-\frac{1}{2}$ we can infer that the z-spin of particle B is $+\frac{1}{2}$.
Popper's argument would be that it is possible (at least
by accident) to measure simultaneously the z-spin of A
and the y-spin of B, and then to infer values for the y-spin

of A and the z-spin of B by conservation of spin. Thus, he would conclude, we can make simultaneously precise *measurements* of both the z-spin and y-spin of A and B, even though spins along orthogonal axes, being conjugate variables, are supposedly not simultaneously measurable. And if we can *measure* both components of spin, we have little reason to deny their reality. For Poppers actual argument, which assumes the original EPR version of the thought experiment, see [13, §*xi].

Simultaneous measurements of the z-spin of A and the y-spin of B are possible. What is questionable about Popper's analysis is, again, the inferences drawn from the results of these measurements. The particular problem in the present case is the uncritical use of conservation arguments. Spin conservation is implied by the quantum formalism only in the following restricted sense: if the *same* component of spin is measured on both decay products, then the probability of the results summing to zero is equal to unity. By itself, the formalism licenses no inferences about spin conservation in the case of measurements of *different* components, which is the case Popper contemplates. In order to invoke conservation arguments in this case, additional assumptions are needed. Specifically, one must *assume* precisely what the possibility of simultaneous measurements are held to *establish*, namely, that each particle actually possesses a definite spin along every axis, and not just along the axis for which the spin is measured. (One must also assume that these definite spins are correlated as required by the conservation principle.) Only thus can a direct measurement of, say, the z-spin of A be held also to constitute a *measurement* of the z-spin of B.

What the foregoing reflections show is that Popper cannot argue for the simultaneous definiteness thesis by appealing to the possibility of simultaneous measurements of conjugate parameters, because his putative examples of the latter are interpretable as simultaneous measurements of the necessary sort only if one assumes simultaneous definiteness. In other words, Popper is arguing in a circle. There would be nothing particularly vicious about this circle if there were independent evidence for the definiteness thesis, but such evidence has not been provided, and none is likely to be forthcoming. Indeed, in this connection Popper's exploitation of the EPR-type thought experiment is especially ironic, because that experiment can be turned into an *empirical* argument *against* both the simultaneous definiteness thesis and a broad class of realistic interpretations of quantum mechanics.

Einstein, Podolsky and Rosen sought to use their thought experiment to prove that quantum mechanics is *incomplete*. Their argument begins with a crucial assumption, namely, a criterion of physical reality:

> If, without in any way disturbing a system, we can predict with certainty (i.e., with probability equal to unity) the value of a physical quantity, then there exists an element of physical reality corresponding to this physical quantity. [44, p. 777]

Then they reason, in effect, that the possibility of indirect, non-disturbing measurements of the kind sketched above, such as a measurement of the z-spin of B carried out by first measuring the z-spin of A and then inferring a value for B by the conservation principle, shows that, without disturbing a system, we can predict with certainty the values of *two* of its conjugate variables. This is because we could have measured either the z-spin or the y-spin of A, and thus could have made a prediction of the value of either component of B's spin, and whichever component we predict, a subsequent direct measurement of it would confirm our prediction. But then it follows, according to the reality criterion, that both the z-spin and y-spin of B are real, and since the orthodox quantum theory rules out the simultaneous definiteness of conjugate variables, one would have to conclude that quantum mechanics is incomplete. In reply, Bohr pointed out that the two indirect measurements cannot be executed simultaneously, since the necessary experimental arrangements are incompatible (for example, the Stern-Gerlach apparatus which measures A's spin cannot be oriented in such a way as to measure simultaneously spin along two different axes), and since he held that the experimental conditions

> constitute an inherent element of the description of any phenomenon to which the term 'physical reality' can be properly attached. [9, p. 700]

he therefore denied that the EPR argument established the *simultaneous* reality of conjugate variables.

Because Bohr makes his own strong assumptions about the circumstances under which we can speak of the *reality* of quantum properties, his reply is hardly a refutation of the EPR argument. But something close to a refutation is possible, or, at least, a very uncomfortable dilemma can be forced upon EPR. Because of the thought experiment's basic symmetry, EPR's premises actually imply the simultaneous reality of the conjugate parameters of both A and B (which is precisely what Popper wants to assert), and this implies, in turn, that each of the systems has its own *separate* physical state. That EPR are committed to this assumption was first suggested in [46]; but for a detailed defense of the attribution one should consult my dissertation [47]. A helpful discussion of these issues may be found in [30, §8-9]. If there is any doubt that Popper agrees with this consequence of the EPR point of view, consider the following remark:

Thus the so-called "paradox" of Einstein, Podolsky, and Rosen is not a paradox but a valid argument, for it established just this: that we must ascribe to particles a precise position *and* momentum, which was denied by Bohr and his school [1, p. 28]

If one believes, in this sense, that the two systems have their own "independent reality" from the time they cease to interact, it would seem to follow that whenever we perform a direct measurement on one of the systems, the result will depend solely on the properties of that system. But while this might appear to be an obvious fact about measurement, it can be shown that the *locality* assumption (i.e., the results of a measurement depend solely on the *local* properties of the measured object) necessarily leads to predictions for a class of correlation measurements, which can be performed with the EPR-type experimental set-up, that differ markedly from the predictions of the quantum theory. Moreover, experiments of this type have been performed, and the results confirm, to a high degree, the quantum mechanical predictions. The basic theorems, which are the work of J. S. Bell, and the experimental results are nicely summarized in [48]. (For an elegant, non-technical treatment, one might look at [49] by David Mermin; his speculations about the philosophical implications of Bell's work should, however, be taken with a grain of salt). What this shows is that one must either give up the realistic thesis that previously interacting systems go into separate physical states (which, of course, entails a denial that they go into separate, *definite* states), or assume the existence of *nonlocal* effects, that is, superluminal signals by means of which the result of a measurement performed on one of the systems can be influenced by our choice of which parameter we will measure on the other system. The hypothetical signals would have to be faster than the speed of light because we can design the experiment in such a way as to insure that there would not be adequate time for a subluminous signal to travel from one Stern-Gerlach apparatus to the other. Nor will it do to argue, if one finds the latter option unsavory, that while the independent reality of the two systems might not be found at the level of the traditional quantum mechanical properties, such as spin, it may exist at a deeper level, say at the level of some hypothetical "hidden parameters". For the negative result is quite general: any technique, which makes measurement results solely dependent on a system's local properties, be they hidden or not, will yield the wrong predictions [50]. Quantum mechanics avoids nonlocal effects precisely by refusing to assign *separate* (let alone *definite*) states to the two systems, describing them instead by means of a single state function for the composite system, $A + B$, a state function which cannot be decomposed as a product or sum of separate state functions for A and B.

These results must be discomfiting to Popper. Since the simultaneous definiteness thesis implies that previously interacting systems have separate physical states, the only way Popper can continue to maintain that thesis, and to assert that it is compatible with quantum mechanics, is to assume the existence of nonlocal effects. While no one has yet provided any independent evidence for such, effects, one may, of course, still assert their existence as a conjecture, but it should be noted that the price for securing compatibility with quantum mechanics is an assumption that is *incompatible* with the relativity theory's assertion that the speed of light is the upper limit for the propagation of physical effects. And, to repeat, this problem is quite general: any attempt to give a realistic interpretation of quantum mechanics will be saddled with nonlocality as long as it attributes an independent reality, at whatever level, to previously interacting systems. The philosophical implications of the work on nonlocality are discussed, with slightly different emphases, in [51,52]. The latter essay should be read with some caution. d'Espagnat argues that the predictions which have been refuted by experiment are actually entailed by three premises: realism, locality and the "free use of induction". Supporters of Popper might take solace from this claim, arguing that realism can be *saved* by denying the third premise, something Popper would have us do anyway. But I think d'Espagnat is simply mistaken in asserting that induction is a necessary premise in the derivation; the crucial extrapolation which he says [52, p. 177] has to be supported by induction could just as well be construed as a highly corroborated conjecture, and not as an inductive extrapolation from limited evidence. Caution is also advised regarding Shimony's rather instrumentalistic reading of Bohr's position.

I have argued that Popper's arguments for the possibility of simultaneously precise *measurements* of conjugate variables are flawed, and that the thesis that quantum systems *possess* simultaneously definite values of conjugate variables (whether measurable or not) can be maintained only at a considerable price. What, finally, of his claim that measurements more precise than those presumably allowed by the indeterminacy relations are needed to *test* those relations? Do the former negative results mean that the indeterminacy relations cannot be put to a test? I think not. If the Heisenberg formulae were scatter relations of the sort Popper claims, then simultaneously precise measurements would be required to test them. But the difficulties with the simultaneous definiteness thesis argue against this interpretation. The indeterminacy formulae are better viewed simply as relating the expected scatter or standard deviation in the *results* of measurements of conjugate observables on *similarly prepared systems*. Nothing requires that the measurements be carried out on the *same* system. Thus, one can measure the position of half the members of an ensemble of similarly prepared systems, and then the momentum of the other half. If, according to ordinary rules of statistical inference, the product of the standard deviations of the two sets of measurements is significantly lower than the minimum posited by the indeterminacy relations, then those relations can be considered falsified; if the product is not significantly lower, then the relations have passed this particular test. No simultaneous measurements are possible, none are required.

5 The propensity interpretation of quantum mechanics

The latest stage in the development of Popper's thinking on the quantum theory is represented by his propensity interpretation of probability and quantum mechanics. In one sense this idea betokens a radical shift by Popper, for while he formerly thought that physicists were pursuing a will-o'-the-wisp in attempting to forge an objective probabilistic theory of individual events, he later believed that the propensity theory accomplishes precisely this goal. But in a more important sense, no shift at all is indicated, for Popper's purpose in introducing the concept of propensity is the same as it was when he advocated a statistical interpretation of quantum mechanics: the defense of objectivity and realism. The propensity interpretation's credentials as a candidate for a realistic interpretation of quantum mechanics are impressive, and its further refinement is an important item on the agenda for research into foundational questions. But there are some confusions in Popper's account of the propensity interpretation, and some of his claims on its behalf are extravagant. Moreover, I will argue that when the negative results of section 4 are taken into account, a genuine rapprochement between the propensity interpretation and Bohr's complementarity interpretation is no longer out of the question.

Popper says that he developed the propensity interpretation from a criticism of the way the relative frequency interpretation handles the probability of individual or *singular* events. (Remember that it was this difficulty which, according to Popper, led to the Copenhagen interpretation's allegedly *subjective* view of quantum mechanical probabilities. Popper gives an account of the development of the propensity interpretation in [1, pp. 30-34] [37] [53]. The latter two treatments emphasize the development from the frequency theory). Popper asks us to consider a very long actual sequence of tosses of a loaded die, for

which the probability of turning up a six is $\frac{1}{4}$, and to imagine interspersed among these tosses, at unknown places, a few tosses of a fair die. The relative frequency of sixes in this sequence will still be virtually indistinguishable from $\frac{1}{4}$, the more so the longer we imagine the sequence to be, and thus there is no obstacle to our continuing to assert that the probability of throwing a six on a toss of the loaded die is $\frac{1}{4}$. What about the probability of getting a six on one of the throws of the fair die? We still want to assert that this probability is $\frac{1}{6}$, even though the toss is a member of an actual sequence in which the relative frequency of sixes is $\frac{1}{4}$. But why? According to Popper, the frequency theorist can only answer that it is because the toss of the fair die, though a member of the aforementioned *actual* sequence, is also a member of a *virtual* sequence of identical tosses, that is, a virtual sequence of tosses of a fair die, and the probability of getting a six on a toss of the fair die is properly assessed by considering the latter sequence. This may appear, at first glance, to be just a slight modification of the frequency theory, specifically, a stipulation that the only admissible sequences for defining probabilities are sequences of repeated similar trials, or, in other words, sequences characterized by repeated realizations of a set of *generating conditions*. But Popper argues that this really represents "a transition from the frequency interpretation to the propensity interpretation," because "probability may now be said to be a property of the generating conditions" [37, p. 34]. His reasoning is as follows:

> ... if the probability is a property of the generating conditions – of the experimental arrangement – and if it is therefore considered as depending upon these conditions, then the answer given by the frequency theorist implies that the virtual frequency must also depend on these conditions. But this means that we have to visualise the conditions as endowed with a tendency, or disposition, or propensity, to produce sequences whose frequencies are equal to the probabilities; which is precisely what the propensity interpretation asserts. [37, p. 35]

A clearer way to put Popper's point would be to note that the frequency interpretation gives us no reason for choosing one virtual sequence over another, say the virtual sequence of tosses of the fair die rather than a virtual sequence of tosses of various dice, some of them loaded. The propensity interpretation gives us a reason by making probability a property of the experimental conditions. In Popper's eyes, the chief virtue of the propensity interpretation, and the characteristic that makes it useful in giving an objective, realistic interpretation of quantum mechanics, is that it enables us to regard probability as a *real physical property of the single physical experiment* or, more precisely, of the *experimental conditions* laid down by the rule that defines the conditions for the (virtual) repetition of the experiment. [1, p. 33]

Various critics have taken Popper to task for obscurities in his formulation of the propensity interpretation, one of the main worries being that it is still not clear how the propensity theory differs from a frequency theory [54, 55]. The likely cause of the critics' worry is Popper's ill-considered claim that the propensity interpretation regards probability statements "as statements about frequencies in *virtual* (infinite) sequences of well characterized experiments," which he thinks is the same as regarding them "as statements about some measure of a property ... of *the whole experimental arrangement*" [1, p. 32]. Indeed, if we take relative frequencies in virtual sequences as measures of propensity, then the distinction between the two interpretations blurs. But Popper's claim is simply a mistake, for while the propensity theory may have been developed through reflection on the frequentist's talk of virtual sequences, nothing requires that we take relative frequences in virtual sequences as the measure of propensities.

One can formulate the propensity interpretation as the thesis that probabilities are theoretical primitives, which are implicitly defined by a suitable set of axioms for the probability calculus, and which we associate with experimental arrangements or generating conditions, either on the basis of experience or as a result of considerations of physical theory, in order to characterize the tendency or propensity of an arrangement to produce outcomes of a certain sort. Probability statements, thus construed, would be tested, as Popper notes, by actual statistical frequencies [1, p. 32], but to say this is not to say that the statements are *interpreted* in terms of frequencies. Interpreting a formal system like the probability calculus is a matter of finding *models* for it. Probabilities might be modeled by relative frequencies; they can just as well be modeled by propensities.

To be sure, Popper has not given a *general*, formal definition of *propensity*, comparable to the ones that can be given for *relative frequency*, but that is not necessary in order to claim that propensities provide a model for the probability calculus. All that is required is an unambiguous rule for assigning numerical values (Popper calls them *weights*) [1, p. 32] to the propensities, a rule which thus allows us to define a function that can be shown to satisfy the probability axioms. Nor is it right to demand one rule and one definition for all domains of inquiry. Why, for example, should we expect biological propensities and physical propensities to be measured in the same way? How they are to be measured is a task for biological

and physical theory to decide.

In the case of quantum mechanics, the rule is provided by Born's statistical interpretation of the state function. More specifically, the quantum mechanical rule may be formulated thus: let $\mathcal{H}$ be the Hilbert space associated with a system, S, whose state is represented by the vector $|\psi\rangle$ in $\mathcal{H}$; let $\hat{A}$ be a Hermitian operator on $\mathcal{H}$ corresponding to the observable of interest; and let $\hat{P}_A(x)$ be the projection operator onto the subspace of $\mathcal{H}$ spanned by the eigenvectors of $\hat{A}$ corresponding to the eigenvalue x. The probability that a measurement of the observable represented by $\hat{A}$ will yield the value x is given by: $\langle\psi|\hat{P}_A(x)|\psi\rangle$. Only if we have no theory to draw upon need we resort to frequencies for an *estimate* of the numerical value of a propensity. But here it is misleading to speak of frequencies as providing a *measure* of the propensity; we can assume that the measure will be provided by some as yet undiscovered theory (or one too complicated to bother with, for practical purposes, as in the case of dice rolling), and that observed, actual frequencies suggest what that measure would be. In no case need we resort to virtual frequencies.

One of the most sensible features of the propensity interpretation is the insistence that propensities be viewed not as properties of individual objects, such as a die by itself, but as relational properties of whole experimental arrangements. As Popper notes, the probability of throwing a six with a loaded die does not depend solely on the die; the weaker the gravitational field, for instance, the more this probability will approach that for a fair die [53, p. 68]. Nor does the probability of a coin's landing heads-up depend solely on the coin; if we toss it on a flat surface, the probability is $\frac{1}{2}$ but if we toss it on a surface containing a number of slots which might catch the coin on edge, the probability of heads will be less than $\frac{1}{2}$ [1, p. 39].

But Popper goes too far when he suggests that relativizing propensities to experimental contexts at last provides a solution to the wave packet reduction problem in quantum mechanics. His argument is that if we view quantum mechanical probabilities as propensitifes, wave packet reduction is seen to be identical with a trivial, non-puzzling, and thoroughly classical phenomenon that can be demonstrated with a pin board [1, pp. 33-36]. Assume that we start with a symmetrical pin board. The more balls we role down, the closer their resulting distribution will approximate a normal distribution, which can be taken to represent the probability distribution, $p(a, e_1)$, for a single ball's reaching a specific place on a specific trial. If we vary the experimental situation by looking only at those balls which happen to hit a specific pin (which Popper says is like performing a position measurement on the balls), we get a different final distribution, which may represent the probability distribution, $p(a, e_2)$, for these balls' reaching a specific place. Popper contends that *exactly* the same kind of thing happens in the case of quantum mechanical measurement, and thus that the wave packet collapse is to be explained as nothing more than the result of a change in the experimental conditions: before a measurement is performed, the theory predicts one probability distribution, after the measurement it predicts another distribution; the difference is that the performance of a measurement changes the experimental situation.

Two comments are in order. First, as Popper's critics have noted [54, pp. 466-467], and as Popper himself, in effect, conceded, the key to this approach to wave packet reduction is not the propensity interpretation, but the recognition that the relevant probability distributions represent *relative* or *conditional* probabilities, rather than *absolute* ones, and this point could just as easily have been made by a frequency theorist [14, p. 73]. The move to conditional probabilities certainly appears more *natural* from the propensity point of view, but nothing precludes the frequentist's offering the same solution.

Second, and more importantly, Popper's suggestion does not go far enough. Even von Neumann's *solution* of the reduction problem by means of the projection postulate recognizes that different probability distributions are called for before and after a measurement. What makes the projection postulate objectionable is that it posits a principle for the evolution of states during observation which differs dramatically from the principle of evolution under other circumstances. As it stands, Popper's analysis of wave packet reduction is open to the same objection. To be sure, we are given a *reason* for picking a new probability distribution after the measurement is performed, But if the propensities which these probability distributions are supposed to represent are real properties (even of the whole experimental arrangement), we ought to be able to trace their continuous temporal evolution, and this Popper has not told us how to do. My guess is that a wholly satisfactory solution will be achieved only when the propensity interpretation's hint that quantum mechanical reality resides in the whole experimental arrangement is taken up in a thoroughgoing reformulation of the quantum formalism, a reformulation that makes the state function itself a property of the whole experimental arrangement. The standard formalism treats the state function as a context-independent property of the system. Popper, himself, seems to be a bit confused on this point. In one place he says "the ψ function describes physical realities", portraying this as the lesson of the propensity interpretation [53, p. 69]. But that interpretation regards *propensities*, which are properties of whole experimental arrangements, as real. The ψ function, as standardly defined, is not a property of whole experimental arrange-

ments, and thus the propensity interpretation does not entail that the ψ function describes reality. One could take an instrumentalistic attitude toward the ψ function and still regard quantum mechanical propensities as real.

Popper quotes Heisenberg saying that the reduction represents a transition from the possible to the actual [1, p. 37], a transition which is completed when "the actual is selected from the possible, which, is done by the 'observer'" [12, p. 23]. What encourages Popper in thinking that the propensity interpretation already provides a complete solution to the wave packet reduction problem is his belief that the problem is solely a result of the Copenhagen interpretation's alleged grant of special status to the observer in quantum mechanics (the idea being that it is the observer's intervention during measurement that brings about the reduction). Showing that it can be explained, instead, merely as the result of a change in experimental conditions surely does away with any need to invoke the observer. But, as I argued in section 3, this is a misreading of the Copenhagen interpretation, or, at least, of Bohr's philosophy. And, in any case, wave packet reduction involves deeper formal issues. Popper's claim that the propensity idea solves the wave packet reduction problem is but part of his larger claim that it helps us to view quantum mechanics as a statistical theory, and thus

> takes the mystery out of quantum theory ... by pointing out that all the apparent mysteries would also involve thrown dice, or tossed pennies – exactly as they do electrons [53, p. 68]

Patrick Suppes has criticized the assertion that quantum mechanics is a statistical theory on the grounds that, in it

> the joint distribution of noncommuting random variables [such as position and momentum] turns out not to be a proper joint distribution in the classical sense of probability [55, p. 771]

Suppes is correct, but it is important to note that his conclusion depends on our calculating the quantum mechanical joint distribution from the standard, context-independent, quantum mechanical state function. I suspect that if we could reformulate quantum mechanics, as per my suggestion, in such a way as to make the state function a property of whole experimental contexts, we would then get classical joint distributions. Thus, I think Popper is wrong in believing that a simple reinterpretation of quantum mechanics reveals it to be a classical, statistical theory. But the propensity interpretation points the way to a reformulation of the theory that might fit this description.

Another claim that Popper makes for the propensity interpretation is that it solves the problem of the wave-particle relationship, and thus presumably, leads us out of "the great quantum muddle". This claim, too, needs to be scrutinized. In general terms, the muddle consists in mistaking a property of a probability distribution function for a property of either the outcomes (such as a toss of the die yielding a six) whose probability that function provides, or of some element of the system (such as the die, itself) to which those outcomes occur. Physicists make this kind of mistake whenever they say things like "an electron is both a particle and a wave", for what is wave-like is not the electron itself, but the electron's state function, $\psi(x)$, and the resulting probability distribution for its position, given by: $p(x) = \psi^*(x)\psi(x) = |\psi(x)|^2$. Others have criticized the claim that electrons are waves in any real physical sense, pointing out, for example, that we are dealing here not with waves in 3-dimensional physical space, but waves in an abstract, multi-dimensional parameter space, as evidenced by the fact that the probability distribution for the electron's momentum is also, usually, wave-like. But Popper has identified the basic confusion more clearly than perhaps any other writer [1, pp. 18-20].

However, as in the case with wave-packet reduction, the propensity interpretation does not seem to be essential to the clarification. A frequency theorist could also warn us not to confuse the shape of the distribution of balls on a pin board with the shape of the balls themselves. A propensity theorist might be less likely to fall into the confusion in the first place, but clear thinking is the real safeguard, not adherence to a particular interpretation of probability.

Popper's own view of the relationship between particles and waves avoids the "quantum muddle", but is flawed in another way. He says that "particles are important *objects* of experimentation" whereas the often wave-shaped probability distributions are, on the propensity interpretation, "properties of the experimental arrangement" [1, p. 39]. What we have, then, is not a relationship between particles and waves, but one between "particles and their statistics" [1, p. 38], and it is misleading to speak of a *duality* between them. Popper is right that talk of *duality* is misleading. But so is talk of the "relationship between particles and their statistics" and this for two reasons. First, as Popper himself has stressed, propensities are properties of whole experimental arrangements, not of the particles investigated by those arrangements, so the crucial relationship is one between *experimental arrangements* and their statistics, not particles and their statistics, for, properly speaking, particles themselves have no statistics. Second, Popper's attitude toward particles is wholly classical, except, of course, for his agreeing that they do not obey deterministic laws. In particular, as was pointed out in section 4, he views particles as real entities endowed at all times with a complete set of simultaneously

definite properties. But, this view of particles is untenable, or, at least, if the statistical predictions of quantum mechanics are correct (and we have no reason to doubt them), Popper's view of particles can only be maintained at the expense of assuming nonlocal effects, which means contradicting the theory of relativity. In sum, particles, as Popper understands them, might not exist at all.

Popper asserts that both particles and propensities are real [1, p. 39]. In light of the difficulties plaguing the simultaneous definiteness thesis, I would urge Popper to give up the claim about the reality of particles, and rest his realistic interpretation of quantum mechanics solely on real propensities. The propensity interpretation does not require us to talk of particles. All we need speak of are experimental arrangements and their propensities to yield certain kinds of outcomes when certain kinds of measurements are conducted with them. This does not entail denying the reality of atomic objects, it simply means that we would consider them real only as constituents of experimental arrangements.

Such a reworking of Popper's propensity interpretation opens up, finally, the possibility of a rapprochement of sorts between Popper and Bohr. Nowhere does one find Bohr speaking of "propensities". The interpretation of the probability calculus is the kind of formal issue in which he displays little interest. And, of course, Bohr's opaque prose style makes it difficult to say exactly what his views are on a number of issues. But in his own way he too emphasizes that solving the problem of objectivity in quantum mechanics requires our according whole experimental arrangements a central role in the interpretation of the theory. He also sees that the statistical character of the theory is a consequence of the fact different outcomes are ordinarily obtained upon repetition of one and the same experiment.

Recall that, for Bohr, *objectivity* is a matter of the unambiguous communicability of experience. What makes quantum mechanics unique is that here an unambiguous account of experience is made more difficult to achieve because of the intimate relationship between measuring instruments and objects of investigation, which means that we have to pay special attention to experimental arrangements. Bohr says:

> While, within the scope of classical physics, the interaction between object and apparatus can be neglected or, if necessary, compensated for, in quantum physics this interaction thus forms an inseparable part of the phenomenon. Accordingly, the unambiguous account of proper quantum phenomena must, in principle, include a description of all relevant features of the experimental arrangement. [34, p. 391]

More specifically, Bohr argues that we should *interpret* quantum mechanics as referring only to what obtains within well-defined experimental arrangements:

> It is certainly far more in accordance with the structure and interpretation of the quantum mechanical symbolism, as well as with elementary epistemological principles, to reserve the word "phenomenon" for the comprehension of the effects observed under given experimental conditions. [33, p. 316]

What makes quantum mechanics an irreducibly probabilistic theory is the fact that even after we have specified as completely as possible what constitutes a repetition of the *same* experiment, we typically find that the *same* experiment yields different results. The next-to-last quotation continues:

> The very fact that repetition of the same experiment, defined on the lines described, in general yields different recordings pertaining to the object, immediately implies that a comprehensive account of experience in this field must be expressed by statistical laws.

And in the same essay, Bohr remarks about the quantum formalism that

> its physical interpretation finds expressions in laws, of an essentially statistical type, pertaining to observations obtained under given experimental conditions. [34, p. 390]

It is but a short step to the conclusion that the *reality* which the quantum theory describes is precisely the tendencies or propensities of experimental arrangements to produce results of a certain sort.

In his intellectual autobiography, Popper makes the following claim about the reception of his propensity interpretation:

> I remember that the theory was not well received to start with, which neither surprised nor depressed me. Things have changed very much since then, and some of the same critics (and defenders of Bohr) who at first dismissed my theory contemptuously as incompatible with quantum mechanics now say that it is all old hat, and in fact identical with Bohr's view. [15, p. 180]

I want to make it clear that I am not claiming that Popper's interpretation of quantum mechanics is identical with Bohr's. There are interesting similarities, which become especially evident when one discounts Popper's

misreading of Bohr's philosophy as a variety of subjectivism. But equally significant are the differences, which center around Bohr's idea of *complementarity*, and derive from a fundamental disagreement over the relationship between *different* experimental arrangements. Popper's simultaneous measurability thesis implies the compatibility of all experimental arrangements. Bohr, on the other hand, argues that the experimental arrangements necessary for the measurement of pairs of conjugate parameters are mutually exclusive, even though both parameters are, in a sense, equally necessary for giving a complete account of atomic objects. To be accurate, Popper effectively concedes the exclusiveness of the experimental arrangements for a certain kind of measurement of conjugate parameters, namely, the kind he designates a "preparation of state". The passage of a beam of particles through a narrow slit would be such a measurement, in as much as it tells us something about the positions of the particles. This kind of measurement corresponds to what was earlier called a "predictive measurement" [1, p. 21]. The mutual exclusiveness of experimental arrangements is what lies at the root of complementarity, and thus it is also the deep reason for Bohr's denial of the simultaneous reality of conjugate parameters. In Bohr's opinion, it is the novel, complementary relationship of quantum mechanical phenomena that makes quantum mechanics unique among physical theories, including other statistical theories [9, 34]. If the propensity interpretation is detached from the simultaneous definiteness thesis, it can become the natural ally of the complementarity interpretation, and one can hope that such an alliance would lead to deeper insight into the implications of quantum mechanics.

Does Popper's propensity interpretation accomplish the original aims of his program in the philosophy of quantum mechanics? In one sense, yes. It can lay claim to being at least the basis for an *objective* interpretation, and it can be argued that in taking propensities to be real properties of experimental arrangements it constitutes a kind of *realistic* interpretation. But let me conclude with a cautionary note. When Popper claims to provide a *realistic* interpretation of quantum mechanics, he is tacitly asserting his allegiance with a long, continuing philosophical tradition, and though he never, to my knowledge, explicitly mentions it, one important assumption of traditional realism is the mutual independence of observer and observed. Yet precisely this assumption is put in question by quantum mechanics. In Bohr's words:

> ...the elucidation of the paradoxes of atomic physics has disclosed the fact that the unavoidable interaction between the objects and the measuring instruments sets an absolute limit to the possibility of speaking of a behaviour

of atomic objects which is independent of the means of observation. [32, p. 25]

If one could assert the simultaneous definiteness of the properties of interacting systems, then the independence of observer and observed could be maintained, for they could then each be assigned a separate state. So Popper's version of the propensity interpretation, incorporating the definiteness thesis, qualifies as a realistic interpretation in the traditional sense. But if definiteness is denied, as I think it ought to be, and the holism of the orthodox quantum mechanical description of interactions is accepted, then it must be understood that, on its own, the propensity interpretation's place in the tradition is open to question.

Author's Note

This paper was originally written in 1981 by invitation for inclusion in a Popper Festschrift. But for reasons not worth recalling now, it was not published in that volume. It is presented here with only minimal editorial changes to update some tenses, reflecting the fact of Popper's since having died and, in only one case, to call attention to one of the author's subsequent papers. These circumstances explain the somewhat dated quality that some readers might note.

References

[1] Popper KR. Quantum Mechanics without 'The Observer'. In: Quantum Theory and Reality, Bunge M (editor), New York: Springer, 1967, pp. 7-43.

[2] Feyerabend PK. On a recent critique of complementarity: part I. Philosophy of Science 1968; 35 (4): 309-331.

[3] Feyerabend PK. On a recent critique of complementarity: part II. Philosophy of Science 1969; 36 (1): 82-105.

[4] Heisenberg W. Über den anschaulichen Inhalt der quantentheoretischen Kinematik und Mechanik. Zeitschrift für Physik A 1927; 43 (3-4): 172-198.

[5] Bohr N. The quantum postulate and the recent development of atomic theory. Nature 1928; 121 (3050): 580-590.

[6] Bohr N. The Quantum Postulate and the Recent Development of Atomic Theory. In: Atti del Congresso Internazionale dei Fisici, Como, 11-20 Settembre 1927 (Bologna: Zanichelli, 1928), vol. 2, pp. 565-588.

[7] Bohr N. The Quantum Postulate and the Recent Development of Atomic Theory. In: Atomic Theory and the Description of Nature, Cambridge: Cambridge University Press, 1934, pp. 52-91.

[8] Jammer M. The Philosophy of Quantum Mechanics: The Interpretations of Quantum Mechanics in Historical Perspective. New York: John Wiley & Sons, 1974.

[9] Bohr N. Can quantum-mechanical description of physical reality be considered complete? Physical Review 1935; 48 (8): 696-702. http://dx.doi.org/10.1103/PhysRev.48.696

[10] Bohr N. Discussion with Einstein on Epistemological Problems in Atomic Physics. In: Albert Einstein: Philosopher-Scientist, Schilpp PA (editor), The Library of Living Philosophers, vol. 7, LaSalle, Illinois: Open Court, 1949, pp. 201-241.

[11] Howard D. Who invented the "Copenhagen Interpretation"? A study in mythology. Philosophy of Science 2004; 71 (5): 669-682. http://dx.doi.org/10.1086/425941

[12] Heisenberg W. The development of the interpretation of quantum theory. In: Niels Bohr and the Development of Physics, Pauli W (editor), Condon: Pergamon, 1955, pp. 12-29.

[13] Popper KR. The Logic of Scientific Discovery. London, New York: Routledge, 2002.

[14] Popper KR. Intellectual Autobiography. In: The Philosophy of Karl Popper, Schilpp PA (editor), The Library of Living Philosophers, vol. 14, LaSalle, Illinois: Open Court, 1974.

[15] Popper KR. Unended Quest: An Intellectual Autobiography. London, New York: Routledge, 2002.

[16] Schrödinger E. Collected Papers on Wave Mechanics. Shearer JF, Deans WM (translators), London: Blackie & Son, 1928.

[17] Scott WT. Erwin Schrödinger: An Introduction to His Writings. Amherst, Massachusetts: University of Massachusetts Press, 1967.

[18] Born M. Zur Quantenmechanik der Stoßvorgänge. Zeitschrift für Physik A 1926; 37 (12): 863-867. http://dx.doi.org/10.1007/BF01397477

[19] Popper KR. Indeterminism in quantum physics and in classical physics: part I. The British Journal for the Philosophy of Science 1950; 1 (2): 117-133. http://dx.doi.org/10.1093/bjps/I.2.117

[20] Popper KR. Indeterminism in quantum physics and in classical physics: part II. The British Journal for the Philosophy of Science 1950; 1 (3): 173-195. http://dx.doi.org/10.1093/bjps/I.3.173

[21] Einstein A. Letter 84, 3 March 1947. In: The Born-Einstein letters: correspondence between Albert Einstein and Max and Hedwig Born from 1916-1955, with commentaries by Max Born. London: Macmillan, 1971. http://archive.org/details/TheBornEinsteinLetters

[22] Heisenberg W. The Physical Principles of the Quantum Theory. Eckart C, Hoyt FC (translators), New York: Dover, 1949.

[23] von Neumann J. Mathematical Foundations of Quantum Mechanics. Beyer RT (translator), Princeton: Princeton University Press, 1955.

[24] Bell JS. On the problem of hidden variables in quantum mechanics. Reviews of Modern Physics 1966; 38 (3): 447-452. http://dx.doi.org/10.1103/RevModPhys.38.447

[25] Popper KR. Probability magic or knowledge out of ignorance. Dialectica 1957; 11 (3-4): 354-374. http://dx.doi.org/10.1111/j.1746-8361.1957.tb01643.x

[26] Schlick M. Die Kausalität in der gegenwärtigen Physik. Naturwissenschaften 1931; 19 (7): 145-162. http://dx.doi.org/10.1007/BF01516406

[27] Jeans JH. The New Background of Science. New York: Macmillan, 1933. http://archive.org/details/newbackgroundofs006677mbp

[28] Heisenberg W. The representation of nature in contemporary physics. Daedalus 1958; 87 (3): 95-108.

[29] Popper KR. Objective Knowledge: An Evolutionary Approach. New York: Oxford University Press, 1972.

[30] d'Espagnat B. Conceptual Foundations of Quantum Mechanics. Reading, Massachusetts: Perseus Books, 1999.

[31] Wigner EP. Remarks on the mind-body question. In: The Scientist Speculates, Good IJ (editor), London: Heinemann, 1961, pp.284-302.

[32] Bohr N. Atomic Physics and Human Knowledge. New York: John Wiley & Sons, 1958. http://archive.org/details/AtomicPhysicsHumanKnowledge

[33] Bohr N. The causality problem in atomic physics. In: Niels Bohr Collected Works, vol.7. Foundations of Quantum Physics II (1933-1958), Kalckar J (editor), Elsevier, 1996, pp.299-322. http://dx.doi.org/10.1016/s1876-0503(08)70376-1

[34] Bohr N. Quantum physics and philosophy - causality and complementarity. In: Niels Bohr Collected Works, vol.7. Foundations of Quantum Physics II (1933-1958), Kalckar J (editor), Elsevier, 1996, pp.385-394. http://dx.doi.org/10.1016/s1876-0503(08)70381-5

[35] Heisenberg W. Physics and Philosophy: The Revolution in Modern Science. Unwin University Books, London: George Allen & Unwin, 1971. http://www.archive.org/details/PhysicsPhilosophy

[36] Carnap R. Logical Foundations of Probability. London: Routledge and Kegan Paul, 1950.

[37] Popper KR. The propensity interpretation of probability. The British Journal for the Philosophy of Science 1959; 10 (37): 25-42. http://dx.doi.org/10.1093/bjps/X.37.25

[38] Bohr N. On the notions of causality and complementarity. Dialectica 1948; 2 (3-4): 312-319. http://dx.doi.org/10.1111/j.1746-8361.1948.tb00703.x

[39] Bohr N. On the notions of causality and complementarity. Science 1950; 111 (2873): 51-54. http://dx.doi.org/10.1126/science.111.2873.51

[40] Bohr N. The Unity of Human Knowledge. In: Niels Bohr Collected Works, vol.10. Complementarity Beyond Physics (1928-1962), David F (editor), Elsevier, 1999, pp.155-160. http://dx.doi.org/10.1016/s1876-0503(08)70208-1

[41] Bohr N. The Rutherford Memorial Lecture 1958: Reminiscences of the Founder of Nuclear Science and of Some Developments Based on his Work. Proceedings of the Physical Society 1961; 78 (6): 1083-1115. http://dx.doi.org/10.1088/0370-1328/78/6/301

[42] Hooker CA. The nature of quantum mechanical reality: Einstein versus Bohr. In: Paradigms & Paradoxes: The Philosophical Challenge of the Quantum Domain, Colodny RG (editor), University of Pittsburgh Series in the Philosophy of Science, vol. 5, Pittsburgh: University of Pittsburgh Press, 1972, pp. 67-302.

[43] Popper KR. Zur Kritik der Ungenauigkeitsrelationen. Naturwissenschaften 1934; 22 (48): 807-808. http://dx.doi.org/10.1007/BF01496543

[44] Einstein A, Podolsky B, Rosen N. Can quantum-mechanical description of physical reality be considered complete? Physical Review 1935; 47 (10): 777-780. http://dx.doi.org/10.1103/PhysRev.47.777

[45] Bohm D. Quantum Theory. New York: Dover Publications, 1989.

[46] Furry WH. Note on the quantum-mechanical theory of measurement. Physical Review 1936; 49 (5): 393-399. http://dx.doi.org/10.1103/PhysRev.49.393

[47] Howard D. Complementarity and Ontology: Niels Bohr and The Problem of Scientific Realism in Quantum Physics. Ph. D. Dissertation, Boston University, 1979.

[48] Clauser JF, Shimony A. Bell's theorem. Experimental tests and implications. Reports on Progress in Physics 1978; 41 (12): 1881-1927. http://dx.doi.org/10.1088/0034-4885/41/12/002

[49] Mermin ND. Quantum mysteries for anyone. The Journal of Philosophy 1981; 78 (7): 397-408. http://www.jstor.org/stable/2026482

[50] Belinfante FJ. A Survey of Hidden Variable Theories. Monographs in Natural Philosophy, New York: Pergamon, 1973.

[51] Shimony A. Metaphysical problems in the foundations of quantum mechanics. International Philosophical Quarterly 1978; 18 (1): 3-17.

[52] d'Espagnat B. The quantum theory and reality. Scientific American 1979; 241 (11): 158-181.

[53] Popper KR. The propensity interpretation of the calculus of probability and the quantum theory. In: Observation and Interpretation: A Symposium of Philosophers and Physicists. Proceedings of the Ninth Symposium of the Colston Research Society held in the University of Bristol, April 1st-April 4th, 1957, Körner S (editor), London: Butterworth Scientific Publications, 1957, pp.65-70.

[54] Sneed JD. Review of "Quantum Theory and Reality" editted by Mario Bunge. Synthese 1968; 18 (4): 464-467. http://dx.doi.org/10.1007/BF00484982

[55] Suppes P. Popper's analysis of probability in quantum mechanics. In: The Philosophy of Karl Popper, vol.2, Schilpp PA (editor), LaSalle, Illinois: Open Court Publishing Company, 1974, pp.760-774.

Permissions

The contributors of this book come from diverse backgrounds, making this book a truly international effort. This book will bring forth new frontiers with its revolutionizing research information and detailed analysis of the nascent developments around the world.

We would like to thank all the contributing authors for lending their expertise to make the book truly unique. They have played a crucial role in the development of this book. Without their invaluable contributions this book wouldn't have been possible. They have made vital efforts to compile up to date information on the varied aspects of this subject to make this book a valuable addition to the collection of many professionals and students.

This book was conceptualized with the vision of imparting up-to-date information and advanced data in this field. To ensure the same, a matchless editorial board was set up. Every individual on the board went through rigorous rounds of assessment to prove their worth. After which they invested a large part of their time researching and compiling the most relevant data for our readers.

The editorial board has been involved in producing this book since its inception. They have spent rigorous hours researching and exploring the diverse topics which have resulted in the successful publishing of this book. They have passed on their knowledge of decades through this book. To expedite this challenging task, the publisher supported the team at every step. A small team of assistant editors was also appointed to further simplify the editing procedure and attain best results for the readers.

Apart from the editorial board, the designing team has also invested a significant amount of their time in understanding the subject and creating the most relevant covers. They scrutinized every image to scout for the most suitable representation of the subject and create an appropriate cover for the book.

The publishing team has been an ardent support to the editorial, designing and production team. Their endless efforts to recruit the best for this project, has resulted in the accomplishment of this book. They are a veteran in the field of academics and their pool of knowledge is as vast as their experience in printing. Their expertise and guidance has proved useful at every step. Their uncompromising quality standards have made this book an exceptional effort. Their encouragement from time to time has been an inspiration for everyone.

The publisher and the editorial board hope that this book will prove to be a valuable piece of knowledge for researchers, students, practitioners and scholars across the globe.

List of Contributors

William M. Shields
Worcester Polytechnic Institute, Worcester, Massachusetts, United States

Chris C. King
Department of Mathematics, University of Auckland, Auckland, New Zealand

Ignazio Licata
ISEM Institute for Scientific Methodology, Palermo, Italy
School of Advanced International Studies on Applied Theoretical and Nonlinear Methodologies in Physics, Bari, Italy

Leonardo Chiatti
AUSL VT Medical Physics Laboratory, Via Enrico Fermi 15, Viterbo, Italy

Tina Bilban
Institute Nova revija, Cankarjeva cesta 10b, SI-1000 Ljubljana, Slovenia

Michael Nauenberg
Department of Physics, University of California, Santa Cruz, United States

Francesco Giacosa
Institute of Physics, Jan Kochanowski University, Kielce, Poland
Institute for Theoretical Physics, Johann Wolfgang Goethe University, Frankfurt am Main, Germany

Yakir Aharonov
School of Physics and Astronomy, Tel Aviv University, Tel Aviv, Israel
Schmid College of Science, Chapman University, Orange, California, USA

Eliahu Cohen
School of Physics and Astronomy, Tel Aviv University, Tel Aviv, Israel
H. H. Wills Physics Laboratory, University of Bristol, Bristol, UK

Tomer Shushi
University of Haifa, Haifa, Israel

Charlyne de Gosson and Maurice A. de Gosson
Numerical Harmonic Analysis Group, Faculty of Mathematics, University of Vienna, Vienna, Austria

Mani L. Bhaumik
Department of Physics and Astronomy, University of California, Los Angeles, California, USA

Peter J. Lewis
Department of Philosophy, University of Miami, Coral Gables, Florida, USA

Michael Nauenberg
Physics Department, University of California, Santa Cruz, United States

Radu Ionicioiu
Department of Theoretical Physics, Horia Hulubei National Institute of Physics and Nuclear Engineering, Bucharest–Măgurele, Romania

John J. O'Connor and Edmund F. Robertson
School of Mathematics and Statistics, University of St Andrews, North Haugh, St Andrews, Fife, Scotland

Otávio Bueno
Department of Philosophy, University of Miami, Coral Gables, FL 33124, USA

N. D. Hari Dass
Tata Institute of Fundamental Research (TIFR), TIFR Centre for Interdisciplinary Sciences (TCIS), Hyderabad, India
Chennai Mathematical Institute, Chennai, India
Centre for Quantum Information and Quantum Computation, Indian Institute of Science, Bangalore, India

John Ashmead
School of Engineering and Applied Science, University of Pennsylvania, Philadelphia, Pennsylvania, United States

Don Howard
Department of Philosophy, University of Notre Dame, Notre Dame, Indiana, United States

Index